AF776818

# PHYSICAL METALLURGY OF REFRACTORY METALS AND ALLOYS

METALLOVEDENIE TUGOPLAVKIKH METALLOV I SPLAVOV

МЕТАЛЛОВЕДЕНИЕ ТУГОПЛАВКИХ МЕТАЛЛОВ И СПЛАВОВ

# PHYSICAL METALLURGY OF REFRACTORY METALS AND ALLOYS

E. M. Savitskii and G. S. Burkhanov
A. A. Baikov Institute of Technology
Academy of Sciences of the USSR
Moscow, USSR

Translated from Russian

CONSULTANTS BUREAU · NEW YORK-LONDON · 1970

*Evgenii Mikhailovich Savitskii* was born in 1912 and completed his studies at the Moscow Institute of Nonferrous Metals and Gold in 1936, specializing in physical metallurgy; from 1937 to 1953 he worked at the Institute of General and Inorganic Chemistry, where in 1940 he was appointed director of the mechanical testing laboratory. Since 1953 he has been associated with the A. A. Baikov Institute of Metallurgy as director of the refractory and rare metal alloys laboratory. In 1953 he was awarded the degree of Doctor of Chemical Sciences, and in 1954 was appointed Professor of Physical Metallurgy. In 1966 Savitskii was elected corresponding member of the Academy of Sciences of the USSR.

*Gennadii Sergeevich Burkhanov* was born in 1932 in Moscow. In 1955 he was graduated from the Moscow Steel Institute, where he specialized in physical metallurgy. After completing graduate studies at the A. A. Baikov Institute, he joined its staff, and since 1963 has been a senior scientific assistant in the refractory and rare metal alloys laboratory.

ISBN-13: 978-1-4684-1574-2 e-ISBN-13: 978-1-4684-1572-8
DOI: 10.1007/978-1-4684-1572-8

The original Russian text, published by Nauka Press in Moscow in 1967 for the A. A. Baikov Institute of Metallurgy of the Academy of Sciences of the USSR, has been corrected by the authors for the present edition. The English translation is published under an agreement with Mezhdunarodnaya Kniga, the Soviet book export agency.

*Евгений Михайлович Савицкий,*
*Геннадий Сергеевич Бурханов*

**Металловедение тугоплавких металлов и сплавов**

Library of Congress Catalog Card Number 69-12519
SBN 306-10841-0


A Division of Plenum Publishing Corporation
227 West 17th Street, New York, N. Y. 10011

United Kingdom edition published by Consultants Bureau, London
A Division of Plenum Publishing Company, Ltd.
Donington House, 30 Norfolk Street, London, W.C. 2, England

# PREFACE TO THE AMERICAN EDITION

The principal reasons which induced the authors to write this book and the features of the book are set forth in the preface to the Russian edition.

That section of the science of metals which in Russian is called "metallovedenie" or the "physical chemistry of metals" is generally referred to in scientific and technical literature published in the English language by the term "physical metallurgy." These concepts are much broader than the term "metallography," used in the scientific and technical literature of various countries, and applied solely to research on the interrelationships of the structure and properties of metals and alloys.

Each science must have its own subject and its own method of research. Certainly, all specialists will agree that metals and alloys, including their solid solutions, mechanical mixtures, and metallic compounds, form the subject of "physical metallurgy" or "physical chemistry of metals." The aim of this science is to produce a theory and to elucidate the experimental relationships which ought finally to make it possible to calculate quantitatively alloys of given properties for any working conditions and parameters.

With regard to the methods of research on metals and alloys, we assume that "physical metallurgy" may properly employ any theoretical and experimental methods which will help to solve its basic problems. The experience of the last few decades has shown that among the experimental methods a leading position is occupied by the physicochemical analysis of metal systems developed by N. S. Kurnakov, that is to say, the systematic investigation of the physical properties of alloys as a function of he variation of their chemical composition and phase composition. This method has also been employed for investigating the interrelationship of the structure and properties of alloys at different temperatures and pressures (see E. M. Savitsky, The Influence of Temperature on the Mechanical Properties of Metals and Alloys, Stanford University Press, 1961). The "physical chemistry of metals" makes extensive use of physicochemical analysis. The science of metals and alloys absorbs all the achievements of the exact fundamental sciences, primarily physics and chemistry. It may even be said more precisely: The "physical chemistry of metals" or "physical metallurgy" is the synthesis of the physics of metals and the chemistry of metal alloys. The distribution and energy of electrons is the basic factor determining the crystal structure and all the properties of metals and alloys. The main problems which have to be solved in the framework of these sciences are the following: the electronic structure and nature of the interatomic bond in metal crystals, solid solutions of metals and metallic compounds, the development of methods for the quantitative measurement of the value of the interatomic forces with variation in the fundamental parameters of thermodynamic equilibrium (composition, temperature, pressure), by the application of electromagnetic, gravitational, and other fields, penetrating radiations, research on the interrelationship of the structure and properties of metals and alloys, the influence on them of impurities, defects and their anisotropy, as well as the influence of specific experimental, technological, or service conditions.

It is clear from this that our science, if only in the interests of its own development, cannot live in an ivory tower of theory, shut off from the technology and the service of metal articles, since these in particular finally decide the suitability of the new metallic materials developed. It is no accident that many of the encyclopedic reviews published in the USA on rare and refractory metals bear the title "Progress in Science and Technology."

There may be different views among specialists regarding the content, limits, and methods of the "physical chemistry of metals" or "physical metallurgy." From the point of view of the writers of this book, in addition to what has been set forth in this preface, its readers may gain some idea after acquaintance with the contents of the book, and particularly with its "Conclusion."

The book was published in Moscow in 1967. Judging by the reviews in periodicals, the effort has been regarded as timely and successful. The Russian edition of the book was sold out in several weeks, indicating the growing demand for books on refractory metals and alloys.

The level of "physical metallurgy" is determined primarily by the requirements of engineering for metals and alloys possessing a definite combination of properties. Originally, therefore, the "physical metallurgy" of iron, steel, and cast iron was first developed, then that of nonferrous and light metals, and in recent years the physical metallurgy of refractory metals and rare metals.

The scientific and technological revolution which is unfolding before our eyes, the appearance of new branches of industry – atomic, jet, electronic, and space technologies – is accompanied by constantly increasing and diverse requirements for metallic materials. These demands may often be satisfied only by the use of rare and refractory metals and their alloys and compounds. The demands of the new technology have also resulted in the appearance of a new branch of the refractory and rare metals which produces, almost all the known elements of Mendeleev's periodic system, and their alloys, and compounds.

The experimental data and theoretical concepts relating to refractory and rare metals have made it possible to reconsider the structure and properties of ordinary industrial ferrous and nonferrous metals and alloys. In this respect, it is possible to include concepts on purity and structural perfection, the part played by interstitial impurities, particularly carbon and oxygen, in the cold brittleness of metals and alloys, information obtained regarding single crystals in the construction of the phase diagrams of alloys of refractory metals, in research on superconducting alloys, the function of rare-earth metals, the "rhenium effect," and the mechanism of hardening by the precipitation of nonmetallic compounds in quenched niobium and molybdenum alloys.

It has become an urgent problem to create a "physical metallurgy" of each refractory and rare metal, as was done earlier for iron, nickel, copper, aluminum, magnesium, and more recently for titanium. The physical metallurgy of niobium, vanadium, rare-earth metals, molybdenum, rhenium, platinum, palladium, lithium, yttrium, and the radioactive metals may be regarded to be already developed, especially at the level of binary systems [see the bibliography appended to the article by E. M. Savitskii in Metallovedenie i termicheskaya obrabotka metallov, No. 10, p. 12 (1967)].

Russian scientists value the achievements of their foreign colleagues highly. We apologize if some of the experimental work carried out in the USA or in other countries is not described in adequate detail in our book, and in connection with a proposed second Russian edition of our book, we should like to reguest our foreign colleagues to send us copies of their publications in the field of the physical metallurgy of refractory metals and alloys.

Unfortunately, a lack of knowledge of the language is often one of the principal difficulties preventing a closer contact between scientists and the exchange of information regarding their researches and new ideas. The translation of books and articles helps to minimize this shortcoming. We are grateful to Plenum Publishing Corporation for their initiative and labor in publishing an English version of our book, and esteem this as a friendly act to the scientists of our country and to us personally.

The A. A. Baikov
Institute of Metallurgy
of the Academy of Sciences
of the USSR, Moscow
January, 1968

E. M. Savitskii
G. S. Burkhanov

# PREFACE TO THE RUSSIAN EDITION

New metals and alloys, new methods of producing and investigating them, new extensive experimental data and fields of application – all this has immeasurably widened the horizons of modern physical metallurgy, metal physics, and the physical chemistry of metals compared with the level of knowledge at the end of the nineteenth and the beginning of the twentieth centuries.

Many earlier data on the properties of metals are now out of date, having been determined on insufficiently pure metals. In recent years, all the refractory transition metals (titanium, zirconium, hafnium, vanadium, niobium, tantalum, rhenium, pure iron, manganese, chromium, molybdenum, tungsten), refractory metals of the platinum group (osmium, ruthenium, iridium, and others), lithium, beryllium, semiconductors (germanium, silicon, boron, high-purity carbon), and the radioactive metals (uranium, thorium, plutonium) have become available for research. The rare-earth metals of the cerium subgroup (lanthanum, cerium, neodymium, praseodymium), metals of the yttrium subgroup, as well as scandium and yttrium are being produced in ever-increasing quantities. They are no longer just rare laboratory specimens, but industrial metals for alloying and for application as alloy bases.

In the 50's and 60's of this century, scientific and production workers created a new combination of the media of the chemical technology of the production of pure substances and vacuum metallurgy: ion exchange chromatography, the iodide method, electrolysis, distillation, vacuum or argon-arc, electron-beam, plasma, and laser melting, vacuum zone refining using different sources of heat, pressure treatment and welding in a vacuum or in an atmosphere of inert gases, vacuum annealing, use of getters, protective envelopes, coatings, and so forth.

The temperature ceiling of physical metallurgical research in particular has been raised. Vacuum metallography, vacuum dilatometry, and the determination of electrophysical properties, high-temperature thermal analysis, neutron diffraction analysis, electron, ion, and proton microscopy with magnifications of hundreds of thousands and even millions of times, all these made their appearance. In addition, research on the structure and special properties of metals and alloys close to the absolute zero of temperature in liquid helium (4.2°K) is being continuously extended, principally in connection with rapid developments in recent years in the physics of metals, physical chemistry, and physical metallurgy of superconducting materials.

This book is based primarily on the experimental work of the Refractory and Rare Metals and Alloys Laboratory of the A. A. Baikov Institute of Metallurgy of the Academy of Sciences of the USSR. Two main circumstances led to the writing of this book. They were the increasing value of refractory metals in the new technology and the complete lack of monographs and textbooks on the physical metallurgy and physical chemistry of refractory metals and alloys (with the exception of titanium and niobium). In the meantime, the need for the publication of reviews of work on these materials had undoubtedly grown, since in recent years, the people who, by the very nature of their work, have to deal with the production, treatment, purification, research on electrophysical and mechanical properties, and, in particular, with the application of semifinished products and components made of refractory metals and alloys, in instruments and apparatus of the new technology (jet aircraft and rockets, satellites and spacecraft, nuclear jets, the special-purpose fleet, devices for converting thermal energy into electrical energy, superconducting magnets and other devices, electrical vacuum instruments, chemical reactors), and also as radioactive isotopes and surgical material in medicine.

Specialists in the field of the production of the devices and appratus mentioned (designers, physicists, chemists, power engineers, technologists, workers in the radioelectronic industry), and finally surgeons, unfortunately do not as a rule receive the necessary grounding in general physical metallurgy and the physical chemistry of metals, and are still less familiar with current literature on research into the structure and properties of refractory metals, published in specialized metallurgical, chemical, and physical journals. Interest in the growth of science itself also calls for periodic reviews of the accumulated experimental data.

It is the hope of the authors that the collection in one book of a considerable proportion of published data on the structure properties and application of alloys will assist the work of the group of specialists mentioned with regard to the use of refractory metals in the new technology, although the authors lay no claim to having exhaustively dealt with all the available information on refractory metals and alloys which, in view of the vigorous development of research in these fields, is constantly being published in the Russian and foreign periodical press.

The universities and technical colleges as yet have no up-to-date textbook on the physical metallurgy and physical chemistry of refractory metals and alloys, despite the fact that the expanding volume of the application of refractory metals in industry is even now calling for suitable specialization in the metallurgical faculties of our universities. In time, such a textbook will be produced, but for the time being, lecturers, graduates, and undergraduates will be able if they so wish to use the corresponding sections of the present monograph.

The planning of the book, the choice of material, and its interpretation have been carried out by the authors from the physicochemical standpoint, since at the present time, as practice has shown, such an approach to the problem of refractory metals and alloys is the most fruitful. The physical chemistry and physical metallurgy of refractory metals have not yet been finally formulated; they are in a state of dynamic development, and a number of cardinal questions, such as the control of plastic properties, the quantitative calculation of the physical properties, and concepts regarding the optimum structure of the metals, have not yet been solved. This is primarily due to the lack of a quantitative theory of metals and inadequate knowledge of the distribution of the electrons in crystals, and of the interatomic bond and the methods of measuring it, and also of the quantitative relationships between the structure and properties of metals, their compounds, and alloys. Of course, all this has affected the contents of the present book. Unfortunately, many questions of the physical metallurgy and physical chemistry of refractory metals (as is furthermore the case with all the other metals) can be considered only on a qualitative level.

Nevertheless, it is essential to insist on drawing the attention of workers in science and production to the need for the creation of a valid physicochemical theory of metals; to take these considerations into account, a special chapter is devoted to the available information on the interatomic bond in refractory metals.

In accordance with the physicochemical bias of this monograph, the main attention is directed to a consideration of the interaction of refractory metals with each other, with other metals, and with nonmetallic impurities, and to the interrelationship of structure and properties in accordance with the position of the refractory metals in the periodic system, as well as to the structure, properties, and application of their alloys. Data on natural resources, methods of purification, melting, and treatment are generally presented as reference material.

In the discussion of structure and properties, the information available on vacuum-melted metals and alloys has been mainly used as being more reliable. The book contains practically no discussion of these questions for cermet materials.

Owing to the intense demand of the new technology for metallic materials, postwar physical metallurgy and the physical chemistry of metals are in a process of vigorous development. A numerous group of rare metals and alloys having become of industrial interest, the range of subjects of physical metallurgy is constantly widening and the overall available knowledge is becoming increasingly difficult to cover in one book, not to mention the considerations associated with publications of considerable bulk. Books are therefore now being produced on the physical metallurgy of individual refractory rare metals. "Nauka" Press has already published monographs on the alloys of niobium and rhenium. As knowledge widens, books will be published on the physical metallurgy of each metal, which of course does not preclude the writing of comprehensive works. Furthermore, in contemporary physical metallurgy, quite new sections are appearing and developing as the result of the discovery in entire groups of metals of special physical properties, for example, the propensity of atomic nuclei to fission under the action of neutrons, the existence of superconducting properties of metals in liquid helium, resulting in the creation of entirely new branches of industry, and branches of physical metallurgy, such as the physical metallurgy of fissile materials and the physical metallurgy of superconductors.

For revealing the true properties of metals, undistorted by impurities and defects, ever-increasing attention is being devoted to their production and investigation in a particularly pure and perfect state in the form of single crystals.

Yet another new branch of physical metallurgy is currently being formed, the physical metallurgy of single crystals, which has come into being through the requirements of the new technology, and a new branch of metallurgical industry has already appeared, the production and treatment of single crystals of refractory and rare metals and alloys.

Taking into consideration the special scientific and practical significance of single crystals of refractory metals and alloys and the rapid development of the experimental investigations of their structure and properties, a specially written lengthy chapter on single crystals has been included in this book for the first time in the literature of physical metallurgy. The foregoing also applies to a considerable degree to the material of Chapter IV, where problems, specific to refractory metals having a body-centered cubic lattice are discussed in connection with methods of combating their cold brittleness.

Considerable attention has also been paid to technological peculiarities, which are very important and specific for refractory metals, and also to available data on the prospects of their industrial utilization.

The authors regard their work as a first attempt to solve a very difficult problem, that of producing a comprehensive printed work on the physical metallurgy and physical chemistry of refractory metals and alloys, useful to all who have to do with the production, investigation, and application of these materials.

The authors will accept with gratitude any comments which readers may have to make on the contents of the book and will endeavor to take them into consideration in a subsequent work.

The authors thank N. V. Ageev, I. V. Burov, V. L. Girshov, K. N. Ivanova, I. A. Karyazin, A. M. Zakharov, Yu. V. Emifov, I. M. Kop'ev, D. V. Ignatov, V. P. Polyakov, and L. F. Myzenkov for reading various chapters of the manuscript and making corrections, as well as M. I. Beloborodov, S. P. Pertsev, E. V. Ottenberg, and I. V. Mal'tsev for assistance in formulating the manuscript.

# CONTENTS

CHAPTER I

# NATURAL RESOURCES AND METHODS OF PRODUCING REFRACTORY METALS

To date there has been no generally accepted definition in scientific literature of those metals which should be regarded as refractory. In this book, refractory metals are considered to be those whose melting point is higher than that of iron (1535°C). They all belong to the group of transition metals having an unfilled d-shell and occurring mainly in the first, second, and third large periods from group IVa to group VIII of the periodic system of elements. The most refractory of them are tungsten (melting point 3380°C), rhenium (3180°C), tantalum (3000°C), and osmium (3045°C).

## Natural Resources and Principal Minerals

Most refractory metals are comparatively rare in nature. Owing to their special physical properties, however, their industrial applications are particularly extensive.

Of the refractory metals having melting points around 1700°C, titanium alone is included in the number of elements which are fairly common in nature (its content in the earth's crust is about 0.6 wt.% [1]). It occupies the tenth place in the elements in order of abundance in the earth's crust. We know that 98.73% of the earth's crust consists of ten elements, oxygen and silicon amounting to 75.13%, while the remaining eight elements (aluminum, iron, calcium, sodium, potassium, magnesium, hydrogen, and titanium) form 23.60%. Approximately 0.5% of the total content of the elements in the earth's crust consists of carbon, chlorine, phosphorus, sulfur, and manganese, while the content of all the other elements is approximately only 0,7%. Attention is drawn in [2] to the regularity, first discovered by Mendeleev, which consists in the general tendency of the clarkes to decrease with increase in the order number of the elements (Fig. 1).

Rhenium is one of the rare elements in the earth's crust. The clarke by weight of rhenium, about $1 \cdot 10^{-7}$%, was determined in 1930 by V. and I. Noddak [3]. According to later data [4-6], the clarke by weight of rhenium is approximately one order higher ($10^{-6}$ wt.%). There are no actual rhenium minerals. It accompanies minerals of molybdenum, copper, lead, zinc, platinium, niobium, and others. The production of rhenium in 1959 was 450-680 kg. According to R. I. Jaffee [7], the possible potential annual production of rhenium was estimated at 9070-13,600 kg. A considerable proportion of the rhenium in the earth's crust is found in molybdenites. In some molybdenites, the rhenium content may attain 0.25-0.30% [6]. Rhenium is also found in copper concentrates [8-10]. There is information concerning the discovery of rhenium in uranium ores [11] and coal deposits [12]. The rhenium content of ores and concentrates from different deposits is low. The rhenium is therefore extracted along with the products of the treatment of these ores and concentrates. The extension of the raw material basis for the industrial production of rhenium and the development of economic, technological systems of extracting it from raw materials is one of the most important problems of geochemistry and metallurgy.

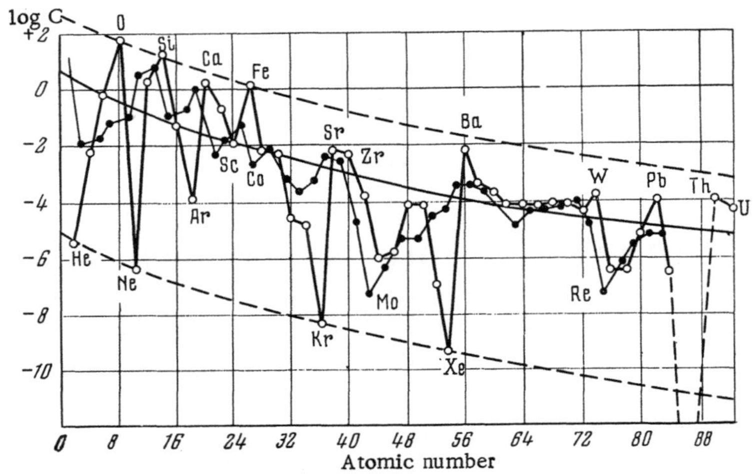


Fig. 1. Logarithms of the clarkes of elements (after Fersman).

Another exceptionally rare refractory metal is osmium, belonging to the family of platinum metals. According to A. P. Vinogradov, the weight clarke of osmium is $5 \cdot 10^{-6}\%$ [1]. According to the authors of [7], osmium is potentially a less available element than rhenium. The annual production of osmium is 45-155 kg [7]. Osmium, ruthenium, rhodium, and iridium often accompany platinum in nature. The metals enumerated are also contained as side products in the purification of nickel [7]. Osmium, one of the most refractory metals in nature, is also found in the form of alloys with iridium (osmious iridium). The weight clarkes of ruthenium, rhodium, and iridium are, respectively, $5 \cdot 10^{-7}$, $1 \cdot 10^{-7}$, and $1 \cdot 10^{-7}\%$ [1].

The platinum content of the earth's crust is $5 \cdot 10^{-7}\%$. Platinum is found in nature mainly in the native state in the form of the pure metal or in the form of natural alloys with other metals. A method of producing spectroscopically pure platinum was developed by I. I. Chernyaev and A.M. Rubenshtein [13]. It should be noted that the raw material reserves of refractory metals of the platinum group (except osmium) are quite adequate for them to find extensive technical application. O. E. Zvyagintseva [14] has dealt with problems of the production and uses of platinum metals.

Tungsten and molybdenum belong to the most important metals for modern technology. Although both metals are considered to be rare metals (they are included by some in what are called the less-common metals), their reserves are relatively large. The tungsten content of rocks is estimated at $1 \cdot 10^{-4}\%$, and the molybdenum content at $3 \cdot 10^{-4}\%$. The most abundant tungsten minerals are wolframite $(FeMn)WO_4$ and scheelite $CaWO_4$. The most important molybdenum mineral is molybdenite $MoS_2$. In every country, there is a tendency to increase the yield of tungsten and molybdenum ores. The world production of tungsten was 60,000 tons in 1965 [15], and of molybdenum (according to concentrate content) in 1965, it was 41,708 tons [37].

The rare metals include the refractory metals of group Va, vanadium, niobium, and tantalum. These metals and their alloys are produced on an industrial scale. Niobium and tantalum are found together in nature. The niobium content of the earth's crust is estimated at $1 \cdot 10^{-3}\%$ and the tantalum content at $2 \cdot 10^{-4}\%$. The principal niobium and tantalum materials are columbite and tantalite; niobium predominates in the first and tantalum in the second. Industrially important tantalum and niobium minerals are also pyrochlore, loparite, coppite, and others; the composition of these minerals includes titanium, zirconium, rare-earth metals, alkali metals, and other metals. Tantalite and columbite contain titanium, tungsten, and lead as impurities. The commonest metal of group Va is vanadium, its weight clarke is $1.5 \cdot 10^{-2}\%$; the principal vanadium minerals are vanadite and carnotite [2, 39].

Of the metals of group IVa, as already mentioned, titanium is one of the most abundant elements in nature. The weight clarkes of zirconium and hafnium are lower ($2 \cdot 10^{-2}$ and $3.2 \cdot 10^{-4}\%$). The content of these metals in the earth's crust is fairly high. In abundance in the earth's crust, zirconium ranks higher than metals such as copper, zinc, tin, lead, and nickel [2].

## Production Methods

The production of highly pure refractory metals is one of the most important problems with which metallurgists are confronted by the new technology. For solving this problem, a whole combination of the means of chemical engineering for the production of pure substances and vacuum metallurgy is employed: ion exchange chromatography, the iodide method, electrolysis, sublimation of fusible impurities, electric arc melting in a vacuum or an inert gas, vacuum electron-beam melting, zone refining, etc. [16]. The initial purity of the metals to a great extent determines their properties and behavior on melting and further treatment to produce semifinished products. Unfortunately, until very recently, questions of the purification of refractory metals (particularly tungsten and molybdenum) in the process of their production were underestimated, nor was it clear which of the impurities was the most harmful.

As is well known, the starting raw material for the production of tungsten is wolframite ($FeWO_4 + MnWO_4$) or scheelite ($CaWO_4$). The ore is roasted and crushed to a powder, which is heated in a solution of caustic soda or potassium hydroxide to produce the tungstate $K_2WO_4$ or $Na_2WO_4$. The latter is crystallized out and separated. In some cases, the tungstate crystals are redissolved in water and recrystallized. By reaction between an aqueous solution of $K_2WO_4$ and hydrochloric acid, tungstic acid $H_2WO_4$ is obtained in the form of a yellow precipitate. The $H_2WO_4$ is dissolved in ammonia water to form ammonium paratungstate $5(NH_4)_2O \cdot 12WO_2 \cdot nH_2O$ in crystal form. On being heated to above 250°C, the crystals are converted to $WO_3$. To remove impurities (iron, silicon, arsenic, manganese, molybdenum, etc.) in a number of cases, the tungstic anhydride is dissolved in KOH, and double (or multiple) precipitation is effected by means of hydrochloric acids. The $WO_3$ is reduced to metallic tungsten in a current of hydrogen at 500-850°C. The tungsten powder is subsequently pressed into rods using a carbonaceous binder followed by sintering and welding in hydrogen or a vacuum at a temperature of up to 3000°C.

The initial raw material for the production of molybdenum is molybdenite with a content of 60-95% $MoS_2$. The principal impurities in molybdenite are $SiO_2$, $Fe_2O_3$, Cu, CaO. By roasting the molybdenite at 600°C, $MoO_3$ is produced which is then dissolved in an aqueous solution of ammonia, the insoluble impurities being removed:

$$MoO_3 + 2NH_4OH = (NH_4)_2MoO_4 + H_2O.$$

On the addition of concentrated nitric acid, molybdic acid is obtained in the precipitate:

$$(NH_4)_2MoO_4 + 2HNO_3 = H_2MoO_4 + 2NH_4NO_3.$$

After washing the precipitate, the latter is redissolved in ammonia water, concentrated, and filtered. As result, ammonium paramolybdate $3(NH_4)_2O \cdot 7MoO_3 \cdot 4H_2O$ is formed. By roasting at 700°C, the crystals are converted to $MoO_2$, which is reduced by hydrogen to metallic molybdenum. The latter is then extruded to form bars which are sintered and welded. Industrial methods of breaking down tungsten and molybdenum minerals and the production of metallic tungsten and molybdenum are described in the literature [17-22].

Considerable progress is being made in research on the development of methods of producing high-purity tungsten and molybdenum to satisfy the demands of the new technology, while being also of considerable scientific interest. The degree of purity of tungsten and molybdenum powder is determined by the purity of the oxides of these metals and the conditions under which these oxides are reduced. In the process of producing compact tungsten and molybdenum, the latter become contaminated with carbon because of the binders used in extrusion (solution of glycerol in alcohol or of paraffin in benzene). Quite recently, it has been found that an increase in carbon content produces brittleness in tungsten [23] and considerably reduces the ductility of molybdenum. Adequately complete removal of carbon by subsequent treatment (arc melting or electron-beam melting) gives rise to exceptional technical difficulties. In the production of high-purity tungsten and molybdenum, therefore, it is very important not to allow contamination by carbon in the technical process of the production of these metals. In the production of rods by noncrucible zone refining using an electron beam, it is obvious that quite an effective means may be to extrude the powders without carbonaceous binder. The experience of the authors' work has shown the effectiveness of this measure for the production of polycrystalline high-purity tungsten and molybdenum, which consequently possess high technological ductility. The production of high-purity tungsten and molybdenum and their subsequent zone refining makes it possible to study more reliably the most important physicochemical properties of these metals, and to solve a number of fundamental scientific problems.

Of the new methods which are being developed for the production of high-purity tungsten and molybdenum, reference should be made to the production of granules of these metals and their halides. In [24], the starting material for the production of high-purity tungsten was a tungsten powder from which tungsten hexafluoride $WF_6$ was produced. This compound was separated from the nonvolatile fluorides and reduced by hydrogen in a "fluidized bed" to tungsten particles acting as seed crystals. The tungsten granules were then subjected to sintering, rolling, forging, and other operations. Data are available on the production of high-purity tungsten and molybdenum by dissociation of halogen compounds [25].

As already mentioned, the starting raw materials for the production of tantalum and niobium are tantalite and columbite. Tantalite and columbite are broken down by fusion with alkalies NaOH, KOH, or decomposition by hydrofluoric acid is used.

According to their mineralogical and chemical compositions, ore concentrates are converted to oxides ($Ta_2O_5$, $Nb_2O_5$), complex fluorides ($K_2TaF_7$ and $K_2NbF_7$), and chlorides. Tantalum and niobium are produced by the reduction of these compounds (the halides are reduced by sodium, magnesium, or calcium; the oxides, by carbon), and electrolysis of molten salts. The production of these metals of high purity gives rise to some difficulty owing to their high chemical activity with respect to hydrogen, oxygen, nitrogen, and carbon. The purest metal may be produced by reduction of the complex fluorides $K_2NbF_7$ or $K_2TaF_7$ by metallic sodium at 900°C [26-29]. Considerable difficulties occur in the separation of niobium and tantalum, owing to the similarity of the properties of these metals and their chemical compounds. The separation of niobium and tantalum is also accompanied by their purification. The principal methods used for the separation of niobium and tantalum compounds are fractional crystallization of the complex fluorides, extraction, rectification of the chlorides, and separation by means of ion-exchange resins [26-29]. Recently, interest has been aroused in niobium pentachloride as starting material for the production of metallic niobium [30].

The starting material for the production of the other group Va metal, vanadium, is vanadium pentoxide which is reduced by aluminum, calcium, or carbon. High-purity vanadium may be produced by thermal dissociation of vanadium iodide $VI_2$.

Questions of the breakdown of minerals of group IVa metals, the separation of zirconium and hafnium and the production of the pure metals are described in detail in [31-33].

Alongside the above-mentioned methods of the treatment of ore concentrates of group Va metals, a start has been made with the direct production of alloys by the reduction of the ore concentrates in the electric arc furnace. Numerous investigations in the USSR and elsewhere have shown that remelting of refractory metals, especially of group IVa and V metals (zirconium, hafnium, vanadium, niobium, tantalum), in the electric arc furnace may remove interstitial impurities, including oxygen, by the evaporation of low oxides in a vacuum.

The question of the high purification of refractory metals (especially group VIa metals) is one of the most important in the problem of refractory metals. Cermet methods do not ensure the high degree of purity of these metals necessary for their use in electronic and other branches of technology. As a method of increasing the purity and ductility of group VIa refractory metals, exceptional interest is presented by electron-beam zone refining [16]. For the production of high-purity tungsten, molybdenum, and other refractory metals, a start has been made in the application of the methods of plasma metallurgy [40, 41], a uniform, fine-grain structure being obtained. According to [34], components of tungsten weighing up to 15 kg, for example, the jet nozzles of spacecraft, are made by the new process.

Information is given in [35] relating to the construction of a high-temperature plasma induction furnace for growing crystals of refractory metals. The apparatus ensures the production of a controllable high-temperature plasma and may be used both for spheroidization and the deposition of coatings from the vapor phase.

The theoretical principles and methods of purifying a large number of metals by the vacuum distillation method are contained in the monograph by V. E. Ivanov et al. [36]. Zone refinement of tungsten anhydride is discussed in [38].

## LITERATURE CITED

1. A. P. Vinogradov, Geochemistry of Rare and Dispersed Chemical Elements in Soils [in Russian] (2nd ed.), Izd. Akad. Nauk (1957).
2. A. A. Saukov, Geochemistry [in Russian], Gosgeolizdat (1950).
3. V. Noddak and I. Noddak, Basic Ideas of Geochemistry [in Russian], Vol. 2, Goskhimtekhizdat, Leningrad (1935).
4. K. B. Lebedev and S. A. Ageev, Izv. Akad. Nauk Kaz.SSR, Seriya Metallurgii, No. 1 (10), p. 48 (1961).
5. V. V. Rodzaevskii, Metallurgicheskaya i Kimicheskaya Promyshlennost' Kazakhstana, No. 3 (7), p. 49 (1960).
6. K. B. Lebedev, Rhenium [in Russian], Metallurgizdat (1963).
7. R. I. Jaffee, D. J. Maykuth, and R. W. Douglass, Rhenium and Refractory Metals of the Platinum Group [Russian translation], IL (1963).
8. S. K. Kalinin, Vestnik Akad. Nauk. Kaz.SSR, No. 9 (18), p. 40 (1946).
9. T. A. Satpaeva, Mineralogical Features of Deposits of the Type of Copper-Bearing Sandstones [in Russian], Izd. Akad. Nauk Kaz.SSR, Alma Ata (1958).
10. K. B. Lebedev. Vestnik Akad. Nauk Kaz.SSR, No. 8, p. 58 (1960).
11. R. G. Petersen, J. C. Hamilton, and A. T. Myers, Econ. Geol., 54(2):254 (1959).
12 V. A. Kuznetsova and A. A. Saukov, Geokhimiya, No. 9, p. 750 (1961).
13. I. I. Chernyaev and A. M. Rubinshtein, Dokl. Akad. Nauk SSSR, 48(5):353 (1945).
14. O. E. Zvyagintseva, Zh. Prikl. Khimiya, 32(9):1889 (1959).
15. P. Grainger, Mining J. Annual Rev., 55 (1966).

16. E. M. Savitskii, G. S. Birkhanov, and Ch. V. Kopetskii, Izv. Akad. Nauk SSSR, Metallurgie i Gornos Delo, No. 6, p. 12 (1963).
17. K. Agte and I. Vatsek, Tungsten and Molybdenum [in Russian], Izd. "Energiya" (1964).
18. C. J. Smithells, Tungsten: A Treatise on its Metallurgy, Properties and Applications (3rd ed.), Chemical Publishing Co., New York (1952).
19. A. N. Zelikan, Metallurgy of Tungsten and Molybdenum [in Russian], Metallurgizdat (1960).
20. G. I. Abashin and G. M. Pogosyan, Technology of Tungsten and Molybdenum Production, Metallurgizdat (1960).
21. L. Northcott, Molybdenum, Academic Press, New York (1956).
22. V. Espe, Technology of Electrovacuum Materials, Vol. 1, Gosenergoizdat (1962).
23. E. M. Savitskii and G. L. Tsarev, Vestnik Akad. Nauk BSSR, Ser. Fiz.-tekhn. Nauk, No. 2, p. 84 (1965).
24. "Pure tungsten granules made by gas reduction," Steel, 153(19):102 (1963).
25. V. S. Emel'yanov, A. I. Evstyukhin, G. A. Leont'ev, and A. N. Semenikhin, in: Metallurgy and Metallography of Pure Metals [in Russian], No. 4, Atomizdat (1963).
26. G. S. Savchenko and I. V. Tananaev, Zh. Prikl. Khim., No. 5, p. 385 (1947).
27. G. A. Meerson, G. L. Zverev, and F. M. Zubkova, Tsvetnye Metally, No. 8, p. 97 (1939).
28. S. M. Chernoborov and N. P. Kolonina, in: Chromatography, Its Theory and Application [in Russian], Izd. Akad. Nauk SSSR (1960).
29. L. A. Nisel'son, Zh. Neorg. Khim., 3(12):2603 (1958).
30. T. R. Ingraham and B. S. P. Whalley, Canad. J. Chem. Engng., 41(6):265–268 (1963).
31. N. P. Sazhin and E. A. Pepelyaeva, in: Researches in the Field of Geology, Chemistry and Metallurgy [in Russian], Izd. Akad. Nauk SSSR, p. 152 (1955).
32. F. Hadsuel and J. K. Hatchen, in: Chemistry of Nuclear Fuel, Metallurgizdat, p. 533 (1956).
33. L. A. Nisel'son, in: Separation of Rare Metals of Similar Properties [in Russian], Metallurgizdat, p. 168 (1962).
34. "Automatisme à l'usine et au bureau, 9(6):15 (1964).
35. Semicond. Prod. and Solid State Technol., 7(9):2 (1964).
36. V. E. Ivanov, I. I. Papirov, G. F. Tikhvinskii, and V. M. Amonenko, Pure and Ultrapure Metals [in Russian], Izd. "Metallurgiya" (1965).
37. Express Information "Tsvetnaya Metallurgiya," No. 29 (Abstract 113) (1966).
38. V. V. Marychev, Dokl. Akad. Nauk SSSR, 159(2):416 (1964).
39. A. N. Zelikman, O. E. Krein, and G. V. Samsonov, Metallurgy of the Rare Metals [in Russian], Izd. "Metallurgiya" (1964).
40. I. D. Kulagin and A. V. Nikolaev, Izv. Akad. Naúk SSSR, OTN, Metallurgiya i Toplivo, No. 2, p. 78 (1959).
41. N. N. Rykalin, A. V. Nikolaev, and I. D. Kulagin, Teplofizika Vysokikh Temperatur, Doklad. Akad. Nauk SSSR, No. 6, p. 871 (1965).

CHAPTER II

# INTERATOMIC BOND, CRYSTAL STRUCTURE, AND PRINCIPAL PHYSICAL PROPERTIES OF REFRACTORY METALS

## Interatomic Bond in Transition Metals

The refractory metals belong to the transition elements, which are characterized by high strength of the interatomic bond and, consequently, by high melting point, mechanical strength, and electrical resistance. As is well known, the nature of the interatomic bond is one of the main factors which determine the crystalline structure and physical properties of metals and alloys.

The character and strength of the interatomic bond in metals and metallic phases are determined by the energy state of the electrons in the crystal lattice. The problem of interatomic cohesion and electron distribution is exceptionally complex and calls for special study. In this monograph, only the principal essential concepts and approaches in estimating the forces of interatomic reaction will be outlined. According to modern classification, four types of interatomic bond are distinguished: metallic, ionic, covalent (or atomic), and the van der Waals bond. The metallic bond occurs in approximately 70% of the elements. Electrons are able to pass from one atom to another to a certain degree. The character of this transition corresponds to a definite type of interatomic bond; where valence electrons are attracted completely from one atom to another an ionic bond is found; the union of the electrons of two atoms in bonding pairs characterizes the covalent bond; in the case of complete detachment of all the valence electrons from the atoms and their distribution in the interionic space, a metallic bond occurs; and, finally, the presence of a slight displacement of the electrons, resulting in the occurrence of weak polar forces, is characteristic of the van der Waals bond [1]. These four extreme types of interatomic bond are rarely found in the pure form; in many substances, including metallic alloys and compounds, the bond is more complex and mixed, one of the types of bond occurring to a greater degree than the others. The strongest bond is the covalent bond, and the weakest is the van der Waals bond.

In transition metals, metallic solid solutions and compounds, in addition to a strongly pronounced metallic bond, there also occurs the covalent bond and sometimes also the ionic bond. The covalent bond, in particular, occurs in metallic compounds, many of which have a high melting point, high electrical resistance, and high hardness. Evidently, the covalent bond also plays an important part in some refractory transition metals. With increase in temperature in metals, particularly those having polymorphic transformations, and in metallic compounds, on account evidently of the destruction of the covalent electron "bridges" by the thermal vibrations of the atoms, the proportion of the metallic bond increases and they become more ductile and electrically conductive.

According to the definition of N. V. Ageev [1], crystals are accumulations of atoms arranged in the form of a space lattice at distances of the order of several angstroms. The conditions here differ sharply from the conditions under which free atoms are situated, but the

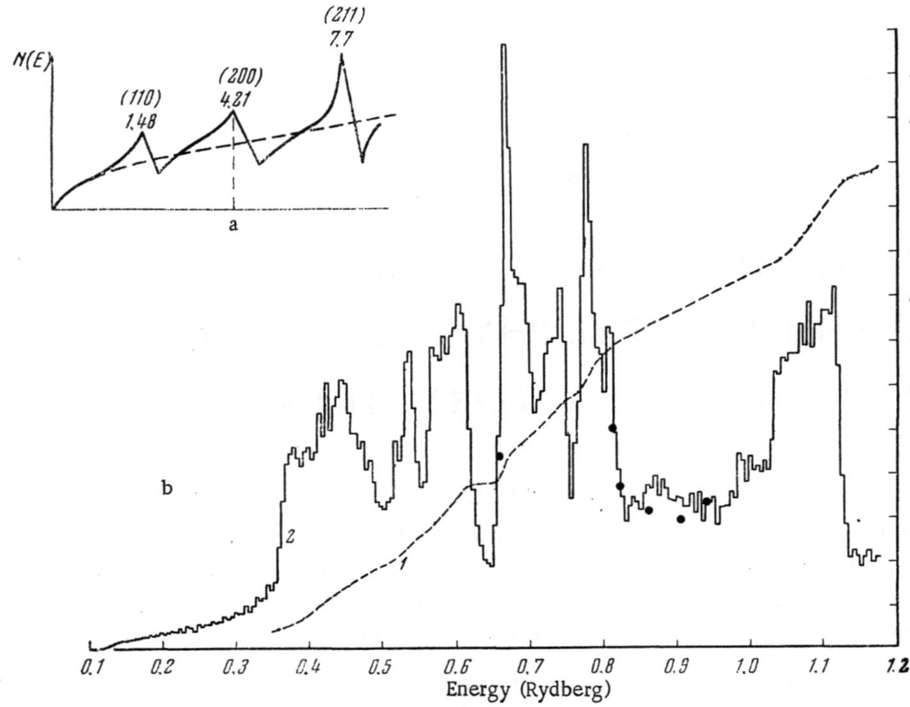


Fig. 2. Diagram of density of states for the body-centered cubic lattice (a), density of electron states for hexagonal rhenium according to Mattias (b).

energy states of the electrons in the periodic field of the crystal may be regarded as resulting from the energy states of the free atom. The discrete, particularly the outer levels of the free atoms on the approach of the latter in the formation of the crystal are spread out in wide energy bands (Brillouin zones). Comparison of the energy spectra of a metal and a nonmetal shows that the electrons in both crystals are distributed according to their energies in bands or zones. The form and character of these bands depend on the crystal structure and may be deduced from considerations of the interaction of the electronic waves with the atomic planes of the crystal. The filling of an energy zone is different for a metal and a nonmetal. In metals, the outer zones are incompletely filled, and may accept additionally a certain number of valence electrons on the introduction into the crystal of atoms of another element of higher valence. At the same time, the energy of the electrons gradually varies owing to the zones being unfilled, or as the result of the overlapping of individual zones for some crystal lattices. In the case of a nonmetal, the individual zones are completely filled with electrons, and any further increase in their number requires a considerable discontinuous increase in energy for filling the next zone. Due to these peculiarities of filling the energy zones metals tend to form wide ranges of solid solution.

Figure 2a shows a diagram of the density of states for the body-centered cubic lattice, characteristic of many refractory metals [2]. The maxima of the curve correspond to the filling of the first, second, and third zones with a concentration of electrons per atom of 1.48, 4.21, 7.7. The energy jumps correspond to the interaction of electrons with the (110), (200), and (211) lattice planes. Figure 2b shows the density of the states for hexagonal rhenium according to data obtained by Mattias.

Among the successes of the zone theory must be included the explanation of the cause of the formation of "electronic compounds" in alloys, the determination of the limits of homogeneous regions in alloys ($\alpha$, $\beta$, $\gamma$, $\varepsilon$-phases), the determination of the existence of a limit number of electrons per cell in filling the Brillouin zones, the explanation of the causes of the formation of imperfect lattices [3]. The conditions for the formation of imperfect lattices in the case of electronic compounds were examined by S. T. Konobeevskii [4, 5]. Subtraction solid solutions (or solid solutions with imperfect lattices) were examined by E. S. Makarov [6]. Owing to the zone theory, it was shown that group VII metals in the majority of metallic compounds, formed by them with monovalent metals, are zerovalent elements, due to the filling of the d-levels by excess electrons. However, the Brillouin zone theory is not universal and cannot give a reply to all questions connected with the nature of the metallic state. As pointed out by S. M. Konobeevskii [3], up till now, this theory has failed to provide any definite bases for understanding the crystal structure of chemical elements, for determining the qualitative connection between structure and valence. The application of the zone theory to the explanation of the nature of the interatomic bond in transition metals has met with considerable difficulty. The zone theory has been found unable to explain the high bond energy and refractory nature of the transition metals, as well as the proneness of some refractory metals to cold brittleness.

There are other approaches to an explanation of the specific characteristics of the transition metals [7]. Thus, for example it is considered that the high bond energy of the transition metals is due to participation in the bonds by d-electrons on hybrid spd-orbitals [8].

The higher values of the electronic specific heat of transition metals is regarded as direct proof of the participation by d-electrons in the collectivized system of electrons in these metals [9]. In the theory of Mott and Jones [2], as a first approximation, division of the s- and d-parts of the hybrid state is assumed. The outer electrons of the elements of the first long period are considered to be in two different but intersecting bands, formed of 3d- and 4s-states of free atoms. In the Mott and Jones theory, the bonding forces are assumed to be due only to 4s-electrons. As pointed out by Hume-Rothery [11], this theory was developed to explain the properties of group VIII elements and it cannot be applied to transition metals of groups IIIa-VIIa, since for metals of these groups, division into separate s- and d-bands is not justified. In the present case, both s- and d-electrons participate in the bonds.

In the Pauling theory [12], it is assumed that bond formation in transition metals is due to electrons in s,p,d-states, and not merely to electrons in s-states. Contrary to the attempts to construct a theory of metals based on the method in which the combination of the electrons is regarded as a whole and each electron cannot be attributed to a definite atom, Pauling advanced the theory of resonant valence bonds. He considered the metallic bond as the result of resonance between different pairs of atoms. According to Pauling, when atoms enter the crystal lattice of a metal their electronic structure is altered, the electrons responsible for the bonds being on hybrid spd orbitals. According to this theory, in metals of the first long period, there are nine orbitals suitable for the formation of hybrid bonds: five 3d, one 4s, and three 4b. There are (according to Pauling) 2.44 atomic d-orbitals and 2.56 d-orbitals, which are bonding orbitals, forming with one s and three p-orbitals 6.56 hybrid bonding spd orbitals. According to Pauling, in the first long period from calcium to vanadium, there is a continuous increase in the number of electrons per atom participating in bonds (from one to five). Pauling assumes that in chromium 5.78 electrons per atom participate in bonds and 0.22 electron per atom is situated on atomic 3d-orbitals; on passing further along the period, the number of electrons participating in the formation of a bond remains constant at 5.7 per atom. A critical analysis of Pauling's theory is given by Konobeevskii [3].

Hume-Rothery [11], analyzing Pauling's views on the nature of the interatomic bond in transition metals, observes that Pauling correctly pointed out the fact that the electrons responsible for the bond are on hybrid spd orbitals, but that Pauling's view concerning the constant number of bonding electrons per atom from group VIa to group VII was not in agreement with the character of the change in physical properties. Furthermore, as pointed out by Hume-Rothery, this theory does not explain the differences in the properties of the transition metals of the first and succeeding long periods. Hume-Rothery, Irving, and Williamson [13] proposed

TABLE 1. Electron Structure of Atoms of the Transition Metals of the First Large Period of the Mendeleev Periodic System

| State of the atom | Electron levels | Number of electrons | | | | | | | |
|---|---|---|---|---|---|---|---|---|---|
| | | 21Sc | 22Ti | 23V | 24Cr | 25Mn | 26Fe | 27Co | 28Ni |
| Isolated | 3 | 2 | 2 | 2 | 2 | 2 | 2 | 2 | 2 |
| | 3 | 6 | 6 | 6 | 6 | 6 | 6 | 6 | 6 |
| | 3 | 1 | 2 | 3 | 5 | 5 | 6 | 7 | 8 |
| | 4 | 2 | 2 | 2 | 1 | 2 | 2 | 2 | 2 |
| In the metal crystal | 3 | 2 | 2 | 2 | 2 | 2 | 2 | 2 | 2 |
| | 3 | 6 | 6 | 6 | 6 | 6 | 6 | 6 | 6 |
| | 3 | 0 | 0 | 0 | 0.2 | — | 2.3 | 8.4 | 9.7 |
| | 4 | 3 | 4 | 5 | 5.8 | — | 5.7 | 0.6 | 0.3 |

another valence system in better agreement with the character of the variation in physical properties of the transition metals according to the three large periods. By valence, these authors understand the number of electrons per atom which participate in covalent bonds. Thus, they take as basis the values of the valences of elements as adopted in inorganic chemistry. The valence here increases to a maximum in group VIa (in agreement with the deductions of Pauling's theory), and then gradually decreases (contrary to Pauling's assumption of constant valence of the elements in groups VIIa and VIII). The valences of transition metals of groups Va, VIa VIIa, and VIII also increase on passing to succeeding long periods. For example, the trivalent chromium compounds are more stable than the hexavalent compounds and a valence of six is more characteristic of tungsten. This valence increase is in agreement with the increase in melting point and other physical properties, characteristic of the interatomic bond.

The electronic structures of the atoms in crystal lattices of the transition metals of the first long period are given in [14]; the maximum interatomic bond is found in chromium, the atoms of which in the crystal lattice have the maximum number of external electrons participating in the bond (Table 1). On passing further along the series a decrease is observed in the number of outer electrons in the atoms of metal crystals, in other words, a decrease in the number of d-electrons participating in the bond, and correspondingly a weakening of the interatomic bond. As will be clear from the above Table 1, the distribution of electrons among the levels in the free atoms of the gas of transition metals differs substantially from the corresponding distribution in the crystal state. The configuration of the outer electron shells of the atoms in the crystal lattice was also determined by Weiss and de Marco [15]. Unfortunately, the electron structures of the atoms (in the metal crystal) of transition metals in the second and third large periods are unknown. It is possible that a knowledge of these structures would in many respects permit an explanation for the increase in the interatomic bond in the transitions vanadium → niobium → tantalum, chromium → molybdenum → tungsten. A detailed review of modern concepts on the electron structure of the transition metals is given in [9] and the most recent approaches (quasiparticles) in [169, 170].

The interatomic bond is produced by electrons, and therefore the problem of determining the strength of the interatomic bond is reduced to a determination of the distribution of the atoms among the atoms in the crystal lattice of the metal or alloy. Direct methods of determining electron distribution include the method of x-ray emission spectra [16], the x-ray determination of electron density (Fourier series) [17], neutron diffraction determination of the spin superstructure, resonance methods, and the experimental determination of the Fermi surface in metals [18-20], while the indirect methods are methods of determining the physical properties. The strength of the interatomic bond is estimated approximately by such physical properties as ionization potential, melting point, recrystallization temperature, Debye temperature, temperature coefficient of linear expansion, elastic modulus, yield point, activation energy of diffusion and self-diffusion, heat of sublimation, etc. The bond strength is considered to increase with increase in each of the above-mentioned values, except the coefficient of linear expansion, which decreases with increase in the interatomic bond.

The ionization potential is characterized by the ease with which the atoms release electrons and may consequently be used as a quantitative characteristic of the metallic properties of the atom. The ionization potential belongs to the periodic functions of the atomic number of an element. For metals, the ionization potential of the detachment of the first electron from an atom is < 10 eV. Lower values correspond to the alkali metals and the highest to the inert gases. Between these extreme limits, the ionization potentials in each period increase gradually.

The properties characterizing the strength of the interatomic bond depend mainly on the structure of the outer electron shells of the atoms of the elements. Since the electron shells of the atoms periodically alter their structure, the physical properties also are in periodical dependence on the atomic number of the element. There exists a definite relationship between the physical properties of the metals; for example Frenkel' [21] showed that the coefficient of linear expansion is inversely proportional to the elastic modulus:

$$\alpha \approx \frac{k}{R_0^3 E},$$

where $R_0$ is the equilibrium distance between the atoms, E is the elastic modulus, $\alpha$ is the coefficient of linear expansion, and k is the Boltzmann constant.

There is a connection between the melting point and the coefficient of linear expansion [22]. For metals with a cubic lattice, such as is the case with many refractory metals, the coefficient of linear expansion is connected to the melting point on the absolute temperature scale by the simple relationship $\alpha \cdot T_{mp} \approx 0.02$.

Curves of the variation of melting point, boiling point, shear modulus, coefficient of linear expansion [56] and coefficient of secondary emission of the transition metals as a function of the position of the metals in the periodic system (Fig. 3) show that the interatomic bond attains a maximum value in group VIa, and then diminishes in groups VIIa and VIII. In the limits of each period, the melting point of the metals varies as a function of the number of electrons situated in the s,p,d-shells of the atoms, and attains a maximum value in group VIa metals: chromium, molybdenum, tungsten, with six electrons in the outer shells. In addition, the melting point of the metals increases when their vertically downward position in the groups of the periodic system is considered; it is evident that this may be due to the partial participation of inner electrons of the atoms in the interatomic bond, and the difference in the electronic structure of the atoms (in the metal crystal) of transition metals of the first, second, and third periods. Approximately the same character is shown by the variation in the elastic modulus and a number of other properties; an inverse function is found in the variation of the coefficient of linear thermal expansion and in the interatomic distances. The relation between the strength of the interatomic bond and interatomic distance is of great importance. The electrostatic forces acting between the atoms is inversely proportional to the square of the distance between them. Thus, the maximum interatomic strength is found in metals having the smallest atomic dimensions. In group VIa, the coefficient of linear expansion and the interatomic distance have the minimum values. There is good correlation between all the properties examined in the course of their variation according to the periodic system. The variation in elastic modulus alone falls somewhat out of the general run; its maximum value does not occur in tungsten

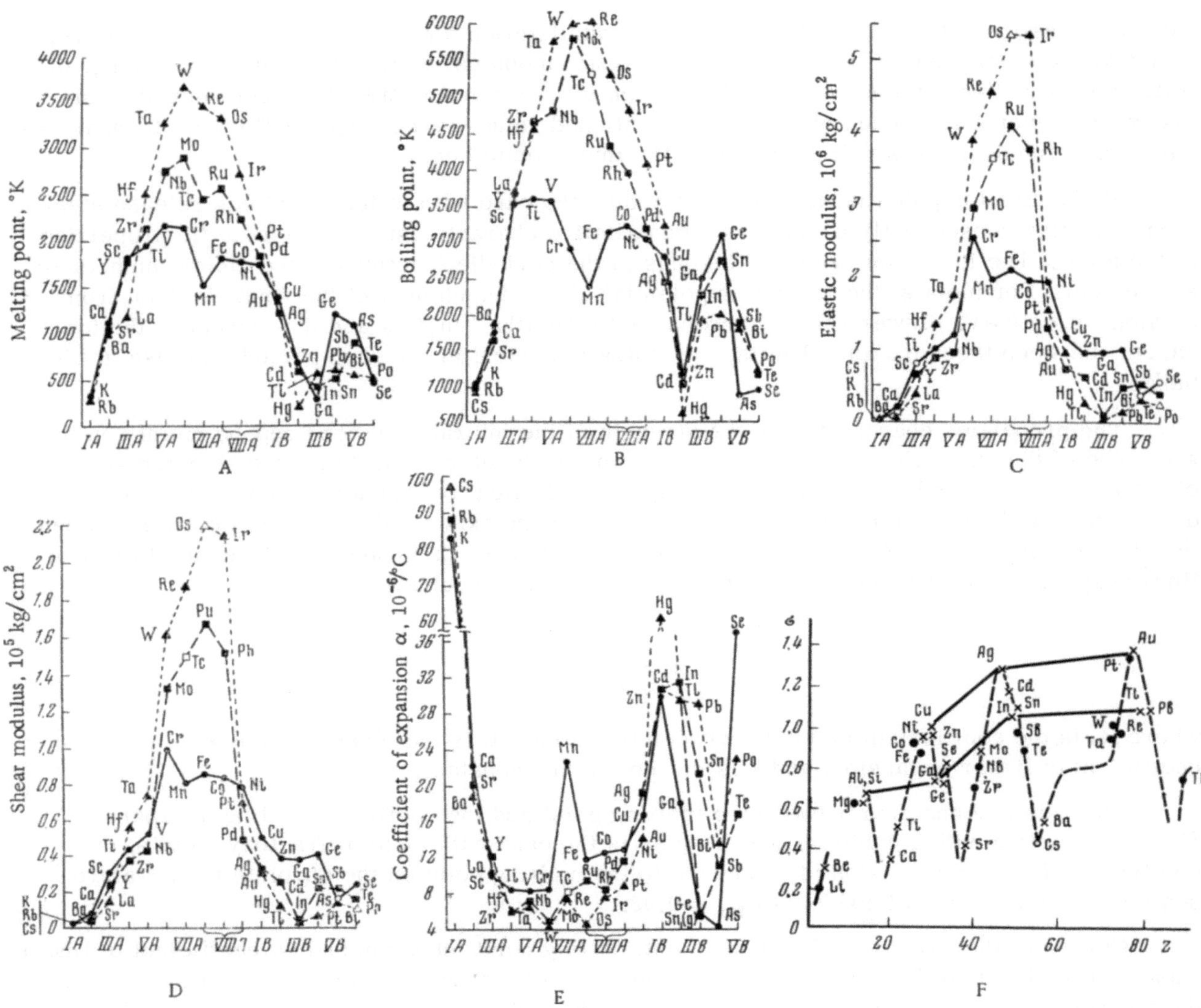

Fig. 3. Variation in melting point, boiling point, elastic modulus, shear modulus, coefficient of linear expansion [56], and coefficient of secondary emission of the transition metals as a function of their position in the periodic system.

(35,000–41,530 kg/mm²), as was to be expected for complete agreement with variation in melting point and coefficient of linear expansion, but in osmium (56,000 kg/mm²).

It may thus be concluded that the maximum of the strength of the atomic bond occurs in metals situated in the middle of the third long period – tantalum, tungsten, rhenium, osmium – having a melting point of 3000°C or over. These and the other transition metals are characterized by maximum values of elastic modulus (yield point), Debye temperature, recrystallization temperature, and minimum values of the coefficient of linear expansion, and also minimum values of the interatomic distances.

Considerable interest is presented by the methods of estimating the strength of the interatomic bond in metals from the variation of some of the physical properties, as discussed by Gilman [23]. The simplest and roughest method is to assume that the strength of the interatomic bond is equal to some part of the elastic modulus in a definite crystallographic direction ($E_{hkl}$). Usually, the bond strength is assumed to be equal to 10% of the value of the elastic modulus (Fig. 4a).

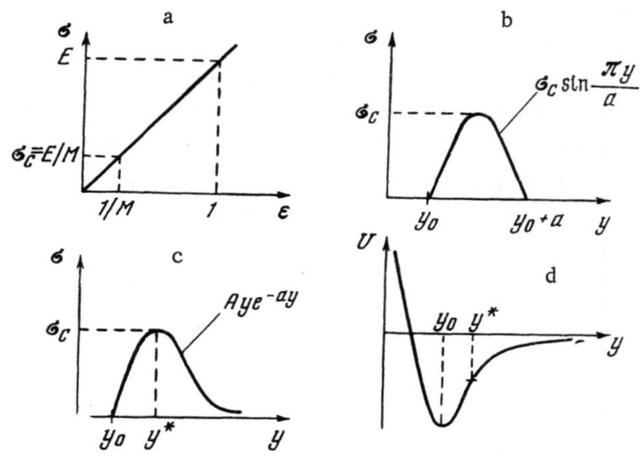

Fig. 4. Diagrams showing strengths of the interatomic bonds as estimated from measurements of various physical properties.

A second method is based on the fact that some resultant force acting on atoms is equal to zero in the absence of externally applied stresses and attains a maximum value with increase in the distance of the atoms from each other, and then falls to zero again as the crystal fractures (Fig. 4b). The existing relation may be expressed by the formula

$$\sigma = \sigma_c \sin \frac{\pi (y - y_0)}{a},$$

where $\sigma_c$ is the direction of the bond, $y_0$ is the initial distance between the atoms, and $a$ is the radius of action of the interatomic forces.

For $y \to y_0$ it may be assumed that $\sigma = \sigma_c \frac{\pi (y - y_0)}{a}$, and from Hookes law $\sigma = E_{hkl} \frac{y - y_0}{y_0}$. Consequently

$$\sigma_c \frac{\pi (y - y_0)}{a} = E_{hkl} \frac{y - y_0}{y_0},$$

whence

$$\sigma_c = \frac{E_{hkl} a}{\pi y_0}.$$

A third method connects the bond strength with the surface energy $\gamma$ of planes along which shear of the crystal occurs under the action of externally applied forces (Fig. 4c):

$$\sigma_c = \frac{1}{e} \sqrt{\frac{2E\gamma}{y_0}} = 0.52 \sqrt{\frac{E\gamma}{y_0}}.$$

A fourth method uses the well-known dependence of the bond energy on the interatomic distance (Fig. 4d). Gilman gives the following final formula, connecting the bond energy with the coefficient of linear expansion:

$$\sigma_c = \frac{3k}{8\alpha_0 V_0},$$

TABLE 2. Crystal Structure of Polymorphic Metals

| Metal | Low-temperature modification | Medium-temperature modification | High-temperature modification |
|---|---|---|---|
| Beryllium | Hexagonal | — | bcc |
| Calcium | fcc | Hexagonal | " |
| Strontium | " | " | " |
| Scandium | Hexagonal | — | " |
| Yttrium | " | — | " |
| Lanthanum | " | fcc | " |
| Cerium | " | " | " |
| Praseodymium | " | — | " |
| Neodymium | " | — | " |
| Samarium | Rhombohedral | — | " |
| Gadolinium | Hexagonal | — | " |
| Terbium | " | — | " |
| Dysprosium | " | — | " |
| Holmium | " | — | " |
| Erbium | " | — | " |
| Thulium | " | — | " |
| Ytterbium | " | — | " |
| Luthenium | " | — | " |
| Thallium | " | — | " |
| Titanium | " | — | " |
| Zirconium | " | — | " |
| Hafnium | " | — | " |
| Tin | Diamond lattice-cubic | — | Tetragonal |
| Uranium | Orthorhombic | Tetragonal | bcc |
| Neptunium | " | " | " |
| Plutonium | Monoclinic | Tetragonal and others | " |
| Polonium | Tetragonal | | fcc |
| Americium | Hexagonal | — | " |
| Manganese | Cubic complex | Cubic complex | bcc |
| Iron | bcc | fcc | " |
| Cobalt | Hexagonal | — | fcc |
| Thorium | fcc | — | bcc |
| Helium (solid) | Hexagonal | — | fcc |

where k is the Boltzmann constant, $\alpha_0$ the coefficient of thermal expansion, and $V_0$ the mean atomic volume.

It should be noted that the method considered reflects the character of interatomic bonding at temperatures above the Debye temperature.

From the results of a comparison of his calculations with the data of other writers, Gilman regards the third method gives the best estimate of the forces of interatomic bonding in a crystal.

For the further study of the nature of the interatomic bond in the transition metals and their alloys, considerable interest attaches to the extensive investigation of the elastic, electric, galvanomagnetic, thermoelectric, emission, and other physical properties of metals and alloys, as well as the study of their interrelationship.

## Crystal Structure

The crystal structure of the transition refractory metals is determined by their electronic structure and is consequently related to the position of these metals in the periodic system. Metals of group Va (vanadium, niobium, tantalum) and group VIa (chromium, molybdenum, tungsten) are monomorphic and have a body-centered cubic lattice, while rhenium, ruthenium, and osmium have a close-packed hexagonal lattice. Thus, metals with maximum strength of interatomic bond have a body-centered cubic or hexagonal lattice, in contrast to "true" metals crystallizing in a face-centered cubic lattice (packing density 74% as against 68% in the body-centered lattice). The cause of this particular spatial arrangement of the atoms of refractory metals has not yet been discovered. It is, however, indisputable that in the present case, the chief causes of the disturbance of the principle of closest packing of spheres (coordination number 12), predominant in inorganic nature are absence of spherical symmetry (pear-shaped form), typical of the outer electron d orbitals of atoms of the transition metals, and the presence of interaction between the internal electrons.

Group IIIa and IVa metals (titanium, zirconium, hafnium) are polymorphic. The low-temperature modification of the metals of these groups has a hexagonal, close-packed lattice, while the high-temperature modification has a body-centered lattice.

The different crystal structures of the modifications of a polymorphic metal result in the existence of different physicomechanical properties in the same metal. Owing to the tendency to diminished energy, the high-temperature modifications of polymorphic metals always have simpler crystal lattices than the low-temperature modifications (Table 2). Such lattices, however, are most favorable for the displacement of atoms relative to each other without disturbance of the bond between them. This enabled E. M. Savitskii [25] to formulate a rule for the plasticity of the high-temperature modifications of polymorphic metals. According to this rule, the high-temperature modifications (close to the melting point) of all polymorphic metals have a body-centered or face-centered crystal structure, owing to which and also the completeness of the metallic bond between the atoms, they possess maximum capacity for plastic deformation, i.e., to a change in shape and dimensions without fracture. The plasticity rule enables us to state that the crystal structure of the highest temperature modification of a polymorphic metal ought to be cubic and predominantly body-centered.

The above-mentioned rule was published in 1950 [25]. The crystal structure of neptunium was then yet unknown, the results of its study being published by Zachariasen in 1952 [26]. It was found that the highest temperature form of neptunium had a cubic structure, while at the ordinary temperature this metal has a complex rhombic structure. What has been said applies fully to the most important metal of atomic engineering, plutonium, the structure of which was published only in 1954 [27, 28]. The existence of the cubic body-centered structure in the high-temperature modification of americium was predicted 10 years before the experimental data were published [29]. With regard to the high-temperature forms of the heavy rare-earth metals of the yttrium subgroup – gadolinium, terbium, dysprosium, holmium, erbium, thalium, lutecium – this was possible 14 years before [30] and for samarium 4 years before [31]. It is also interesting that helium, produced in the solid state under pressure, is not exception to this rule [32].

The plasticity rule may be confidently extended also to the high-temperature forms of all transuranic (and rare-earth) polymorphic metals, which so far have been obtained only in microquantities, and whose crystal structure has not yet been determined experimentally (curium, berkelium, californium, einsteinium, fermium, mendelevium, jolium, lawrencium, promethium). At the same time, there are many reasons for believing that all (or almost all) of the rare-earth and transuranic elements will be found to be polymorphic. The plasticity rule for the high-temperature modifications of metals has a deep physical significance. It obviously points to a close connection between polymorphism and the variation in the electronic structure of atoms of polymorphic metals with temperature [33].

The refractory metals situated on the right of group VIa have a hexagonal, close-packed or cubic face-centered lattice. Thus, when the d-shells are more than half-filled with electrons, modifications with closest packings are formed; when the d-shells are half-filled or rather less than half-filled with electrons, a body-centered lattice results. The influence of d-electrons on the formation of one structure or another is underlined in [34]. Table 2 which we have compiled on the basis of information [27-31, 35-39] published toward the commencement of 1966, gives a list of polymorphic metals possessing temperature polymorphism. Polymorphism under the influence of pressure was not considered, although it is possessed by a number of metals: bismuth, antimony, cadmium, lithium, cesium, rubidium, mercury, barium, gallium, zinc. Some metals, including refractory ones (tungsten, chromium, and others), may show an uncharacteristic (for them) type of crystal lattice when produced by electrolysis. More often, the hexagonal type of lattice is found [35]. For this reason, evidently, tungsten and chromium are sometimes considered to be polymorphic. There is reason to suppose that some of these crystal lattices actually correspond to the structures of hydrogen compounds of the metals.

Table 2 shows 32 polymorphic metals, but their number is increasing, mainly on account of the rare-earth and transuranium metals as and when each of them is produced in pure form, and data on the crystal structure at different temperatures refined. Thus, about half of all the known metals possess polymorphism.

The phenomenon of polymorphism imposes definite features on the behavior of metals in their treatment under pressure or in service under certain conditions. It is therefore essential to take it into consideration in selecting the best working conditions for plastic deformation, as well as in the development of alloys based on polymorphic metals.

Polymorphic metals – titanium, zirconium, hafnium, scandium, iron, manganese, and some others – having a relatively high melting point, do not possess high-temperature strength, since at the temperatures of the polymorphic transformations, the crystal lattice of the metal is rearranged and abruptly loses the capacity for resisting external forces. This explains why heat-resistant alloys based on monomorphic nickel operate for a long time at temperatures above 900°C, while alloys based on the above-mentioned metals which, although more refractory, are polymorphic, are suitable for work only at 600-700°C.

## Physical Properties

We shall examine the fundamental physical properties of refractory metals, the most important properties from the point of view of their application in the new fields of technology as particularly heat-resistant alloys and alloys with special physical properties. It should be noted that many values of the physical properties of refractory metals quoted in scientific literature have not been determined for sufficiently pure metals and are therefore quite approximate. Further work is necessary on the refinement of the values of physical properties using high-purity refractory metals. For example, it has recently been found that the melting point of tungsten is not 3410 but 3380°C. The previously accepted value of 2750°C for the melting point of osmium has been found to be inaccurate; repeated determinations show a value of 3045 ± 30°C [40]. The elastic, electrical, emission, and other properties are also being refined. The mechanical properties of refractory metals depend very considerably on the degree of purity, as has been shown convincingly in the case of metals of groups Va, VIa, VIIa [10], as well as the rare-earth metals [41]. We shall examine more particularly the question of the influence of different impurities on the properties of refractory metals.

### 1. Density

As is well known, the density of elements is a complex periodic function of the atomic number. As is clear from Table 3, titanium (4.5 g/cm$^3$) is the lightest refractory metal. The

TABLE 3. Density of Refractory Metals

| Metal | Group of periodic system | Density, $g/cm^3$ |
|---|---|---|
| Titanium | IVa | 4.5 |
| Zirconium | | 6.5 |
| Hafnium | | 13.09 |
| Vanadium | Va | 6.1 |
| Niobium | | 8.57 |
| Tantalum | | 16.6 |
| Chromium | VIa | 7.19 |
| Molybdenum | | 10.2 |
| Tungsten | | 19.3 |
| Rhenium | VIIa | 21.02 |
| Ruthenium | VIII | 12.3 |
| Rhodium | | 12.5 |
| Osmium | | 22.5 |
| Iridium | | 22.4 |
| Platinum | | 21.5 |

majority of refractory metals, however, have a density higher than 10 $g/cm^3$ and are included in the category of heavy metals (tungsten, molybdenum, rhenium, tantalum, osmium, and others). Osmium is the heaviest metal, its density is 22.5 $g/cm^3$.

## 2. Thermodynamic Properties

For calculations of thermodynamic reactions and working out the optimum technological melting conditions for refractory metals, the following thermodynamic characteristics are important: melting and boiling points, heats of fusion and evaporation, rate of evaporation, vapor pressure of the metals, free energy of formation of the oxides, nitrides, carbides, etc. These properties are also very important from the point of view of the utilization of refractory metals in various apparatus and constructions, for example, when used in electrovacuum instruments, as constructional heat-resistant materials, etc. We have already examined the variation in the melting points of refractory metals as a function of their position in the periodic system. The following have the highest melting points: tungsten 3380°C, rhenium 3180°C, osmium 3045°C, tantalum 3000°C. These metals also have the highest boiling points (Table 4). The vapor pressure of many refractory metals has been studied in satisfactory detail over a wide temperature range. The temperature dependence of the vapor pressure of refractory metals in a vacuum is shown in Fig. 5 [48]. The lowest vapor pressure compared with the other metals is shown by tungsten. A low vapor pressure is also characteristic of other refractory metals (tantalum, molybdenum, niobium), an exception being chromium, which has high values of vapor pressure and consequently is distinguished by high volatility [49].

TABLE 4. Melting and Boiling Points, Heats of Fusion and Vaporization of Refractory Metals [40-47, 56]

| Metal | Melting point, °C | Boiling point, °C | Heat of fusion, cal/g | Heat of evaporation, cal/g |
|---|---|---|---|---|
| Titanium | 1668 | 3260 | 104 | 2350 |
| Zirconium | 1852 | 3580 | 60.3 | 1360 |
| Hafnium | 2222 | 5400 | 29.1 | 885 |
| Vanadium | 1919 | 3350 | 82.5 | 2150 |
| Niobium | 2468 | 4927 | 69 | 1782 |
| Tantalum | 3000 | 5427 | 41.5 | 995 |
| Chromium | 1875 | 2199 | 61.5 | 1474 |
| Molybdenum | 2610 | 5560 | 69.8 | 1222 |
| Tungsten | 3380 | 5900 | 46 | 1038 |
| Rhenium | 3180 | 5630 | 42.4 | 815 |
| Ruthenium | 2250 | 4900 | 60.3 | 1340 |
| Rhodium | 1960 | 4500 | 50.5 | 1150 |
| Osmium | 3045 | 5500 | 36.9 | 790 |
| Iridium | 2445 | 5300 | 32.6 | 790 |
| Platinum | 1769 | 4530 | 24.1 | 625 |

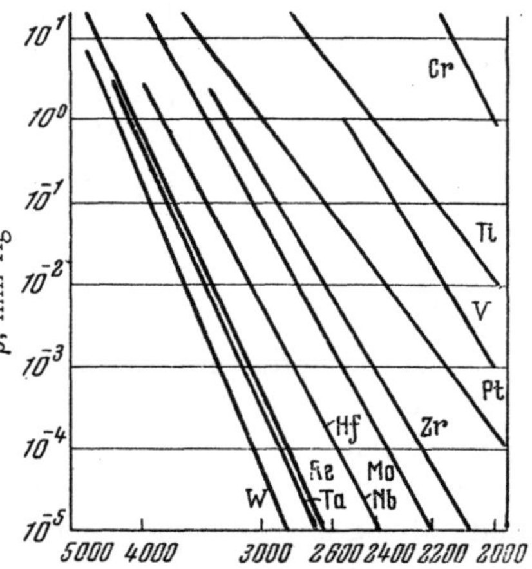


Fig. 5. Temperature dependence of vapor pressure of refractory metals in a vacuum.

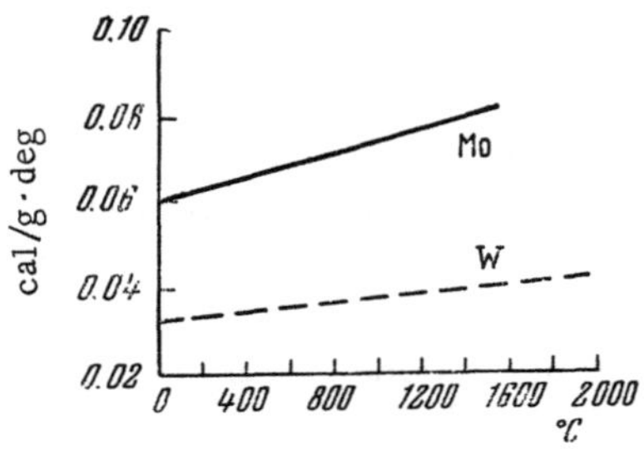


Fig. 6. Variation of the specific heat of tungsten and molybdenum as a function of the temperature.

Vapor pressure is related to the rate of evaporation by the following equation of Langmuir [50]:

$$v = P\sqrt{\frac{M}{2\pi RT}},$$

where v is the rate of evaporation, g/cm$^2$·sec; P is the vapor pressure, mm Hg; M is the molecular weight; R is the universal gas constant; and T is the absolute temperature, °K.

One of the most important thermodynamic characteristics is the specific heat, the variation of which is influenced considerably by the temperature. The nature of this influence varies in different temperature ranges. For metals in the solid state in the temperature range from absolute zero to room temperature, the specific heat varies according to the equation of a cubic parabola. At temperature close to room temperature, the specific heat of most metals assumes a value determined by the Dulong and Petit rule. The specific heat of refractory metals at room temperature is characterized by the following data [43] (in cal/g·deg):

| | | | |
|---|---|---|---|
| Titanium | 0.125 | Tungsten | 0.032 |
| Zirconium | 0.0659 | Rhenium | 0.03262 |
| Hafnium | 0.0352 | Ruthenium | 0.057 |
| Vanadium | 0.120* | Rhodium | 0.059 |
| Niobium | 0.0642 | Osmium | 0.039 |
| Tantalum | 0.034* | Iridium | 0.032 |
| Chromium | 0.1068 | Platinum | 0.0314 |
| Molybdenum | 0.066* | | |

With increase in temperature, some increase in the specific heat is observed, which may be described by the empirical equation $c = \alpha + \beta T$. The increase in specific heat is the result of the increase in volume and the associated increase in the potential energy of the interatomic forces, and also the result of the increase in the collectivized energy of the electrons. In transition metals, the electronic component exercises the most appreciable effect on the specific heat, this component including the specific heat of both the s- and d,f-electrons.

The specific heat of molybdenum increases more rapidly with increase in temperature than does the specific heat of tungsten (Fig. 6) [50].

As pointed out by A. N. Krestovnikov [51], for metals in the liquid state, an increase in temperature up to the boiling point has no effect on the specific heat. This statement has been confirmed experimentally for a number of metals by the use of accurate calorimetric measurements of the specific heat. Data on the specific heat of some systems are to be found in [171].

## 3. Thermal Conductivity

The refractory transition metals are characterized by low value of the thermal conductivity as compared with other metals. Particularly low values of the thermal conductivity (in cal/cm·sec·deg) is found in the transition metals of group IVa, as is shown by the following

*At 0°C.

data [43, 56]:

| | |
|---|---|
| Titanium . . . . . . . . . . | 0.041 |
| Zirconium . . . . . . . . . | 0.04 |
| Hafnium . . . . . . . . . . | 0.0533 (50°C) |
| Vanadium . . . . . . . . . | 0.074 (100°C) |
| Niobium . . . . . . . . . . | 0.124 (0°C) |
| Tantalum . . . . . . . . . . | 0.13 (20°C) |
| Chromium . . . . . . . . . | 0.16 (20°C) |
| Molybdenum . . . . . . . . . . | 0.298 (204°C) |
| Tungsten . . . . . . . . . . . . | 0.31 (20°C) |
| Rhenium. . . . . . . . . . . . . | 0.17 |
| Ruthenium . . . . . . . . . . . | 0.17 |
| Rhodium. . . . . . . . . . . . . | 0.36 |
| Iridium . . . . . . . . . . . . . | 0.35 |
| Platinum . . . . . . . . . . . . | 0.17 |

The thermal conductivity of refractory metals depends to a considerable extent on their purity (Table 5), and also on the degree of deformation of the metals and the grain size.

Figure 7 [50] shows the variation in thermal conductivity of tungsten and molybdenum as a function of temperature. The thermal conductivity of tungsten and molybdenum is an important factor in calculations of the thermal balance of details of electric vacuum apparatus.

## 4. Thermal Expansion

It has already been mentioned that the refractory metals of group Va - VIIa have the lowest values of the coefficient of linear expansion compared with other elements of the periodic system, tungsten being found to have the lowest coefficient ($4.6 \cdot 10^{-6}$). The following values of the coefficient of thermal expansion ($\alpha \cdot 10^{-6}$) relate to the temperature range 0-100°C [56]:

| | |
|---|---|
| Titanium . . . . . . . . . . | 8.35 ± 0.15 |
| Zirconium . . . . . . . . . | 5.78 ± 0.07 |
| Hafnium . . . . . . . . . . | 6.01 ± 0.06 |
| Vanadium . . . . . . . . . | 8.3 |
| Niobium . . . . . . . . . . | 7.07 ± 0.05 |
| Tantalum . . . . . . . . . . | 6.55 ± 0.05 |
| Chromium . . . . . . . . . | 8.4 |
| Molybdenum . . . . . . . . | 4.98 ± 0.15 |
| Tungsten . . . . . . . . . . | 4.59 ± 0.03 |
| Rhenium . . . . . . . . . . | 6.63 ± 0.06 |
| Ruthenium . . . . . . . . . | 9.36 ± 0.27 |
| Rhodium . . . . . . . . . . | 8.40 ± 0.10 |
| Osmium . . . . . . . . . . | 4.7 ± 0.1 |
| Iridium . . . . . . . . . . . | 6.63 ± 0.12 |
| Platinum . . . . . . . . . . | 14.1 ± 0.1 |

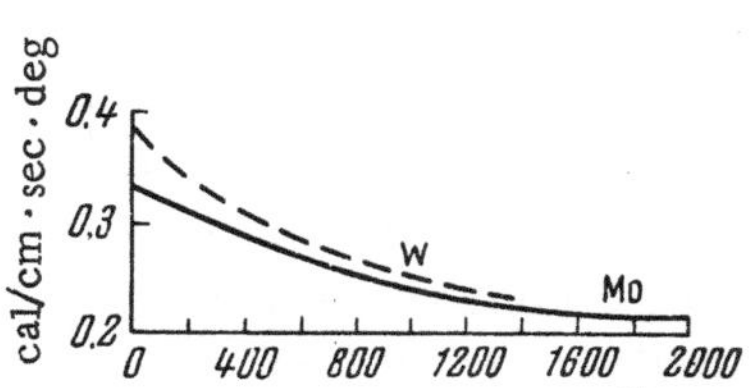


Fig. 7. Variation in thermal conductivity of tungsten and molybdenum as a function of temperature.

The degree of purity of a metal appreciably affects the value of the coefficient of linear expansion. This relationship may be illustrated by the example of niobium (Table 6).

The resistance to high-temperature heat cycles is connected with the resistance to thermal fatigue, which is a direct function of the coefficient of thermal expansion.

The coefficient of thermal expansion is very important in estimating the stresses in a weld between two different materials.

TABLE 5. Thermal Conductivity of Niobium of Different Degrees of Purity at Temperatures 0-600°C [52, 53]

| Purity of the metal, % | Thermal conductivity, cal/cm · sec · deg | | | | | | |
|---|---|---|---|---|---|---|---|
| | 0° | 100° | 200° | 300° | 400° | 500° | 600° |
| 99.90 | 0.125 | 0.130 | 0.135 | 0.140 | 0.145 | 0.151 | 0.156 |
| 99.92 | — | — | 0.131 | 0.134 | 0.137 | 0.140 | 0.143 |
| 99.8 | — | — | 0.139 | 0.134 | 0.137 | 0.140 | 0.143 |

TABLE 6. Coefficient of Linear Expansion of Niobium ($\alpha \cdot 10^{-6}$) in the Range 100-1000°C as a Function of the Degree of Purity [52, 53]

| Purity of metal, % | 100° | 200° | 300° | 400° | 500° | 600° | 700° | 800° | 900° | 1000° |
|---|---|---|---|---|---|---|---|---|---|---|
| 99.0 | 7.10 | — | — | — | — | — | — | — | — | — |
| 99.90 | — | — | 7.38 | 7.54 | 7.61 | 7.87 | 8.03 | 8.20 | 8.37 | 8.52 |
| 99.92 | 7.19 | 7.40 | 7.62 | 7.83 | 8.04 | 8.25 | 8.46 | 8.63 | 8.89 | — |

## 5. Electrical Properties

Most refractory metals have high electrical resistance. As the d-shells become filled with electrons, the electrical resistance decreases regularly along the large periods (Fig. 8). Metals of group VIIa (manganese, technicium, rhenium) are exceptions to the general rule, since they have an anomalously high resistivity compared with transition metals of neighboring groups. According to the data of [36, 43, 51, 161], refractory metals at room temperature have the following resistivity values (in $\mu\Omega \cdot$cm):

| | | | | | |
|---|---|---|---|---|---|
| Titanium | 42-47 | Tantalum | 12.4-13.6 | Ruthenium | 7.16-7.6 |
| Zirconium | 40-44 | Chromium | 12.8-14 | Rhodium | 4.7 |
| Hafnium | 33-40 | Molybdenum | 5.78 | Osmium | 9.5 |
| Vanadium | 24.8 | Tungsten | 5.5 | Iridium | 5.3 |
| Niobium | 12.7-14.6 | Rhenium | 19.14 | Platinum | 10.58 |

The high electrical resistance of the refractory metals is due to the peculiarities of their electronic structure and is explained on the basis of the zone theory. In the transition metals, the electrons are known to be in states formed from the s- and d-states of the free atoms, the density of the states in the d-band being the highest. As in the other metals, an electric current occurs in the case of the transition metals on account of the s-electrons. Mott [57] points out that the probability of the transition of a conductivity electron from an initial state on the Fermi surface to some final state is proportional to the final density of the states on that surface. Since the density of the states in the d-band is greater than in the s-band, the probability of s-d transitions in interactions with lattice vibrations is greater than that of s-s transitions in ordinary scattering. The additional scattering of electrons due to s-d transitions also explains the high resistivity of the transition metals. Electrons, which have jumped from an s- to a d-band on scattering do not remain in it; the number of s-d transitions is equal to the number of d-s transitions and the resultant effect is s-s transitions; the d-state is an intermediate one [58].

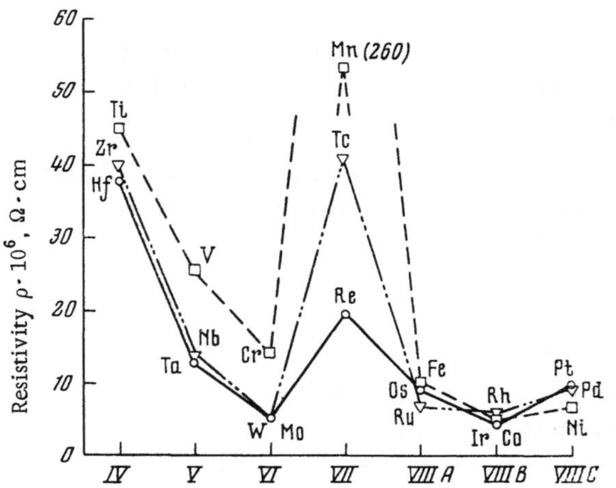


Fig. 8. Variation in resistivity of refractory metals as a function of their position in the periodic system.

Maximum electrical resistance is found in the rare-earth metals, the atoms of which have only one electron in the 3d-shell. For example, the resistivity of erbium at room temperature is 120 $\mu\Omega \cdot$cm. Electron scattering in the case of the rare-earth metals has a more complex character than that of the other metals, due to the presence in them of unfilled d- and f-shells.

TABLE 7. Temperature of Transition of Metals to the Superconducting State [60]

| Metal | Transition temperature, °K |
|---|---|
| Aluminum | 1.20 |
| Titanium | 0.53 |
| Vanadium | 5.10 |
| Zinc | 0.91 |
| Gallium | 1.10 |
| Zirconium | 0.7 |
| Niobium | 9.17 |
| Molybdenum | 0.9–0.98 |
| Technicium | 8.24 |
| Ruthenium | 0.47 |
| Cadmium | 0.56 |
| Indium | 3.37 |
| Tin | 3.70 |
| Lanthanum | 5.00 |
| Hafnium | 0.35 |
| Tantalum | 4.40 |
| Rhenium | 1.70 |
| Osmium | 0.71 |
| Iridium | 0.14 |
| Mercury | 4.15 |
| Thallium | 2.38 |
| Lead | 7.22 |
| Thorium | 1.39 |
| Uranium | 0.80 |
| Tungsten | 0.05 |

Resistivity is a structure-sensitive property. Therefore, any structural changes in metals and alloys under the influence of external factors (temperature, pressure, etc.) produce an appreciable change in resistivity. By the action of work hardening, it is known that there is an increase in resistivity of metals. It is true that for many metals, this increase is only a few percent. Tungsten is an exception in this respect; the resistivity of a stretched tungsten wire depends considerably on the degree of work hardening, as is shown in Fig. 9 [50]. The increase in resistivity is accompanied by a decrease in the temperature coefficient of electrical resistance. The product of these two values, according to Matthiessen's rule, is constant. Similar relationships are found for molybdenum and other refractory metals. Annealing of the work-hardened metal results in a resistivity of the initial value, the value of the temperature coefficient increasing at the same time.

The electrical resistivity of refractory transition metals also increases in the range of elastic deformation, and increases with the temperature. At high temperatures when $T \gg \theta$ ($\theta$ is the Debye temperature), the electrical resistance increases in direct proportion to the temperature; at low temperatures ($T \ll \theta$), Bloch's theory gives the relationship

$$\rho = aT^5.$$

Experimentally, for different metals there are found different types of the dependence of $\rho$ on T, which may be represented by the formula $\rho = aT^n$, where n varies from 2 to 6 for different metals [59].

In the case of melting accompanied by an increase in volume, the resistivity of the liquid metal increases approximately by a factor of 1.5-2, due to the abrupt disturbance in the order of arrangement of the atoms, and also to some changes in the interatomic bond. In elements having a strongly pronounced covalent bond, a decrease in electrical resistance on melting is observed, due to the breakdown of the covalent bond and a transition to the metallic character of the bond on melting.

The high resistivity of the refractory metals enables them to be used in the production of high-ohmic alloys. As basis for high-ohmic alloys for work at elevated temperatures, the noble refractory metals are widely used because of their high technical properties, resistance to corrosion, and the stability of their properties. When palladium and platinum are alloyed with refractory metals, metals of group VIa (chromium, molybdenum, tungsten) increase their electrical resistance most effectively.

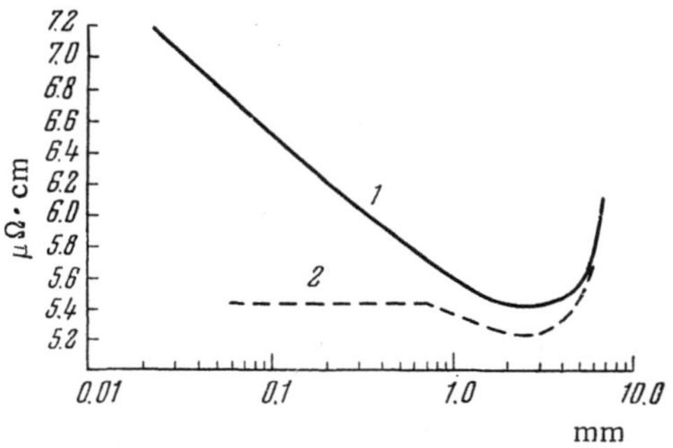


Fig. 9. Dependence of resistivity of tungsten wire on diameter. 1) Stretched; 2) annealed.

## 6. Superconductivity

The phenomenon of superconductivity has been found in 25 pure metals and in a number of alloys and compounds. Many of the refractory metals are superconductors; these include titanium, zirconium, hafnium, vanadium, niobium, tantalum, molybdenum, tungsten, technicium, rhenium, ruthenium, osmium, and iridium.

Niobium has the maximum temperature of transition to the superconducting state, and tungsten has the minimum temperature (Table 7).

It is to be expected that with improvement in methods of purifying metals, superconducting properties will be found in a number of other metals, including refractory metals, and the transition temperatures of known superconducting metals will also be determined more exactly. The assumption is made in [61] that in a state of high purity, platinum and gold will be found to be superconductors. Much higher superconducting materials (called hard superconductors) are found among alloys of refractory metals, in many cases the superconducting characteristics of the alloys being higher than in the metals of which they are composed. Currently, more than 1000 superconducting alloys are known, and their number is constantly increasing.

Theory is as yet unable to explain why the electrical resistance of some metals remains equal to zero at a temperature which is slightly above absolute zero, and then suddenly increases at a certain critical temperature ($T_c$). The presence of superconductivity does not interfere with the distortion of the crystal lattice. This shows that, in the present case, the conductivity electrons do not interact with the crystal lattice, and are in a special energy state [34]. The current theory of superconductivity [62, 63], produced for ideal, isotropic superconductors, cannot explain the processes occurring in real metal crystals, nor indicate from what elements and in what combinations alloys with the necessary critical parameters should be made. Here, as in other directions of the physics and physical chemistry of the solid state, experimental investigations play a leading part.

Superconductivity is a phenomenon inherent only in the metallic state of matter. The transition temperature depends on the individual nature of the superconductor, but at the present time, no superconductors have been found having a transition temperature above 18°K, although it would be very attractive to find superconductors for hydrogen, neon, and nitrogen temperatures. The nature and mechanism of the superconducting state have not yet been elucidated; it is assumed that it is a special thermodynamic state of matter, and that the superconducting transition is analogous to second-order phase transitions. It is not impossible that metastable metal phases possess superconductivity. It is important to emphasize that the superconducting state is destroyed not only by heating the superconductor above the critical temperature, but also by the application of a sufficiently powerful (critical) magnetic field or by passing a large (critical) current through the superconductor. All these three critical characteristics of superconductors are interrelated and depend on the electronic and crystal structure of metals and alloys, the methods of their production, and thermomechanical treatment.

Metallurgical factors have a great influence on the properties of superconducting materials and their phase composition. It suffices to say that the production of superconducting magnets became possible only after the development of methods of melting and treatment of refractory metals and their alloys in a vacuum and protective atmospheres.

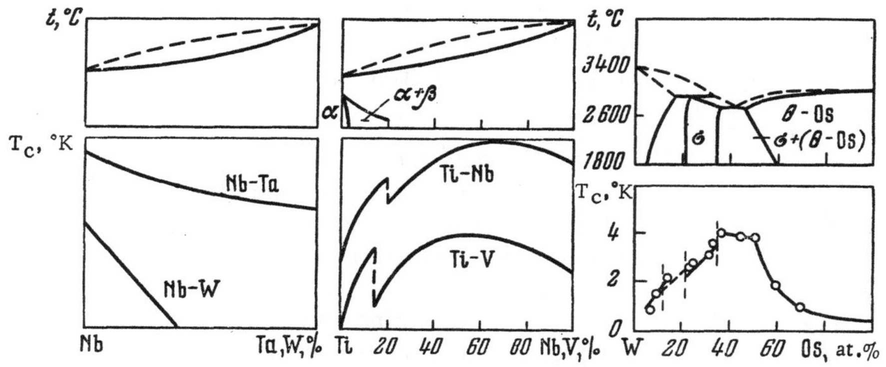


Fig. 10. Dependence of critical temperature of transition to the superconducting state $T_c$ on alloy composition in binary systems.

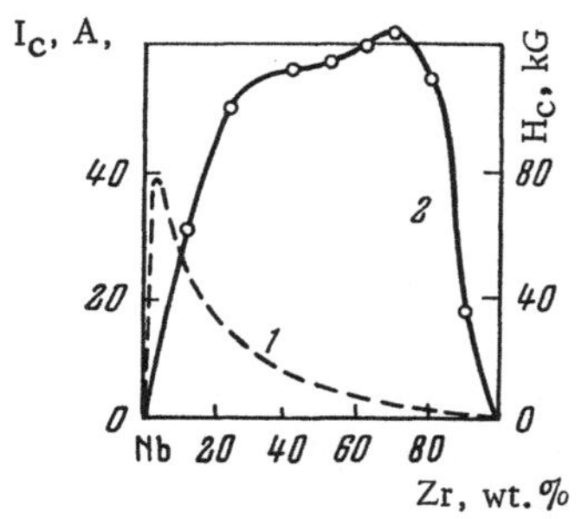


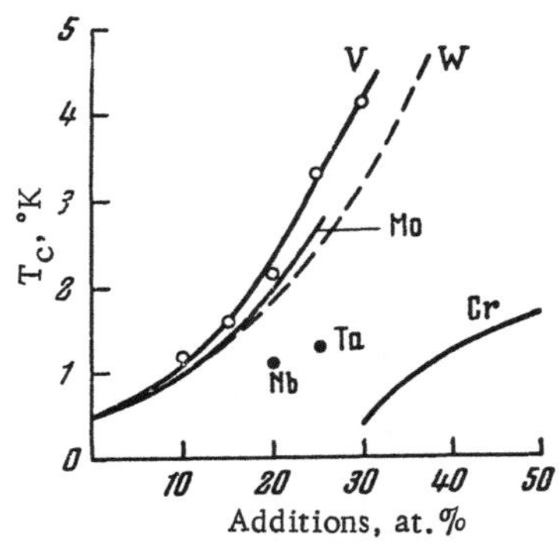


Fig. 11. Dependence of value of critical current (1) and critical magnetic field (2) on composition of alloys of the niobium–zirconium system.

Fig. 12. Dependence of superconductivity transition temperature $T_c$ on composition of hexagonal solid solutions based on ruthenium.

It has been found that by the methods of thermomechanical treatment (application of different degrees of deformation, different annealing conditions), it is possible to vary by a factor of two or three such an important superconducting characteristic as the critical current density [64].

The study of superconducting alloys, however, is still in its infancy. Little or no systematic study has been made of alloys of binary systems, the relationships of the dependence of superconducting characteristics on purity, composition, and thermomechanical treatment have not yet been definitely established. Still less has been done on multicomponent alloys.

The establishment of the relationships in the change of structure and properties of superconductors, the extension of physicochemical analysis to the range of helium temperatures, the development of metallurgical criteria for the production of superconducting alloys and the theoretical bases for the production of semimanufactured products and components made from them in our opinion are problems of a new section of physical metallurgy, superconductor metallurgy, which is currently being created.

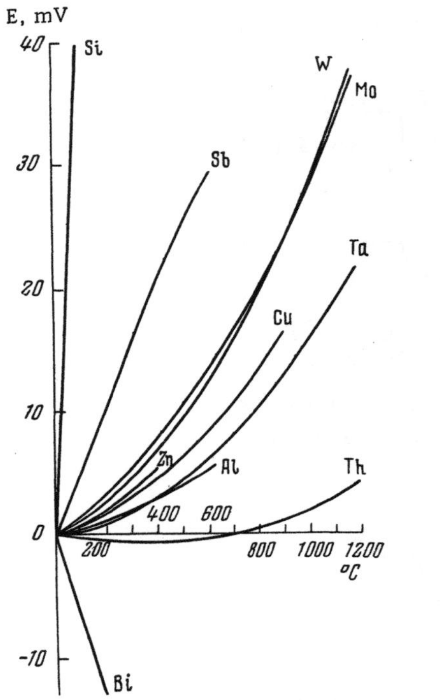


Fig. 13. Thermoelectromotive force of pure metals paired with platinum.

A review of the current position of the development of new superconducting materials having high critical parameters (transition temperature, critical magnetic field, and critical current) is to be found in [64, 66, 67].

The superconducting characteristics of the alloys follow the laws of physicochemical analysis like other physical properties of the systems. In particular, the transition temperature of alloys consisting of substitutional solid solutions of two superconducting metals vary almost additively, depending on the composition. This enables us to predict the variation of $T_c$ of any metal solid solutions on the basis of the data of phase diagrams, which is also in agreement with experimental data (with the exception of Nb–Mo).

Figure 10 shows the variation in the superconductivity transition temperature of alloys of the niobium–tantalum, niobium–tungsten, titanium–vanadium, titanium–niobium, and tungsten–osmium systems [64].

The superconducting characteristics of alloys of the niobium–zirconium system have been investigated in particular detail [64-66]. Figure 11 shows the dependence of the value of the critical current and critical magnetic field on the composition of alloys of the niobium–zirconium system. The principal merit of these superconductors is that they have good technical properties and high values of superconducting characteristics.

Many solid solutions, based on metals of the platinum group, are superconductors. Figure 12 shows the dependence of the critical transition temperature of hexagonal solid solutions based on ruthenium on the concentration of the alloying addition [61].

The critical transition temperature may be considerably increased by the formation of metallic compounds; for example, the $\sigma$-phases of tungsten–rhenium and tungsten–osmium alloys vary the transition temperature by 8 and 4°K, respectively. Currently, more than 900 superconductor metallic compounds are known.

The highest transition temperature is found in $Nb_3Sn$ (18.07°K). The compounds $Nb_3Sn$ and $V_3Ga$ have phenomenal values of 250 and 210 kG, respectively, of the critical magnetic fields (fields destroying the superconducting state). As a rule, the maximum values of all three superconducting characteristics are found in metallic compounds of the type $Nb(V)_3X$ (where X is tin, silicon, gallium) with lattice of the type of $Cr_3Si(\beta\text{-}W)$ [67]. Superconducting properties are also found in compounds of metals of the platinum group with elements of groups IIIb-VIb, forming a structure of the type of $\beta$-W, $\alpha$-Mn, Laves phases, $\sigma$-phases, and also a hexagonal close-packed lattice. The density of states on the Fermi surface for the compound $Nb_3Sn$ is 4.75 el/at [172]. The applicability of this criterion, however, to strongly alloyed alloys is not clear.

The development of compositions of superconducting alloys and compounds based on refractory metals, and also of methods for their thermomechanical treatment is one of the most important problems for the immediate future in the field of superconductor research. Superconducting materials have extensive application prospects in various fields of advanced technology: thermonuclear reactors, powerful accelerators, magnetohydrodynamic generators, lasers, masers, traveling wave tubes, computers, gyroscopes, frictionless bearings, modern telescopes, etc.

## 7. Thermoelectric Properties

As is well known, the thermoelectric properties of the elements are characterized by physical quantities such as integral and differential thermoelectric emf, absolute thermoelectric emf, Peltier coefficient, etc. Many refractory metals are characterized by high values of the integral thermoelectric emf and are used in high-temperature pyrometry. Most of them are electropositive to platinum.

Tungsten and molybdenum have the highest thermoelectric emf of the refractory metals.

Figure 13 shows the curves of variation of thermoelectric emf of some refractory metals paired with platinum [68]. Owing to their high melting points, low vapor pressure and low diffusion mobility, tungsten, molybdenum, and rhenium are used for stable, high-temperature thermocouples, enabling measurements to be made up to 2500°C. Because of their high oxidizability, however, their use is possible only in neutral atmospheres or in a vacuum.

## 8. Thermionic Emission Properties

One of the most important fields of application of refractory metals (tungsten, molybdenum, rhenium, and tantalum) comprises the various constructions of electrovacuum instruments and tubes. In recent years, another field of application of refractory metals has made its appearance, i.e., thermionic power converters. The emission properties of refractory metals have therefore been studied in some detail. With regard to alloys based on refractory metals, interest in the study of their emission properties has recently arisen.

Much of the information on the emission properties of refractory metals has been obtained on insufficiently pure cermet specimens. It is not beyond question that in a number of cases, a study of these properties on high-purity, vacuum-melted materials would give completely new results.

It is known that saturation current density of thermionic emission is determined from the formula [69]

$$i = AT^2e^{-\frac{\varphi}{kT}},$$

where i is the thermionic current density, $A/cm^2$; T is the absolute temperature; A is Richardson's constant, $A/cm^2 \cdot deg$; $\varphi$ is the work function, eV; and k is the Boltzmann constant.

The temperature dependence of the work function appears at high temperatures (approximately by two orders higher than room temperature), when it is not the quantum effects which make themselves felt but also the Maxwellian distribution of electrons over different levels:

$$f = e^{-\frac{\varepsilon}{kT}},$$

where f is the average number of electrons on a given energy level and $\varepsilon$ is the energy of the electrons.

The work function of refractory metals in a vacuum is characterized by the following

TABLE 8. Thermionic Emission Properties of Tungsten [70, 71]

| Temperature, °K | Rate of evaporation, $g/cm^2 \cdot sec$ | Thermionic emission current, $A/cm^2$ | Effectivity of thermionic emission, A/W |
|---|---|---|---|
| 1000 | — | $1.07 \cdot 10^{-15}$ | $1.77 \cdot 10^{-15}$ |
| 1200 | — | $9.73 \cdot 10^{-12}$ | $5.88 \cdot 10^{-12}$ |
| 1400 | $1.69 \cdot 10^{-22}$ | $6.62 \cdot 10^{-9}$ | $1.73 \cdot 10^{-9}$ |
| 1600 | $1.69 \cdot 10^{-20}$ | $9.27 \cdot 10^{-7}$ | $1.20 \cdot 10^{-7}$ |
| 1800 | $3.61 \cdot 10^{-17}$ | $4.47 \cdot 10^{-5}$ | $3.15 \cdot 10^{-6}$ |
| 2000 | $1.47 \cdot 10^{-14}$ | $1.00 \cdot 10^{-3}$ | $4.18 \cdot 10^{-5}$ |
| 2200 | $2.09 \cdot 10^{-12}$ | $1.33 \cdot 10^{-2}$ | $3.48 \cdot 10^{-4}$ |
| 2400 | $1.28 \cdot 10^{-10}$ | 0.116 | $2.01 \cdot 10^{-3}$ |
| 2600 | $3.92 \cdot 10^{-9}$ | 0.716 | $8.55 \cdot 10^{-3}$ |
| 2800 | $7.40 \cdot 10^{-8}$ | 3.54 | $3.01 \cdot 10^{-2}$ |
| 3000 | $9.47 \cdot 10^{-7}$ | 14.15 | $8.82 \cdot 10^{-2}$ |
| 3200 | $8.85 \cdot 10^{-6}$ | 27.84 | 0.244 |
| 3400 | $6.35 \cdot 10^{-5}$ | 142.1 | 0.507 |
| 3600 | — | 383.5 | 1.04 |

TABLE 9. Thermionic Emission Properties of Molybdenum [50]

| Temperature, °K | Rate of evaporation, $g/cm^2 \cdot sec$ | Pressure, mm Hg | Thermionic emission current, $A/cm^2$ |
|---|---|---|---|
| 1000 | $1.37 \cdot 10^{-24}$ | $1.01 \cdot 10^{-21}$ | — |
| 1200 | $1.23 \cdot 10^{-20}$ | $7.47 \cdot 10^{-19}$ | $4.44 \cdot 10^{-11}$ |
| 1400 | $1.26 \cdot 10^{-16}$ | $7.88 \cdot 10^{-15}$ | $2.25 \cdot 10^{-8}$ |
| 1600 | $1.23 \cdot 10^{-13}$ | $8.63 \cdot 10^{-12}$ | $2.90 \cdot 10^{-6}$ |
| 1800 | $2.82 \cdot 10^{-11}$ | $2.10 \cdot 10^{-9}$ | $1.33 \cdot 10^{-4}$ |
| 2000 | $2.07 \cdot 10^{-9}$ | $1.62 \cdot 10^{-7}$ | $2.88 \cdot 10^{-3}$ |
| 2100 | $7.30 \cdot 10^{-8}$ | $6.07 \cdot 10^{-6}$ | $3.58 \cdot 10^{-2}$ |
| 2200 | $1.40 \cdot 10^{-6}$ | $1.20 \cdot 10^{-4}$ | $3.16 \cdot 10^{-1}$ |
| 2300 | $1.79 \cdot 10^{-4}$ | — | $8.52 \cdot 10^{-1}$ |

data (in eV) [69]:

| | | |
|---|---|---|
| Titanium. . . . . . 3.95 | Molybdenum . . . . . 4.29 | Tungsten . . . . . 4.51 |
| Vanadium . . . . . 4.13 | Ruthenium . . . . . . 4.52 | Rhenium . . . . . 4.99 |
| Chromium. . . . . 4.59 | Rhodium . . . . . . . 4.75 | Osmium. . . . . . 4.70 |
| Zirconium. . . . . 4.16 | Hafnium . . . . . . . . 3.53 | Iridium . . . . . . 4.57 (5.3) |
| Niobium . . . . . . 3.99 | Tantalum . . . . . . . 4.10 | Platinum . . . . . 5.30 |

On analyzing these data, some regularity will be found in the variation in the work function of refractory metals: It increases along the large periods with increase in the change of the nucleus; in the third large period, osmium and iridium are exceptions to the general rule; in their case, the work function decreases compared with rhenium. Evidently, the degree of purity, surface phenomena and closeness of packing of the electrons in a definite direction affect the thermionic emission. These factors interfere to a considerable degree with the appearance of a more definite dependence of the work function on the position of the refractory metals in the periodic system.

The emission properties of tungsten and molybdenum in a vacuum are given in Tables 8 and 9.

The current density of electron emission, as shown by the tables, increases rapidly with temperature. The rapid increase in thermionic current at the melting point of the metals is of

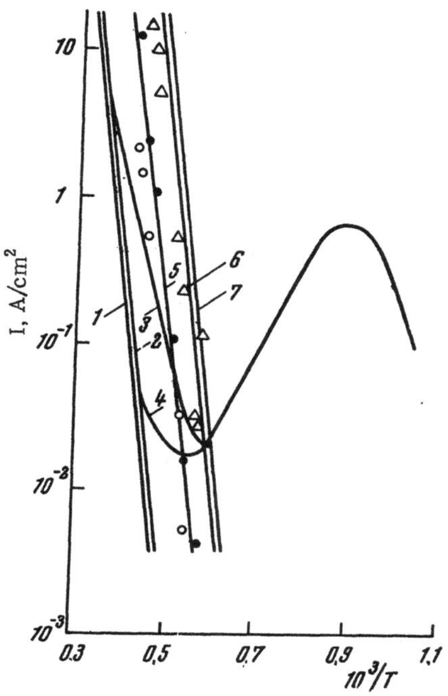


Fig. 14. Thermionic emission properties of some refractory metals and metal compounds in cesium vapor.

| Curve | Cathode | p, mm Hg | Curve | Cathode | p, mm Hg |
|---|---|---|---|---|---|
| 1 | W | 0 | 5○ | ZrC | 0.002 |
| 2 | Ta | 0 | 5● | | 0.08 |
| 3 | Ta | 0.4 | 6 | $(UC)_{0.2}(ZrC)_{0.8}$ | 0 |
| 4 | W | 0.05 | 6△ | | 0.3 |
| 5 | ZrC | 0 | 7 | UC | 0 |

great theoretical and practical interest. So far, the emission properties of molten metals has been studied very little. In foreign literature, there is a reference to an attempt to use liquid tungsten cathodes for use in special instruments and constructions, in which the surface of the tungsten cathode is melted by laser beams [72].

In the single-crystal state, metals possess appreciable anisotropic emission properties along different crystallographic directions. Anisotropy of the emission properties is well explained by the different density of packing of the atoms on different faces of the metal crystal [73]. Of the refractory metals, anisotropy of the thermoemission properties of tungsten has been studied most fully [74-76]. The emission properties of tungsten and molybdenum single crystals, produced by electron-beam zone refining in a vacuum will be considered later (in Chapter V).

A much higher current density and lower work function may be obtained if the refractory metals are coated with a thin film of alkali metals or alkaline earth metals, and the characteristics are not measured in a vacuum but in the vapors of the corresponding metals. In this case, the work function is substantially reduced (for example, for cesium on tungsten 1.36 eV may be obtained). The effect which has been discovered opens up for refractory metals and their alloys wide prospects for their use as thermocathodes in various kinds of converters for transforming heat energy into electrical energy.

The refractory metals may be arranged in the following order of work function in cesium: iridium (lowest work function), rhenium, tungsten, molybdenum, zirconium, tantalum, niobium, hafnium. Other conditions remaining the same, a higher work function of refractory metals in a vacuum thus corresponds to a less effective work function in cesium vapor. This is due to the better adsorptive capacity for cesium vapor of materials having a high work function [163, 164].

Figure 14 shows the results of an investigation of the emission properties of some refractory metals and refractory metal compounds in cesium plasma. As follows from these data, the anode current density was found to be much higher when measured in cesium plasma than when measured in a vacuum [74].

The thermionic emission of tungsten, molybdenum, tantalum, niobium, rhenium paired with alkali metals has been studied [75-78] by means of thermoelectronic projectors. Face-centered metals paired with cesium give the highest emission at high cathode temperatures with the closest-packed (110) surface, which in the pure metal has the highest work function [75].

The variation of the electronic and ionic current densities with the reciprocal temperature for different crystallographic faces of tungsten in cesium vapor is shown in Fig. 15 [75].

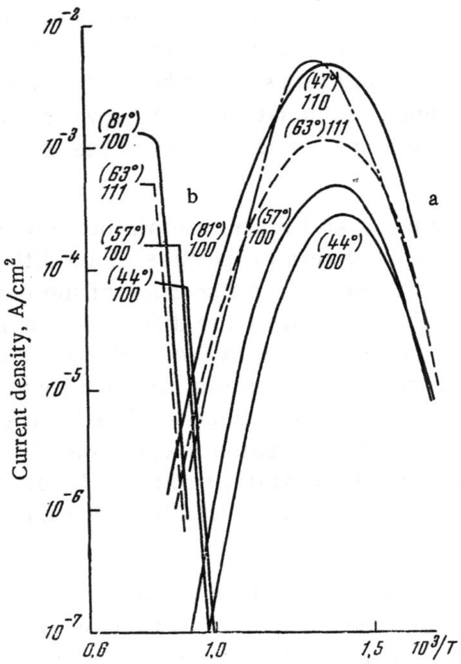


Fig. 15

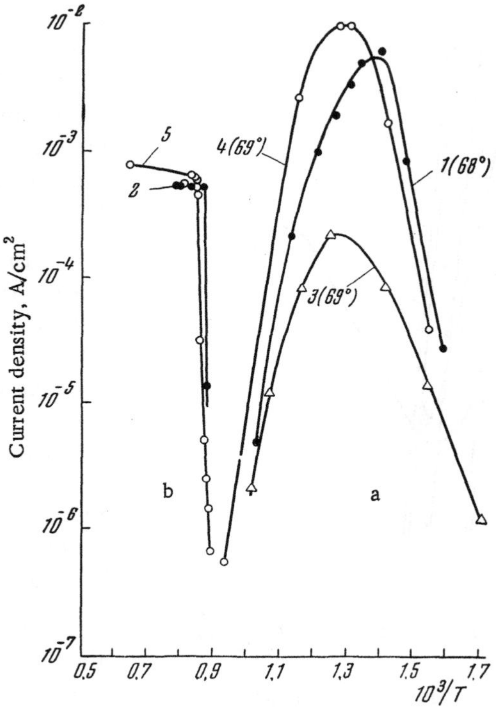


Fig. 16

Fig. 15. Electronic (a) and ionic (b) current density versus temperature for different crystallographic planes of tungsten (temperature of the cesium in the ampoule is shown in parentheses).

Fig. 16. Electronic (a) and ionic (b) current density versus temperature of molybdenum (the temperature of the cesium in the ampoule is shown in parentheses). 1,2) (110) plane; 3) (112) plane before contamination with fluorine and chlorine; 4,6) the same, after contamination.

In the case of face-centered metals, the highest emission in cesium vapor is also found to be from the closest-packed (111) plane having the highest work function in a vacuum. Hexagonal rhenium in cesium vapor gives a strong emission in the three close-packed planes. Maximum thermoelectronic current from the closest-packed plane of the crystal is found in a temperature range in which the metal is partially coated with cesium. In the range of more complete cesium coating, such a relationship has not been found [75].

In all cases, the work function on close-packed surfaces in cesium vapor is greater than the ionization potential of cesium which is 3.89 eV. The emission from other crystallographic surfaces is insignificant. The emission pattern obtained on molybdenum surfaces in potassium vapor is quite different from the emission pattern in cesium vapor. In the case of potassium vapor, spots have also been observed from other crystallographic planes having high indices. In a vacuum, such planes have a low work function. The ionization potential of potassium is 4.32 eV.

In a number of investigations [75-78], it has been shown that when the work function of a plane of the base metal is higher than the ionization potential of the vaporized alkali metal, emission occurs at high temperatures only from planes having a high work function. In the case of commensurability of these values, a complex emission pattern is found.

Correlation of the work function and ionization potential is determined to a considerable extent by adsorption of the alkali metals on the surface of the emitters. The adsorbed layers are usually polarized, and the dipole moment is added to or subtracted from the work function [75, 79].

Impurities reaching the surface of emitters may have a considerable influence on the electronic emission in alkali metal vapor. The influence of impurities in particular is used to explain the considerable increase in emission from the (112) plane for metals having a body-centered lattice (tungsten and molybdenum) in the process of their work as emitters. Webster [75] makes the assumption that in the process of the electron bombardment, atoms of fluorine and chlorine are liberated from glass, and enter the atomic packing voids of the (112) plane, making these planes close-packed with a high work function, leading to a change in their adsorptive capacity. Figure 16 shows the electronic emission from the (112) plane before and after contamination, and also the electronic emission from the (110) plane in cesium vapor for molybdenum. It is evident that the influence of impurities on thermionic emission may also be due to chemical reaction between emitter impurities and the alkali metals.

In the case of tantalum, strong emission in cesium vapor after lengthy operation has been found only from the (110) plane [76, 78]. A similar result has been obtained for niobium [77].

The application of a thin film of zirconium on tungsten also leads to a sharp reduction in the value of the work function [80]. The reduction in the work function is explained in this case by assuming that the zirconium film is responsible for the emission. It could equally well be assumed that an alloy of tungsten and zirconium is formed, resulting in a reduction in the work function.

Until recently, the influence of alloying on the emissive properties of refractory metals was not given serious attention. Some data on this question are to be found in [81].

The thermionic emission properties of alloys of the hafnium–scandium and molybdenum–niobium systems (a continuous series of solid solutions) have been studied [82, 83]. In the case of a multiphase system, according to [84] the work function of the mixture of phases is determined by the equation:

$$\varphi = \sum_{i-1}^{N} f_i \varphi_i,$$

where $f_i$ is the proportion of current given off by spots of i type, and $\varphi_i$ is the work function of spots of the same type.

The variation in work function in the case of solid solutions (scandium–hafnium, molybdenum–niobium systems) is evidently determined to a considerable extent by the variation in the crystal lattice parameters and, due to this, the variation in the potential barrier at the metal boundary.

According to [85], the potential barrier at the metal boundary is connected to the crystal lattice parameter by the following relationship

$$W_a = C \frac{e^2}{2a},$$

where $a$ is the lattice parameter.

The emission of cathodes is appreciably diminished by the action of oxygen. However, for metals such as titanium, zirconium, and hafnium, this influence is slight, since the corresponding oxides are rapidly removed from the surface. Oxygen exerts a considerable influence on the emission of tungsten.

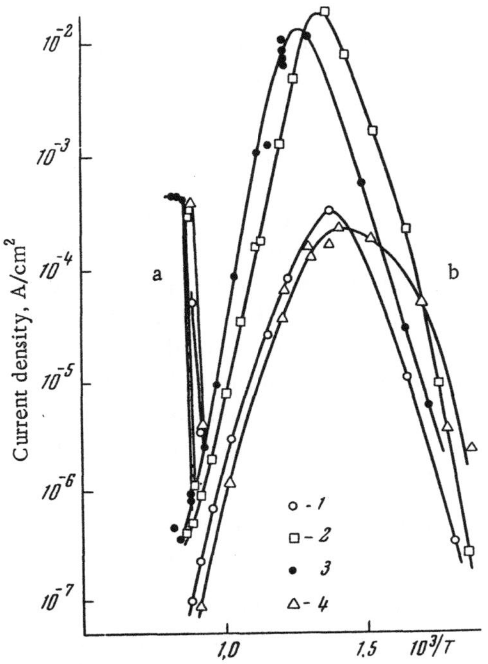


Fig. 17. Electronic (a) and ionic (b) current density in cesium plasma for emitting surfaces of a rhenium single crystal. 1) (10$\bar{1}$0) plane; 2) (1000) plane; 3) (122$\bar{1}$) plane; 4) (123$\bar{1}$) plane.

The influence of alloying on the thermionic emission of refractory metals has already been discussed. It would be of decided interest to investigate the influence of alloying on the thermionic emission of refractory metals in the vapor of alkali metals and alkaline earth metals. Work in this direction has only just begun to be developed. For obtaining reliable results in this case, the method of physicochemical analysis and the establishment of a correlation between thermionic emission and other physical properties of the metals and alloys may be found to be particularly fruitful.

The work of cathodes in thermionic emission converters in the presence of cesium vapor presents a series problem. The high working temperature of the cathodes gives rise to the necessity of investigating the evaporation of refractory metals in cesium vapor and other atmospheres. Maximum interest is presented by an investigation into the volatility of tungsten, rhenium, tantalum, molybdenum, and niobium.

A study of the evaporation of tungsten in the presence of various inert gases [86, 87] has shown a reduction in the rate of evaporation due to the formation of a diffusion barrier. The influence of cesium vapor on volatility has been studied [88]. Cesium vapor considerably reduces the rate of evaporation of molybdenum compared with the rate of evaporation in a vacuum. The electronic and ionic current densities of a rhenium single crystal in cesium vapor have been measured (Fig. 17) [103].

## 9. Nuclear Properties

In connection with the use of refractory metals as constructional materials in atomic engineering, the coefficient of thermal neutrons by these metals is of considerable importance. These data are given below (in barns/atom) for a neutron velocity of 2200 m/sec [43]:

| | | | | | |
|---|---|---|---|---|---|
| Zirconium | 0.180 | Vanadium | 4.98 | Tantalum | 21 |
| Niobium | 1.15 | Titanium | 5.8 | Rhenium | 86 |
| Ruthenium | 2.56 | Platinum | 8.8 | Hafnium | 115 |
| Molybdenum | 2.7 | Osmium | 15.3 | Rhodium | 156 |
| Chromium | 3.1 | Tungsten | 19.2 | Iridium | 440 |

We have already discussed some properties of refractory metals in relation to their position in the periodic system. A number of properties of metals and alloys are determined not only by the nature of the metal or (in the case of an alloy) by the nature of the metal phases composing it (i.e., by the peculiarities of the atomic structure, character of the chemical bond, type of crystal lattice, etc.), but they also depend on the form, size, and character of the arrangement of the structural components and other structural factors (structurally sensitive properties).

The structurally insensitive properties are principally a function only of the chemical composition of the alloys. Some structurally sensitive and structurally insensitive properties are given below according to their categories as follows [173]:

| | Structurally insensitive | Structurally sensitive |
|---|---|---|
| Mechanical | Density | Resistance to fracture |
| | Elastic modulus | Plasticity (Ductility) |
| Thermal | Thermal expansion | |
| | Melting point | |
| | Thermal conductivity | |
| | Specific heat | |
| | Emittance | |
| Electrical | Resistance (metallic) | Resistance (semiconductor and at low temperatures) |
| | Electrochemical potential | |
| | Thermoelectric properties | |
| Magnetic | Paramagnetic and diamagnetic properties | Ferromagnetic properties (including magnetostriction) |
| Optical | Reflectivity | |
| Nuclear | Radiation absorption | |

In conclusion, Tables 10 and 11 give the basic physical properties of the refractory metals of groups Va and VIa.

## Compatibility of Refractory Metals and Alloys with Various Materials

The use of refractory metals in space, rocket, electronic, nuclear, and other fields of engineering has given rise to the problem of the compatibility of refractory metals with various materials (ceramics, fusible liquid metals, vapors and the plasma of alkali and alkaline-earth metals, nuclear fuel, etc.). One material should not spoil the working properties of another material. If metals are to prove serviceable for several thousands of hours at high temperatures in contact with ceramics, vapors, or molten metals, it is essential to clarify the question of chemical interaction with these materials, which are selected primarily on the basis of structural, technological, and economic considerations. The impurity content of these materials has a definite influence on the nature and velocity of the chemical reaction of refractory metals with different materials in contact with them. In an investigation of compatibility, therefore, it is very important to ascertain the permissible content of the individual impurities in a material. The compatibility of certain physical properties of the contacting materials, for example equality of the coefficients of thermal expansion in bonding a ceramic with a refractory metal, is also of considerable importance.

The compatibility of niobium, tantalum, niobium–zirconium alloys, and tungsten–rhenium alloys with the oxides $Al_2O_3$, MgO, BeO, $Y_2O_3$, and $ThO_2$ has been studied [89], using commercially pure and high-purity oxides.

The study of the interaction of the materials enumerated was carried out by metallographic, x-ray diffraction, and electron microprobe methods. The results of this study showed $ThO_2$, $Y_2O_3$, BeO, $Al_2O_3$, and MgO of commercial purity to be compatible with niobium, tantalum, and alloys of tungsten with 5 and 25% Re at temperatures of up to 1540°C and holding times of up to 1000 h.

TABLE 10. Physical Properties of Group Va Metals [161]

| Property | Vanadium | Niobium | Tantalum |
|---|---|---|---|
| **Atomic properties and specific gravity** | | | |
| Atomic number | 23 | 41 | 73 |
| Atomic weight | 50.95 | 92.91 | 180.95 |
| Interatomic distance, Å | 2.632 | 2.859 | 2.859 |
| Atomic radius, kX | 1.36 | 1.45 | 1.46 |
| Atomic volume, $cm^3/g$-atom | 8.35 | 10.8 | 10.9 |
| Structural type | A2 | A2 | A2 |
| Specific gravity | 6.14 (20°) | 8.58 | 16.654 |
| Natural isotopes | 50 (0.24%), 51 (99.76%) | 93 (100%) | 180 (0.01%), 181 (99.99%) |
| Neutron capture cross section (barns/atom) | 4.5 ± 0.9 | 1.1 | 21.3 |
| Self-diffusion activation energy, kcal/mole | – | 100-105 | 110 |
| **Thermal properties** | | | |
| Melting point, °C | 1900 ± 25 | 2468 ± 10 | 3000 |
| Boiling point, °C | 3350-3000 | 5127-4927 | 5300-6030 |
| Vapor pressure, mm Hg | $10^{-8}$ (1200°)<br>$10^{-4}$ (1600°)<br>$10^{-2}$ (1800°)<br>$10^{-1}$ (2000°)<br>10 (2500°) | $10^{-11}$ (1680°)<br>$10^{-5}$ (2194°)<br>$10^{-4}$ (2355°)<br>$10^{-3}$ (2539°) | $9 \cdot 10^{-11}$ (1737°)<br>$10^{-5}$ (2407°)<br>$10^{-4}$ (2599°)<br>$10^{-3}$ (2820°)<br>$5 \cdot 10^{-3}$ ($T_{pl}$) |
| Rate of evaporation, $g/cm^2 \cdot sec$ | – | $1.63 \cdot 10^{-12}$ (2000°K)<br>$6.79 \cdot 10^{-6}$ (3000°K)<br>$6.8 \cdot 10^{-5}$ (3269°) | $1.16 \cdot 10^{-7}$ (2467°K)<br>$1.08 \cdot 10^{-6}$ (2628°K)<br>$1.06 \cdot 10^{-5}$ (2812°K) |
| Coefficient of thermal expansion, $\times 10^{-6}$ | 9.7 ± 0.3 (20-720°)<br>8.95 (200-1000°)<br>8.3 (20-100°)<br>9.6 (20-500°)<br>10.4 (20-900°)<br>10.9 (20-1100°) | 7.1 (20°)<br>7.56 (18-300°)<br>7.88 (300-1000°)<br>7.95 (up to 900°)<br>7.2 (20-100°)<br>7.4 (20-200°)<br>7.5 (20-300°)<br>19.8 (20-2200°) | – |
| Thermal conductivity, cal/cm · sec · deg | 0.074 (100°)<br>0.088 (500°) | 0.125 (20°)<br>0.130 (100°)<br>0.131-0.139 (200°)<br>0.134-0.140 (300°)<br>0.140-0.151 (500°)<br>0.143-0.156 (600°) | 0.130 (20°)<br>0.174 (1430°)<br>0.186 (1630°)<br>0.198 (1830°) |
| Specific heat, cal/g · deg | 0.119 (0-100°) | 0.065 (100°)<br>0.0797 (1400°)<br>0.0643 (0°)<br>0.08267 (1600°) | 0.036 (0°)<br>0.034 (100°)<br>0.038 (1000°)<br>0.044 (1970°) |
| Molecular heat, cal/mole · deg | 5.40 (298-1900°K) | 6.012 (0°C)<br>7.730 (1600°C)<br>5.95 (298°K)<br>7.39 (1800°K)<br>8.00 (3000°K) | 6.024 (0°C)<br>7.401 (1600°C)<br>6.08 (298°K)<br>7.02 (1800°K)<br>7.53 (3000°K) |
| Entropy, cal/mole · deg | 7.0 ± 0.1 (290°K) | 8.73 (298°K)<br>20.35 (1800°K)<br>26.67 (3000°K) | 9.90 (298°K)<br>21.66 (1800°K)<br>25.37 (3000°K) |

TABLE 10 (continued)

| Property | Vanadium | Niobium | Tantalum |
|---|---|---|---|
| Heat of fusion, kcal/mole | 4.0 | 6.4 | 5.9-7.5 |
| Heat of evaporation, kcal/g-atom | 119.9 ± 6 (1900°)<br>121. ± 7 (25) | 166.5 | 180 |
| Thermal radiation, W/cm$^2$ | – | 6.4 (1600°K)<br>18.5 (2000°K)<br>45.3 (2400°K)<br>130.6 (1800°K) | 7.36 (1600°K)<br>21.6 (2000°K)<br>105.5 (2800°K)<br>144.4 (3000°K)<br>214.5 (3270°K) |
| | Optical and electronic properties | | |
| Index of refraction | – | 1.80 | 2.05 |
| Ionization potential | 6.8 | 6.77 | 7.3 ± 0.3 |
| Work function, eV | 3.79 | 4.01 | 4.1 |
| Electronic emission, mA/cm$^2$ | – | $2.19 \cdot 10^{-5}$ (1600°K)<br>$1.16 \cdot 10^{-2}$ (2000°K)<br>0.8 (2400°K)<br>60.67 (2800°K) | $9.1 \cdot 10^{-6}$ (1600°K)<br>$6.2 \cdot 10^{-3}$ (2000°K)<br>0.5 (2400°K)<br>12.53 (2800°K)<br>46.60 (3000°K) |
| Secondary emission, mA/cm$^2$ | – | 1.18 | 1.35 |
| Ionic emission | – | 5.52 | 10.0 |
| | Magnetic properties | | |
| Magnetic permeability (CGS system) | $5.50 \cdot 10^{-6}$ (25°)<br>$4.57 \cdot 10^{-6}$ (1700°) | $2.28 \cdot 10^{-6}$ (298°K)<br>$2.34 \cdot 10^{-6}$ (20.4°K)<br>$1.79 \cdot 10^{-6}$ (1575°C) | $0.93 \cdot 10^{-6}$ (20°C) |
| | Electrical properties | | |
| Resistivity, μΩ·cm | 24.8 (20°) | 12.7 (20°) | 12.4 (18°)<br>54 (1000°)<br>71 (1500°) |
| Temperature coefficient of resistance, $10^{-3}$/deg | 3.40 (0-100°)<br>3.30 (0-200°)<br>2.80 (0-100°) | 3.95 (0-100°) | 3.82 (0-100°) |
| Temperature of transition to superconducting state, °K | 5.13 | 9.22 | 4.38 |
| Electrical equivalent (for valence 5) | 0.1056 | 0.1926 | 0.3749 |

An alloy of niobium with 1% Zr reacted with commercially pure $ThO_2$, $Y_2O_3$, and BeO since they contained small amounts of silicon; the electron microprobe method showed the presence, close to the contact zone, of inclusions of $Nb_5Si$. For the same alloy, in contact with spectrally pure $ThO_2$, $Y_2O_3$, and BeO, no traces of reaction were found. The investigations [89] showed that incompatibility was due not only to the contamination of the ceramic by silicon, but also to the presence in the alloy of zirconium, which combined with the oxygen forming $ZrO_2$, with reduction of $Al_2O_3$ and MgO. Therefore, on contact of the alloy with even high-purity alumina together with $Nb_5Si$ and $ZrO_2$, the presence of the intermetallic compounds $Nb_5Al$ and $Nb_2Al$ was observed; the compound $Nb_5Si_3$, as was to be expected, was absent. Figure 18 shows the distribution of aluminum in this alloy, in contact with $Al_2O_3$, as a function of the distance from the adjacent surface.

TABLE 11. Physical Properties of Group VIa Metals [36, 38, 43, 56]

| Property | Chromium | Molybdenum | Tungsten |
|---|---|---|---|
| Atomic properties and specific gravity | | | |
| Atomic number | 24 | 42 | 74 |
| Atomic weight | 52.01 | 95.94 | 183.85 |
| Interatomic distance Å | 2.87 | 3.1403 | 3.1583 |
| Atomic volume, $cm^3/g$-atom | 7.23 | 9.41 | 9.53 |
| Structural type | A2 | A2 | A2 |
| Specific gravity, $g/cm^3$ | 7.19 | 10.2 | 19.35 |
| Isotope composition | 50.52, 53.54 | 92 (15.86%)<br>94 (9.12%)<br>95 (15.7%)<br>96 (16.5%)<br>97 (9.45%)<br>98 (23.75%)<br>100 (9.62%) | 180, 182, 183,<br>184, 186 |
| Arrangement of valence electrons (d + s) | $3d^5 4s^1$ | $4d^5 5s^1$ | $5d^4 6s^2$ |
| Neutron capture cross section, barns/atom | 3.1 | 2.7 | 19.2 |
| Thermal properties | | | |
| Melting point, °C | 1875 | 2610 | 3380 |
| Boiling point, °C | 2199 | 5560 | 5900 |
| Vapor pressure, mm Hg | – | $7.5 \cdot 10^{-10}$ (1800°)<br>$2.2 \cdot 10^{-6}$ (2100°)<br>$1.5 \cdot 10^{-4}$ (2400°) | $10^{-5}$ (2554°)<br>$10^{-4}$ (2767°)<br>$10^{-3}$ (3016°)<br>$10^{-2}$ (3309°) |
| Rate of evaporation, $g/cm^2 \cdot sec$ | – | $1.06 \cdot 10^{-10}$ (1800°)<br>$2.82 \cdot 10^{-8}$ (2100°)<br>$1.30 \cdot 10^{-7}$ (2200°)<br>$1.80 \cdot 10^{-6}$ (2400°)<br>$1.57 \cdot 10^{-5}$ (2600°) | $1.47 \cdot 10^{-7}$ (2554°)<br>$1.46 \cdot 10^{-6}$ (2767°)<br>$1.45 \cdot 10^{-5}$ (3016°)<br>$1.43 \cdot 10^{-4}$ (3309°) |
| Coefficient of thermal expansion, $\times 10^{-6}$ | 6.2 (20°) | 5.44 (20°) | 4.45 (0-500°) |
| Thermal conductivity, cal/cm·sec·deg | 0.16 (20°) | 0.298 (204°) | 0.31 (20°) |
| Specific heat, cal/g·deg | 0.1068 (20°) | 0.061 (0-100°)<br>0.072 (20-550°) | 0.357 (from −185 to +20°)<br>0.0320 (0-20°)<br>0.0324 (0-100°)<br>0.0375 (15-423°)<br>0.0479 (2095-2212°) |
| Heat of fusion, kcal/mole | 61.5 | 69.8 | 46 |
| Optical and electronic properties | | | |
| Emission coefficient | – | 0.55 (730°)<br>6.3 (1330°) | 0.9 (80°)<br>18 (1600°)<br>64 (2200°)<br>153 (2700°) |
| Ionization potential | – | 7.18 | – |
| Work function, eV | 4.59 | 4.29 | 4.51 |
| Electrical properties | | | |
| Resistivity, μΩ·cm | 12.8 (20°) | 5.78 (27°) | 5.5 (20°) |
| Temperature coefficient of electrical resistance, $10^{-3}$/deg | 2.5 (0-100°) | 4.7 (20-100°) | 4.82 (20-100°) |
| Temperature of transition to superconducting state, °K | – | 0.9-0.98 | 0.05 |

TABLE 12. Free Energy of Formation of Oxides at Different Temperatures

| Oxide | $-\Delta F$, kcal/g-atom $O_2$ | | |
|---|---|---|---|
| | 1000° C | 1310° C | 1540° C |
| NbO | 70 | 66 | 62 |
| $ZrO_2$ | 100 | 95 | 90 |
| $Al_2O_3$ | 99 | 93 | 88 |
| $SiO_2$ | 76 | 70 | 66 |
| MgO | 108 | 97 | 100 |
| $Y_2O_3$ | 111 | 107 | 103 |
| $ThO_2$ | 116 | 111 | 106 |

Unalloyed niobium and alloys of niobium with 1% Zr and with 5% Zr were tested in contact with different sorts of $Al_2O_3$ at 1310°C for 1000 h. In the case of pure niobium, no reaction was observed in contact with $Al_2O_3$ of commercial and spectroscopical purity. An alloy of niobium with 5% Zr in contact with $Al_2O_3$ + 0.5% Si gave a reaction zone with the formation of $Nb_2Al$, $Nb_3Al$, and inclusions of $Nb_5Si_3$. In [89] the occurrence of the above-mentioned reaction products, resulting from the interaction of niobium–zirconium alloys with $Al_2O_3$ + Si, is given a thermodynamic basis. To produce the reaction products, continuous diffusion of the metals to the adjacent alloy–ceramic surface is essential. Metals which form intermetallic compounds with niobium are obtained as the result of the reduction of oxides by the more thermodynamically active zirconium contained in the alloy. On the basis of the values of the free energy of formation of the oxides (Table 12), at the test temperatures, niobium is unable to react with $SiO_2$ and $Al_2O_3$, while zirconium, present as a component of the alloy, is capable of reducing both these oxides with the formation of $ZrO_2$. However, $ZrO_2$ was not found in the zone of formation of the intermetallic compounds or ceramics, but in the alloy below the reaction zone. This led the authors of [89] to assume that the direct reduction of $Al_2O_3$ by zirconium is very unlikely.

In this connection the following mechanism for the reaction of a niobium–zirconium alloy with the ceramic ($Al_2O_3$) has been proposed:

$$Al_2O_3 + 6Nb \rightleftarrows 2Nb_3Al + \frac{3}{2}O_2.$$

This reaction does not take place in the presence of unalloyed niobium. The authors of [89] assume that the small equilibrium constant of this reaction and the presence of an element capable of forming stable oxides leads to the displacement of the equilibrium to the right. With the removal of oxygen ($Zr + O_2 \rightarrow ZrO_2$), the reaction continues with the formation of $Nb_3Al$. The mechanism of formation of $Nb_5Si_3$ is evidently similar.

The problem of the production of stable bonds of metals with a ceramic assumes considerable importance in the development of nuclear, isotope, solar, and thermionic emission converters (TEC) of heat energy into electric energy, as well as of magnetohydrodynamic (MHD) generators, where refractory metals and alloys may be used as cathode materials. In addition to high chemical stability of the cathode materials to the insulating ceramic, important properties of the contacting materials are the coefficient of thermal expansion, melting point, vapor tension, thermal and electrical conductivity, etc.

Data on the compatibility of some refractory metals and alloys with oxides of the rare-earth metals (for use of the latter as cathode insulators) are given in [90], in which not only is chemical stability considered, but also the compatibility of a number of physical properties, in particular the coefficient of linear expansion.

Pure oxides of the rare-earth metals and their alloys (alloys of oxides of erbium, yttrium, and holmium with thorium oxide) were used as insulating ceramic materials. Attractive properties for the use of oxides of the rare-earth metals as TEC ceramics are the low coefficient of thermal expansion and the stability of the oxides of these metals to thermal shock; thorium oxide is characterized by a low vapor tension and high chemical stability (high value of the free energy of formation of the oxide), but it is very sensitive to thermal shock.

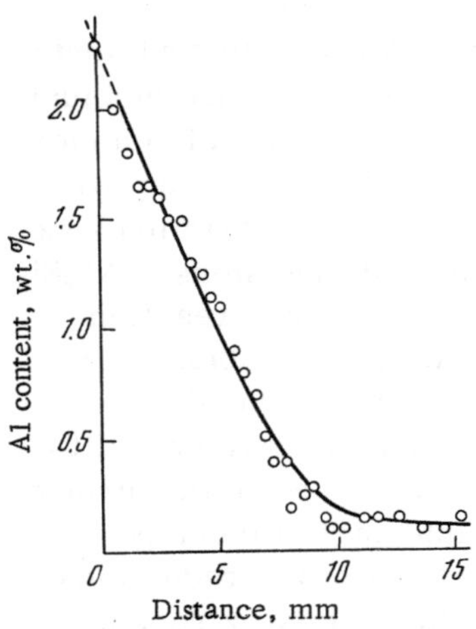


Fig. 18. Distribution of aluminum in an alloy of niobium with 1% Zr, in contact with $Al_2O_3$ as a function of the distance from the adjacent surface.

The plasma torch was used for preparing $ThO_2$–$Y_2O_3$ alloys in the proportions 25 : 75 and 50 : 50. Compatibility tests (at 2000°C) of the selected oxides and their alloys with tantalum, niobium, molybdenum, tungsten, and rhenium showed that reactivity increased in the following order: thorium oxide, hafnium oxide, yttrium oxide, aluminum oxide, erbium oxide, and holmium oxide. The refractory metals were arranged in the following order (in order of their increasing reactivity): tungsten, rhenium, niobium, molybdenum, tantalum [90].

The tests were made in a vacuum of $10^{-4}$-$10^{-6}$ mm Hg. At temperatures of up to 2000°C, tungsten and molybdenum did not react with thorium and yttrium oxides and spectroscopically pure hafnium oxide. At temperatures above 2000°C it was found [90] that the metals became saturated with oxygen in contact with the oxides of the rare-earth metals. The existence of an intermediate phase was found in the rhenium–yttrium oxide system.

Hafnium oxide of spectroscopic purity has high chemical stability and low vapor tension.

The question of the nature of the interaction of oxides and metals has not yet been studied adequately. The use of alloys based on refractory metals will evidently show promise of better compatibility of the physical properties. By alloying, alloys may be produced having the necessary coefficient of expansion, but at the same time the chemical stability of the alloy with regard to ceramics should not be impaired.

Considerable interest is attached to the use of metallic compounds as cathode materials. Tests which have been made so far have not been met with success. The most promising materials will evidently be solid solutions based on refractory metals.

Table 13 gives some materials recommended for use in TEC's at 2000°C [90].

The study of questions of compatibility of refractory metals (used as TEC electrodes) with ambient vapor of alkali and alkaline earth metals, in particular cesium, is of considerable interest.

There is little published information on the corrosion of refractory metals in cesium vapor. Experiments made on these lines as yet bear a qualitative character. Results described in [91] show that tungsten, molybdenum, and tantalum are compatible with cesium vapor. Rhenium also has high stability in liquid cesium and cesium vapor [92]. High stability in cesium

TABLE 13. Some Promising Thermionic Emission Converter Materials for Use at 2000°C

| Composition | Content, wt.% | Coefficient of thermal expansion, $10^{-6}$/deg | Composition | Content, wt.% | Coefficient of thermal expansion, $10^{-6}$/deg |
|---|---|---|---|---|---|
| Cathodes | | | Insulators | | |
| Tungsten . . . . . . . . . . . | Pure | 4.5 | $ThO_2$. . . . . . . . . . | Pure | 10.4 |
| Rhenium . . . . . . . . . . . | Pure | 6.7 | $Y_2O_3$. . . . . . . . . . | Pure | 9.3 |
| Tungsten–rhenium . . . . . | 75-25 | 5.6 | $ThO_2$–$Y_2O_3$. . . . . . | 25-75 | 9.5 |
| Ruthenium–molybdenum* | 89-11 | 8.5 | $ThO_2$–$Er_2O_3$ . . . . . | 50-50 | 10 |

*Cathode at 1500°C

TABLE 14. Compatibility of Nuclear Fuel with Cathodes of Refractory Metals for Thermionic Emission Converters [93]

| Fuel | Temperature, °C | Tungsten | Molybdenum | Niobium | Tantalum |
|---|---|---|---|---|---|
| UC | 1800—1900 | Compatible | * | * | * |
| $UC_2$ | 1800 | † | * | * | * |
| | 1200 | Compatible | Compatible | Compatible | Compatible |
| 10 molar % UC + 90 molar % ZrC | 1800 | Compatible | Compatible | ‡ | ‡ |
| $UO_2$ | 1800 | Compatible | § | § | § |
| | 2000 | » | § | § | § |
| | 1200 | » | Compatible | Compatible | Compatible |

*Formation of a liquid phase.
†Formation of a carbide layer and penetration along the grain surfaces.
‡Diffusion of uranium and zirconium at a high rate through the metal.
§Diffusion of fuel components along the grain boundaries (as a result, the cathode behavior is similar to that of a disperse cathode).

and also the considerable increase in current density, specific power, and efficiency of the TEC's renders rhenium and its alloys promising materials for TEC cathodes.

The interaction of refractory metals with uranium-containing nuclear fuel has been investigated [93]. This was done in connection with the problem of producing thermionic emission converters of nuclear energy into electrical energy. The interaction of tungsten, molybdenum, tantalum, and niobium with uranium carbide (UC) was studied at temperatures of 1800-2000°C. Tungsten was found to be the most stable metal with respect to uranium carbide; at 1800°C no visible traces were found of any reaction when these materials were in contact for 50 h. Under the same conditions, contact of molybdenum, niobium, and tantalum with UC led to the formation of carbides of these metals and the occurrence of a liquid uranium phase. Slight traces of reaction of UC with tungsten were found at 2000°C. The interaction of the above-mentioned materials may be described by the following reaction: UC + metal $\rightleftharpoons$ carbide of the metal + uranium (liquid). Tungsten also does not react with $UO_2$ at 1800 and 2000°C. In the case of the other metals, penetration of $UO_2$ components along the grain boundaries is found.

According to the results given in [93], the reaction of metals with uranium carbide may be suppressed by the introduction of an excess of carbon in the nuclear fuel ($UC_2$ + UC) or by stabilization of the uranium carbide (by the addition of zirconium).

Table 14 summarizes data on the compatibility of nuclear fuel with some refractory metals, based on a study of diffusion.

Indisputable interest attaches to the investigation of the compatibility of rhenium (in particular, tungsten–rhenium and molybdenum–rhenium alloys) with carbide nuclear fuel.

Investigations in recent years have shown the promising nature of the use of a number of low-melting metals and eutectics as heat carriers in power systems, due to the high values of the coefficients of heat transfer, and also the possibility of using these materials at high temperatures owing to the comparatively large temperature interval during which they remain in the liquid state. In this connection, the choice of a constructional material, which is inert to liquid metal media, becomes of considerable importance. Reaction between the solid and liquid metals may occur by mutual solubility (solution of the solid metal in the liquid metal or the formation of solid solutions in the surface layers of the constructional material), the formation of metallic compounds or intergranular penetration of the liquid metal, and also as result of reaction with the oxygen in the material. In connection with this last-mentioned circumstance, the use of refractory metals in single-crystal form, where the size of the components permits, offers some interest. Research has shown that niobium is the most inert of the refractory metals with regard to solution by lithium, sodium, potassium, and some other fusible metals and alloys [94]. The absence of reaction is due to the insignificantly low solubility of niobium in these liquid metals, and also to the absence of intermediate intermetallic phases in the corresponding systems. This, however, refers to heat carriers, well purified from oxygen; if the heat carriers are contaminated by oxygen, niobium reacts with them and becomes embrittled. Filters of various constructions are used for reducing the oxygen content in a stream of metallic heat carriers. For example, for reducing the oxygen content in a stream of sodium (or sodium–potassium alloys), the principle of the action of the filter is based on the low solubility of $Na_2O$ at temperatures slightly above the melting point of sodium [95].

The stability of molybdenum and some of the alloys based on it with regard to liquid sodium and potassium has been pointed out [96]. On this account, molybdenum is beginning to be used in reactor systems with sodium cooling; as in the case of niobium, special attention is paid to low oxygen content of liquid metal coolant [96]. Rhenium shows high stability in liquid potassium and sodium [91]; it also does not react with liquid gallium at 800°C, and under these conditions may be used as protective coating for molybdenun and nickel alloys and steel [97, 98]. Corrosion of rhenium is observed in molten lithium [99].

It has been found that niobium and molybdenum are most suitable for use in liquid lithium at temperatures above 1000°C; the solubility of these metals in lithium at 1000°C is $10^{-4}$%; vanadium and chromium are not stable at this temperature [100]. Investigations have shown that niobium and molybdenum are capable of resisting the action of molten uranium and its alloys [96, 101]. Molybdenum and niobium may be used as envelope for fuel of molten uranium and bismuth or plutonium and bismuth; the inertness of these metals to uranium alloys is also based on their insignificant mutual solubility and the impossibility of the formation of intermetallic compounds. Absence of reaction with liquid uranium is also found in yttrium [102].

The compatibility of tungsten, rhenium, tantalum, molybdenum, and niobium with nuclear fuel – a solid solution of carbides of uranium and zirconium (zirconium carbide diminishes the tendency to spalling of the uranium carbide in cyclic heating) – has been investigated [104]. It was shown that at working temperatures of about 2000°C, tungsten, tantalum, and rhenium are most promising materials for continuous use in contact with UC + ZrC; above 2000°C the reaction rate of the contacting materials rapidly increases. The compatibility of refractory metals with plutonium carbide was also investigated, this being in connection with the problem of producing a compact power source, using spontaneously decomposing isotopes, for example $Pu^{238}$. It was found that none of the refractory metals was compatible with PuC at temperatures above 1200°C. It was considered that the application of protective layers would extend the temperature range of the use of these materials in contact with PuC to higher temperatures [104].

Rhenium, in view of the absence of reaction with carbon, may be used as isolating layer between graphite (or a carbonaceous medium) and the carbide-forming refractory metal of the basic construction [163].

TABLE 15. Mechanical Properties of Refractory Metals of Commercial Purity Subject to Tension

| Metal | Tensile strength, kg/mm² | Elongation, % | Reduction in area, % |
|---|---|---|---|
| Titanium | 20–30 | 40–72 | — |
| Zirconium | 25–35 | 35–40 | — |
| Hafnium | 41–61 | 35–43 | 40 |
| Vanadium | 27 | 25 | — |
| Niobium | 35 | 50 | 80 |
| Tantalum | 38 | 70 | 95 |
| Molybdenum | 80–90 | 10–15 | — |
| Tungsten | 80–110 | 0 | — |
| Rhenium | 115–120 | 25–28 | — |
| Rhodium | 56 | 45 | — |
| Iridium | 40 | — | 10 |
| Platinum | 14–15 | 31 | 92 |

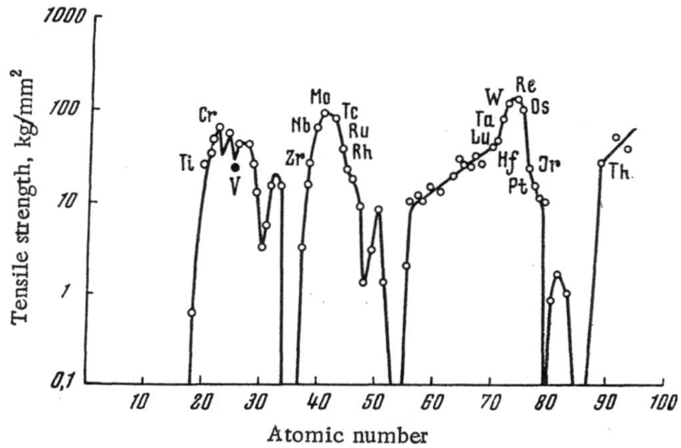


Fig. 19. Variation in tensile strength of refractory metals with their position in the periodic system.

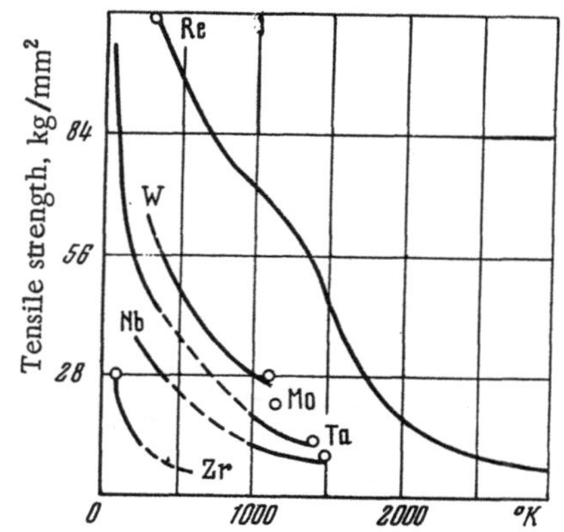


Fig. 20. Variation in tensile strength of some refractory metals with temperature [34].

The possibility of using tungsten–rhenium alloys as constructional materials for the capsules of isotope fuel blocks with $Po^{210}$ in a nuclear jet engine for spacecraft is indicated in [162]. Working temperature of the capsule is 2200-1600°C; time of operation, 3 months.

## Mechanical Properties

The high energy of the interatomic bond of metals of groups Va, VIa, and VIII, including rhenium (group VIIa), determines their high strength at room and elevated temperatures. Available information on the mechanical properties of refractory metals is quite approximate and to a great extent depends on the purity of the metal, and dimensions, method of preparation, and test conditions of the specimens. Values of the yield point of refractory metals at room temperature are given in Table 15. The character of the variation of tensile strength of the elements as a function of the atomic number is approximately similar to the variation in melting point (Fig. 19) [105]. The existence of such a correlation is particularly clear when curves showing yield point versus the corresponding temperature are plotted [106]. As is clear from Figs. 19 and 20, rhenium with a hexagonal structure has the maximum strength. To a considerable degree, this is due to the exceptionally high capacity of rhenium for work hardening. According to [107, 108], rhenium is work hardened 3.5 times faster than tungsten and molybdenum.

The high strengthening on hardening of rhenium is associated with its tendency to twinning, and the increase in the area of the intergranular boundaries is explained by the boundaries of the twins [109]. In addition to rhenium, two other refractory monomorphic metals, osmium and ruthenium, have a hexagonal lattice. The poor deformability of osmium and ruthenium of commercial purity at low temperatures does not permit their work-hardening capacity to be determined accurately over a wide range of reduction. Evidently the coefficient of hardening of these metals is close to that of rhenium.

In addition to hexagonal rhenium, osmium, and ruthenium, iridium and rhodium, metals having a face-centered cubic lattice, possess a high work-hardening capacity. The work-hardening capacity of the metals enumerated above may also be judged by the variation in hardness with the reduction in rolling (Fig. 21) [47]. Attention should be drawn to the fact that metals which are more prone to strengthening are characterized by high elastic properties. According to the authors of [110], the elastic constants constitute a most important factor affecting the rate of hardening of rhenium; at the same time, polycrystalline rhenium hardens more intensively than single-crystal rhenium, owing to the interaction between adjacent grains, stimulating repeated slip. This article also shows that the rate of hardening of a single crystal of

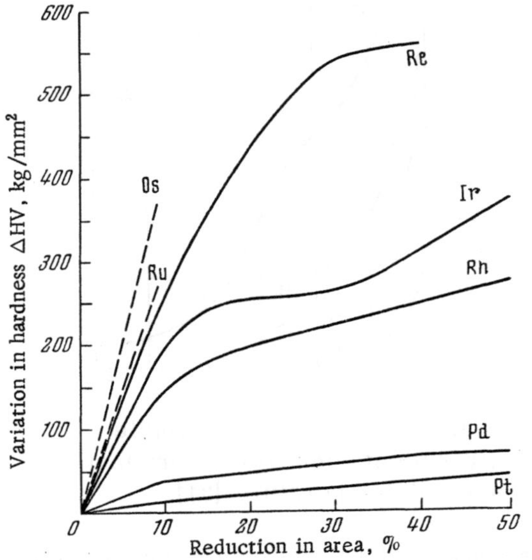


Fig. 21. Variation in hardness of some refractory metals after rolling with reduction in area.

rhenium depends on the initial crystallographic orientation; in the deformation of polycrystalline specimens, this leads to uneven work hardening of the individual grains and the formation of tensile stresses, which are the cause of intergranular cracking.

The high melting points and the anomalously high values of the modulus of elasticity of ruthenium and osmium are evidence of their high-strength properties. Similar strength properties are possessed by other refractory transition metals in the three large periods of the Mendeleev system, including all the metals of the platinum group.

The influence of temperature on the mechanical properties of monomorphic and polymorphic metals and their alloys and compounds, and the physicochemical processes occurring on heating have been examined in detail by E. M. Savitskii [39]. As the result of the researches made [39], certain fundamentally important relationships were found in the temperature variations of the mechanical properties of metallic materials with the type of crystal structure, character of the interatomic bond, and the nature of the physicochemical interaction of the components. The exponential dependence of the strength properties of pure monomorphic metals on temperature was shown, since in these metals heating does not produce any qualitative variations in the crystal structure. At the same time, plasticity increases steadily with temperature.

In metals of the hexagonal system, which at room temperature show a higher resistance to deformation than do metals of the cubic system, a sharp increase in plasticity is observed on heating in certain temperature ranges. This applies equally to fusible hexagonal metals, zinc, cadmium, magnesium and to refractory metals, rhenium, ruthenium, osmium (heating of the latter being in an inert atmosphere).

For polymorphic metals, a characteristic feature is the presence of discontinuities in the variation of the mechanical properties, due to the occurrence of new crystal modifications, transition to a high-temperature modification being accompanied by a sharp increase in plasticity [25, 39]. The existence of discontinuities is particularly clearly shown if the temperature dependence of the mechanical properties is plotted in semilogarithmic coordinates. The determination of the relationship in the variation of mechanical properties with temperature for polymorphic metals made it possible to formulate the plasticity rule already mentioned.

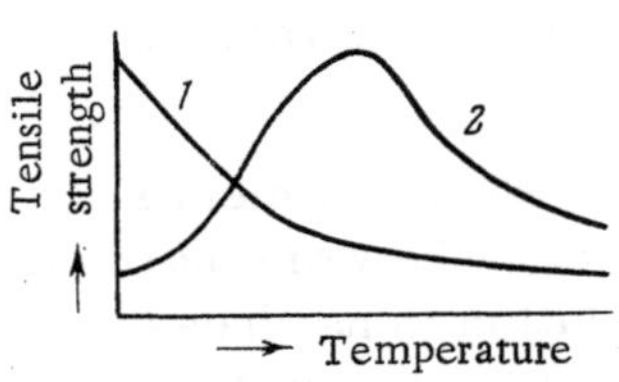


Fig. 22. Diagram showing influence of temperature on strength of plastic (1) and brittle (2) metallic materials.

The character of the variation of mechanical properties with temperature of alloys of different types of metallic systems has also been studied [39].

As shown in [39], brittle metallic materials, like semiconductors (graphite, silicon, germanium), metallic compounds, and certain metals of commercial purity (chromium), when heated show a deviation from the exponential temperature function of the

TABLE 16. Mechanical Properties of Niobium Subjected to Tensile Stress at Elevated Temperatures [113]

| Test temperature °C | Yield point, kg/mm² | Tensile strength kg/mm² | Elongation (on a length of 25.4 mm), % | | Reduction in area, % |
|---|---|---|---|---|---|
| | | | uniform | total | |
| 20 | 25.3 | 34 | 18.1 | 48.0 | 63.5 |
| 315 | 18.8 | 37.8 | 17.5 | 35.4 | 78.9 |
| 480 | 13.2 | 29.7 | 15.3 | 24.8 | 82.2 |
| 650 | 7.0 | 15.5 | 15.5 | 43.0 | 72 |
| 870 | 6.6 | 10.7 | 13.8 | 47 | 96 |
| 1095 | 5.7 | 7.0 | 8.8 | 34.0 | 100.0 |
| 1205 | 5.3 | 6.5 | 5.2 | 21.0 | 73.2 |
| 1315 | 2.8 | 3.9 | 17.2 | 70.0 | 81.8 |
| 1370 | 1.9 | 2.7 | 12.6 | 97.4 | 100.0 |

TABLE 17. Mechanical Properties of Tantalum at Various Temperatures [114, 118]

| Temperature, °C | Yield point, kg/mm² | Tensile strength, kg/mm² | Elongation (on a length of 50.8 mm), % | Strain-hardening coefficient, (m)* | Coefficient of sensitivity to rate of strain, (n)† |
|---|---|---|---|---|---|
| —195 | 104.0 | 104.0 | 3.7 | 0.018 | 0.0075 |
| —75 | 51.4 | 51 5 | 23.7 | 0.099 | 0.0307 |
| 25 | 40.3 | 47.2 | 25.3 | 0.164 | 0.0285 |
| 100 | 29.9 | 41.7 | 24.6 | 0.243 | 0.0175 |
| 205 | 24.9 | 39.4 | 12.7 | 0.326 | —0.0012 |
| 320 | 26.6 | 52.0 | 18.0 | 0.298 | 0.006 |
| 430 | 23.4 | 45.8 | 24.3 | 0.316 | 0.0240 |
| 540 | 18.3 | 42.1 | 16.2 | 0.356 | 0.0176 |
| 655 | 13.3 | 31.4 | 17.3 | 0.266 | 0.0220 |
| 760 | 11.9 | 21.3 | 23.1 | 0.205 | 0.0324 |
| 870 | 8.5 | 15.6 | 32.8 | 0.212 | 0.0374 |
| 980 | 8.7 | 15.2 | 33.1 | 0.181 | 0.0461 |
| 1095 | 5.7 | 11.8 | 43.2 | 0.217 | 0.0506 |
| 1205 | 5.3 | 10.3 | 47.5 | 0.130 | 0.0581 |

*$m = (\partial \log \sigma / \partial \log E)/\varepsilon T$, where $\sigma$ is the yield stress, E is the true strain, $\varepsilon$ is the rate of strain, T is the temperature.

†$n = (\partial \log \sigma / \partial \log E)/\varepsilon T$, calculated for $\varepsilon$ varying from 0.009 to 0.09 mm/min for E = 0.05.

strength properties [39], and increase their strength in a certain temperature range (0.7 to 0.9 $T_{mp}$), and then return to the exponential curve (Fig. 22). In its quantitative value, the effect of the increase in strength on heating is very considerable, amounting to hundreds percent for some substances. It should be taken into account in the development of refractory materials and the analysis of their behavior under working conditions. The physicochemical causes of this effect are not yet sufficiently clear; it is, however, indisputable, that this effect is typical of substances having a mixed interatomic bond and depends on the increase in the proportion of metallic bond when they are heated. More exactly, this effect is determined by the ratio of the strength and plasticity of materials which are brittle at room temperature. The increase in the proportion of metallic bond and the increase in mobility of the atoms on heating results in an increase in plasticity. Since the maximum manifestation of strength requires the occurrence in materials of some plasticity, the increase in the capacity for plastic deformation when brittle materials are heated is accompanied by an increase in strength up to some maximum value. The resulting microcracks, which under normal conditions lead to the fracture of the

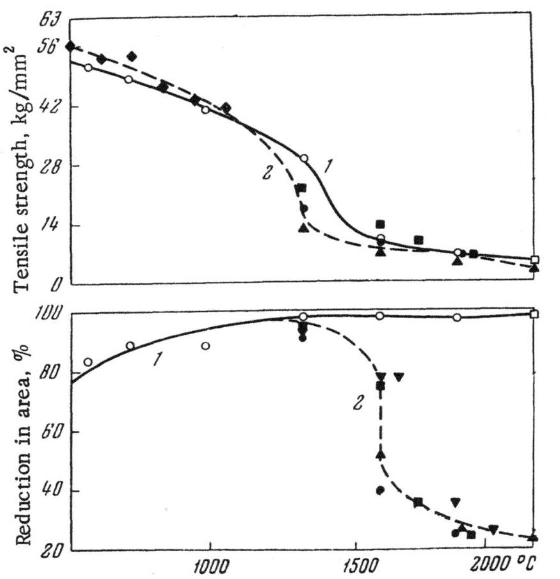


Fig. 23. Mechanical properties of tungsten produced by electric-arc vacuum melting (1) and powder metallurgy methods (2).

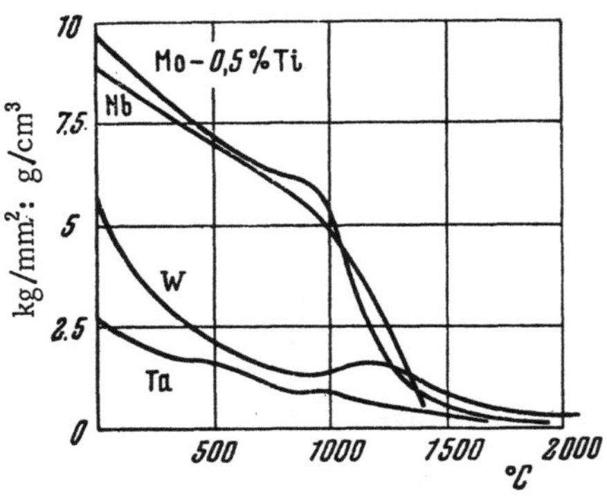


Fig. 24. Dependence of specific (ratio of yield point to density) of refractory metals on temperature.

specimen, are healed owing to the high diffusion mobility of the atoms. When maximum strength is attained with further increase in plasticity and increase in the rate of strength-reducing processes in proportion to the degree of heating, the strength diminishes as in all plastic materials.

The character of the influence of temperature on the mechanical properties of refractory metals under tensile stress depends largely on the initial purity and structure of the metal (recrystallized structure, deformation texture, etc.), i.e., finally on the production history of the specimens (methods and conditions of melting and mechanical and heat treatment).

The tensile properties of niobium of different degrees of purity was studied in detail in [112-115]. Table 16 presents the mechanical properties of vacuum-arc-melted niobium. To eliminate the action of the atmosphere, the tests were carried out in a vacuum. As shown by the table, at a temperature of 315°, some work hardening of niobium was observed, this being the result of strain aging. This question will be examined in more detail later. According to [114], strain aging was also found in electron beam melted niobium at a temperature of 500°C.

The mechanical properties of tantalum at elevated temperatures were studied in [114-120]. The properties of tantalum according to [118, 121] are given in Table 17. For tantalum, maximum strain aging effect under tension was found at 320°C.

A review of the mechanical properties of tungsten of different degrees of purity in a wide temperature range are given in [122]. The mechanical properties of deformed tungsten having a fibrous structure and tungsten subjected to recrystallizing anneal were investigated. As in the case of niobium and tantalum, an increase in strength was also found for tungsten, due to strain aging in the range 500-600°C. At temperatures above 1650°C, a considerable scatter was found in the plastic properties of tungsten.

Figure 23 compares the temperature dependence of the mechanical properties of tungsten obtained by electric-arc vacuum melting and by powder metallurgy methods [12]. At high temperatures, a pronounced difference in plasticity is observed; tungsten produced by electric-arc melting retains high plasticity at temperatures above 1650°C; the plasticity of cermet tungsten, characterized by a higher impurity content, falls considerably above 1650°C.

The influence of temperature on the mechanical properties of noble refractory metals is examined in [47]. The high content of interstitial impurities in rhodium and iridium produced

TABLE 18. Hardness (HB) of Refractory Metals*

| Metal | HB, $kg/mm^2$ |
|---|---|
| Titanium | 60 |
| Zirconium | 67 |
| Hafnium | 160 |
| Vanadium | 80 |
| Niobium | 80 |
| Tantalum | 125 |
| Chromium | 90-100 |
| Molybdenum | 150 |
| Tungsten | 286 |
| Rhenium | 274 |
| Ruthenium | 194 |
| Rhodium | 55 |
| Osmium | 350-600 |
| Iridium | 163 |
| Platinum | 30.5 |

*Hardness depends largely on the degree of purity of the metal; the table gives the approximate values of the hardness for metals of commercial purity.

by the ordinary methods results in an increase in the strength characteristics at low temperatures and in their reduction at elevated temperatures.

Owing to the use of refractory metals and alloys in aircraft construction, there is considerable practical interest in the assessment of the specific strength of these materials. Figure 24 shows the influence of temperature on the specific strength of tungsten, tantalum, molybdenum, and niobium [123]. At temperatures of up to 1370°C, niobium and molybdenum have considerable advantages compared with tungsten and tantalum with regard to specific strength; above 1370°C tungsten has the highest specific strength.

Hardness is one of the criteria of the strength properties of metals. For niobium and alloys based on it, the following relationships apply: Tensile strength = 0.28 HV (room temperature) and tensile strength = 0.30 HV (elevated temperature) [124-126]. The atomic number-strength and atomic number-hardness curves are similar in character. Among the refractory metals, tungsten, rhenium, and osmium have maximum Brinell hardness at room temperature (Table 18).

Some of the refractory metals (especially the polymorphic metals titanium, zirconium, hafnium) lose their hardness with relatively small rise in temperature. Other refractory metals (tungsten, molybdenum, rhenium, osmium) retain their high hardness up to temperatures close to the melting point. The ability of a metal to retain its hardness with increase in temperature increases with the atomic number and melting point.

Affiliated with the strength properties of metals are the elastic properties, to which reference has already been repeatedly made in connection with the nature of the interatomic bond in refractory metals. The value of the elastic modulus of metals and alloys is largely determined by crystal structure and interatomic binding forces, which in turn depend on the electronic structure of the atom. As is well known, the elastic modulus may be calculated from the natural frequency of vibration, as determined by the resonance method

$$E = 1.6388 \cdot 10^{-8} \left(\frac{l}{d}\right)^4 \frac{Q}{l} f^2 \ [\mathrm{kg/mm^2}],$$

where $l$ is the length of the specimen, cm; d is the diameter of the specimen, cm; Q is the weight of the specimen, g; and $f$ is the natural frequency of vibration of the specimen, cps.

The curve of elastic modulus versus atomic number is in many respects identical with the curve of tensile strength versus atomic number; both these properties are in periodic dependence on the atomic number of the element. Tungsten, rhenium, iridium, and osmium (in increasing order) have the highest values of elastic modulus, exceeding 40,000 $kg/mm^2$. The elastic modulus is connected approximately to the atomic volume and melting point by the following relationship:

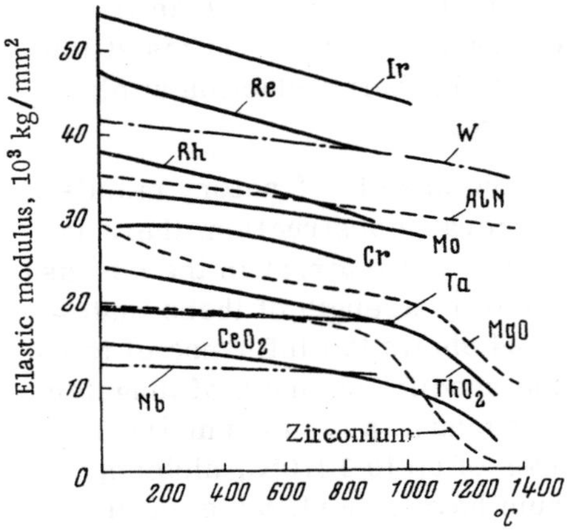


Fig. 25. Elastic modulus of some refractory materials as a function of temperature.

$$E = K\frac{T^2_{mp}}{V^3_{at}},$$

where $T_{mp}$ is the melting point in °K, $V_{at}$ is the atomic volume, and K is a constant.

In addition to the refractory metals, metals having a low atomic volume value (for example beryllium) also have a high elastic modulus. The temperature dependence of the elastic modulus of some refractory metals and oxides is shown in Fig. 25 [123].

Important characteristics of the elastic properties are compressibility and shear modulus. Compressibility is a measure of the interatomic forces opposing a change in volume, and shear modulus is a measure of the interatomic forces opposing a change in shape with constant volume. Compressibility and shear modulus are functions of the atomic number of the element. Table 19 presents data on the elastic properties of some refractory metals.

The creep strength of metals at high temperatures increases with increase in their melting point. This is because the creep process takes place by dislocation climb, as the result of diffusion of the atoms. The creep rate varies with temperature according to an equation proposed by Dorn [127],

$$\varepsilon = kE^{-Q/RT},$$

where ε is the creep rate, R is the gas constant, and Q is the activation energy of the creep process.

Maximum creep strength is possessed by chromium, molybdenum, niobium, tantalum, tungsten, and rhenium. Creep of the two most refractory metals, tungsten and rhenium, at different temperatures is represented graphically in Fig. 26. The tests were carried out in an argon or hydrogen atmosphere [165].

TABLE 19. Elastic Properties of Refractory Metals [43, 56]

| Metal | Compressibility coefficient $\times 10^3$, cm/kg | Normal elastic modulus, kg/mm$^2$ | Shear modulus, kg/mm$^2$ | Poisson ratio |
|---|---|---|---|---|
| Titanium | — | 9,000-10,000 | — | — |
| Zirconium | 1.097 | 8,960 | 3,330 | 0.35 |
| Hafnium | — | 9,800-14,060 | — | — |
| Vanadium | — | 13,500 | — | — |
| Niobium | — | 9,080 | 8,820 | 0.39 |
| Tantalum | 0.52 | 18,830 | 7,000 | 0.35 |
| Chromium | — | 25,000 | — | — |
| Molybdenum | 0.347 | 33,630 | 12,200 | 0.31 |
| Tungsten | 0.293 | 41,500 | 15,140 | 0.30 |
| Rhenium | — | 47,000 | — | — |
| Ruthenium | — | 42,000 | — | — |
| Rhodium | — | 28,640 | — | — |
| Osmium | — | 57,000 | — | — |
| Iridium | — | 53,830 | — | — |
| Platinum | — | 15,320 | — | — |

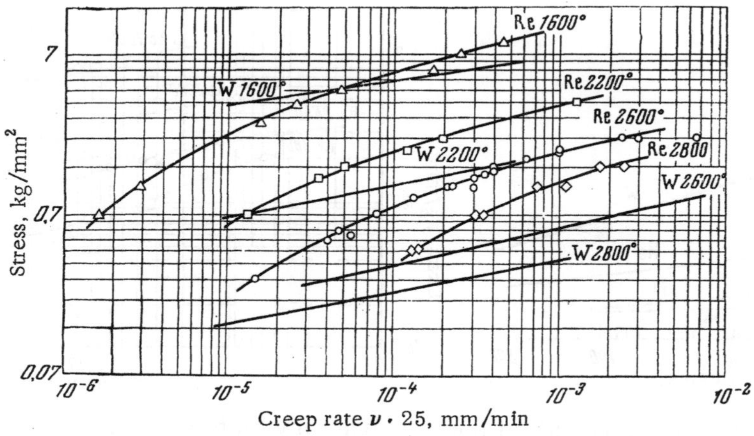


Fig. 26. Creep rate $\nu$ of rhenium and tungsten.

Allen and Carrington [128] investigated compressive creep of refractory metals of the platinum group (face-centered cubic lattice) and compared it with the creep of molybdenum and tungsten (body-centered lattice). The stress producing a deformation of 1% in 24 h at 1000°C was (in kg/mm$^2$) for platinum 0.14, for rhodium 4.6, for iridium 9.2, for molybdenum 4.6 to 6.16, and for tungsten 9.2.

Thus, according to [128], rhodium and iridium have approximately the same compressive creep strength as molybdenum and tungsten. According to the same data, the compression creep strength of tantalum is 4.4-6.3 kg/mm$^2$.

The results of creep tests on niobium (99.2% Nb) are given in Table 20.

Figure 27 compares the stress-rupture strength values of some refractory metals at 1095°C [131]. According to [132], the 100-hour stress-rupture strength of recrystallized molybdenum (arc vacuum-melted) is 8.4 kg/mm$^2$, and the endurance strength of recrystallized tungsten (100 h at 1095°C) is approximately 15.5 kg/mm$^2$.

The variation of the stress-rupture strength of some refractory metals for a 10-h test was studied in [133-135].

Figure 28 presents data on the variation in stress-rupture strength of tungsten and rhenium at different temperatures [165]. Above 1000°C, the stress-rupture strength of rhenium is slightly higher than that of tungsten.

It has already been pointed out that the mechanical properties of refractory metals is largely determined by their degree of purity. Impurities which form interstitial solid solutions with refractory metals have a particularly strong influence. Tables 21 to 23 illustrate the influence of interstitial impurities on the mechanical properties of niobium and tantalum.

TABLE 20. Results of Creep Tests on Niobium [129, 130]

| Condition | Temperature, °C | Stress, kg/mm$^2$ | Time taken to reach the given deformation (%), h | | |
|---|---|---|---|---|---|
| | | | 0.05 | 0.1 | 0.2 |
| Annealed at 1150°C . | 500 | 12.6 | 85 | 140 | 390 |
| Annealed at 1100°C . | 600 | 6.4 | 27 | 110 | 470 |
| The same . . . . . . . | 600 | 12.6 | 15 | 60 | — |

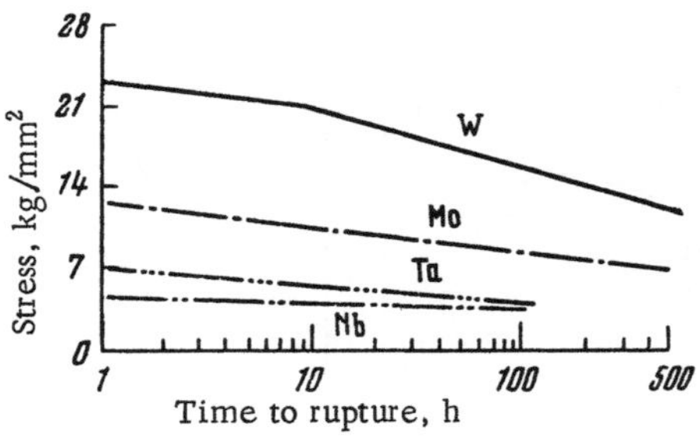


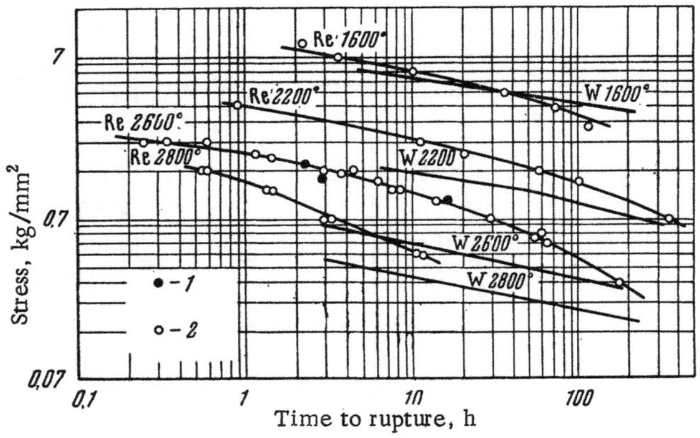


**Fig. 27.** Stress-rupture strength of some refractory metals (vacuum-melted) at 1000°C.

**Fig. 28.** Stress-rupture strength of tungsten and rhenium at different temperatures in an atmosphere of argon (1) and hydrogen (2).

**TABLE 21. Influence of Interstitial Impurities on the Mechanical Properties of Tantalum at Room Temperature [136**

| Composition of metal, wt.% | Tensile strength, kg/mm² | Yield point, kg/mm² | Elongation, % |
|---|---|---|---|
| Unalloyed tantalum . . . . . . . . . . | 20.4 | 18.3 | 36 |
| Tantalum – 0.0560% oxygen . . . . | 42.0 | 35.1 | 26 |
| Tantalum – 0.0225% nitrogen . . . | 51.3 | 47.8 | 25 |
| Tantalum – 0.0955% carbon . . . . | 50.6 | 34.4 | 20 |

**TABLE 22. Influence of Oxygen on the Mechanical Properties of Niobium under Tension [137]**

| Oxygen content, wt.% | Proportionality limit, kg/mm² | Tensile strength, kg/mm² | Elongation, % |
|---|---|---|---|
| 0.03 | 19.8 | 28.8 | 29.3 |
| 0.161 | 42.8 | 53.7 | 16.9 |
| 0.208 | 48.7 | 64.2 | 17.7 |
| 0.279 | 54.5 | 69.7 | 20.7 |
| 0.315 | 68.5 | 95.9 | 20.5 |
| 0.371 | 74.5 | 96.0 | 10.4 |
| 0.410 | 75.7 | 92.2 | 9.8 |
| 0.565 | Specimen cracked before test | | |

**TABLE 23. Influence of Interstitial Impurities on the Hardness of Tantalum [138]**

| Impurity content, wt.% | | | Brinell hardness (P = 500 kg), kg/mm² |
|---|---|---|---|
| carbon | oxygen | nitrogen | |
| 0.0033 | 0.077 | 0.027 | 90–105 |
| 0.001 | 0.044 | 0.010 | 73–75 |
| 0.006 | 0.006 | 0.007 | 60–64 |
| 0.001 | 0.006 | 0.009 | 56–60 |
| 0.002 | 0.004 | 0.005 | 55–57 |
| 0.002 | 0.006 | 0.001 | 53–55 |

TABLE 24. Mechanical Properties of Niobium and Tantalum in the Peened and Recrystallized Conditions [139, 140]

| Metal | Condition | Tensile strength, $kg/mm^2$ | Elongation, % | Reduction in area, % |
|---|---|---|---|---|
| Niobium | Peened* | 68 | 8 | 30 |
| | Recrystallized (annealed at 1200°C) | 38 | 23 | 81 |
| Tantalum | Peened* | 91-95 | 4-7 | – |
| | Recrystallized (annealed at 1300°C) | 60-63.5 | 15-22 | – |

*Degree of deformation 90%

As will be seen from the tables, in metals of group Va, the permissible impurity content is quite high. No sharp drop in plasticity is observed at the impurity content levels shown in the tables. This circumstance is due to the high solubility of interstitial impurities of group Va metals compared with group VIa metals and rhenium. The more detailed influence of interstitial impurities and the change in plasticity of refractory metals will be discussed in Chapter IV.

## Recrystallization of Refractory Metals

The recrystallization temperature may serve as a measure of the interatomic bond with metals and alloys. As a rule, the higher is the recrystallization temperature of a metal or alloy, the stronger is the interatomic bond and the higher is the strength at room and elevated temperatures. Therefore, in the development of heat-resistant alloys, primary interest lies with alloying elements which raise the recrystallization temperature.

When a metal or alloy is heated above the recrystallization temperature, a considerable reduction in strength occurs, accompanied by a sharp increase in plasticity. By way of example, Table 24 gives the mechanical properties of niobium and tantalum (arc vacuum-melted) in the peened and recrystallized states.

The recrystallization annealing of chromium, molybdenum, and tungsten of commercial purity results in their embrittlement, which is due to the redistribution of impurities and their precipitation on the grain boundaries (see Chapter IV). In a state of high purity with respect to interstitial impurities, the plasticity of these metals (like all the rest) is increased as the result of recrystallization.

A peculiarity in the behavior of rhenium during recrystallization annealing should be noted. Unlike other metals, rhenium shows no sharp increase in plasticity at the commencement of recrystallization, but only at higher temperatures (above 2000°C), when recrystallization is fully terminated [103].

### 1. Recrystallization Diagrams

It is well known that in the physical process of recrystallization, the atoms of the deformed metal overcome the atomic bonds in the distorted crystal lattice with the formation of nuclei of new, undistorted grains and their subsequent growth as the result of the transition of atoms from distorted crystallites to thermodynamically more stable, undistorted crystallites. As the result of recrystallization, the lattice distortions are eliminated and all the internal stresses are relieved. The first stage in recrystallization, consisting in the replacement of deformed grains by new undeformed grains, is called *treatment recrystallization*. The second stage is *collective recrystallization*, as the result of which growth occurs of the new, undistorted grains which have come into contact with each other. The rate of

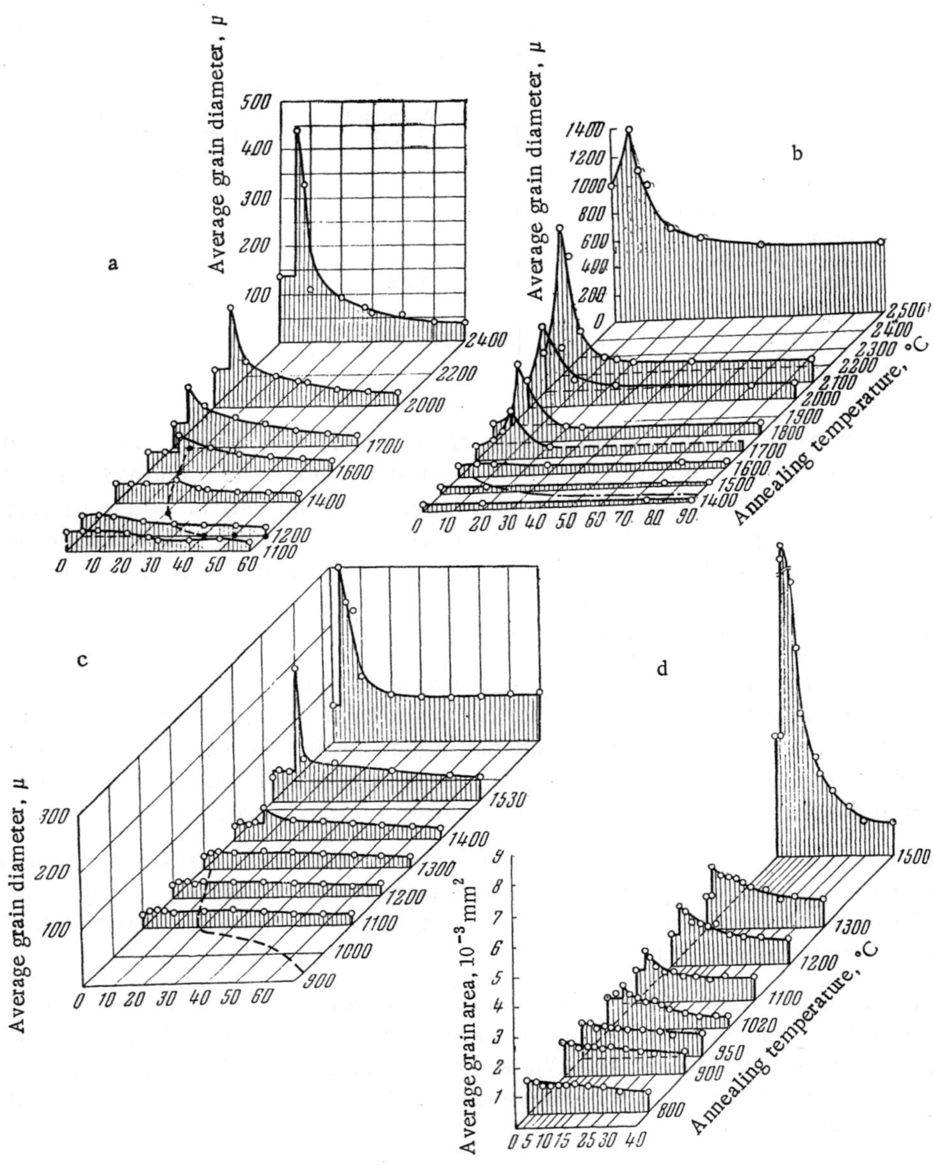


Fig. 29. Recrystallization diagrams of rhenium (a), tungsten (b), molybdenum (c), and chromium (d).

collective recrystallization is much lower than that of treatment recrystallization. In the final result, collective recrystallization ought to lead to the formation of a single-crystal structure of the metal, which thermodynamically is the most stable state. In practice, however, this final state (without special control) is never reached. According to the concepts of G. Tammann [141] and A. A. Bochvar [142], the retardation in grain growth in collective recrystallization is caused by various inclusions enveloping the grains. In high-purity metals, attainment of the single-crystal state is possible by collective recrystallization, and is used for the growth of single crystals of metals which do not possess a particularly high melting point and recrystallization temperature.

Published data on the temperatures of the commencement of recrystallization of different metals relate largely to metals of commercial purity, subjected to considerable cold deformation. The following are the temperatures (°C) of the commencement of recrystallization of re-

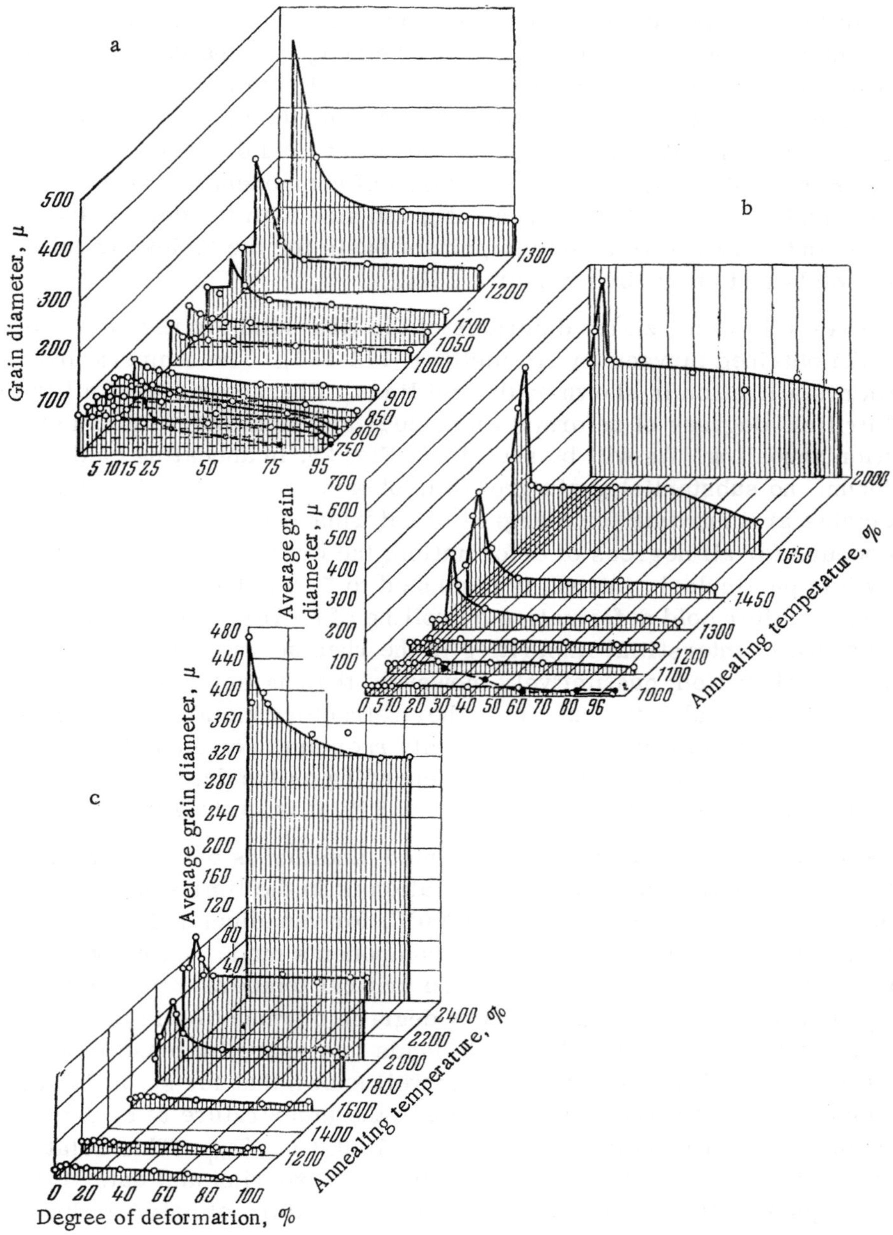


**Fig. 30. Recrystallization diagrams of vanadium (a), niobium (b), and tantalum (c).**

fractory metals (degree of deformation of more than 99%):

| | | | | | |
|---|---|---|---|---|---|
| Titanium | 500 | Tantalum | 1250 | Rhodium | 650* |
| Zirconium | 550 | Chromium | 950 | Ruthenium | 1175† |
| Vanadium | 800 | Tungsten | 1250 | Iridium | 1200* |
| Niobium | 940 | Rhenium | 1500 | Platinum | 525* |

*Degree of deformation 50%.
†Degree of deformation 85%.

The temperature of the commencement of recrystallization and the size of the grain, grown as the result of recrystallization annealing, depend very strongly on the degree of previous cold deformation. This relationship is expressed by a recrystallization diagram of the first kind in the coordinates grain size–degree of deformation–annealing temperature. Such diagrams may be used for selecting the correct annealing conditions after cold deformation, and also for determining the upper limit of the service temperatures of components. The recrystallization diagram of the second kind shows the variation of grain size with the degree and temperature of hot deformation, and is therefore a very important basis for the technological conditions in the treatment of metals and alloys by pressure [151].

Figure 29 gives recrystallization diagrams of the first kind for rhenium and refractory metals of group VIa obtained by vacuum arc melting [143-150]. The temperature of the commencement of recrystallization is shown by dashed lines. Characteristic of all the metals is the existence of critical degrees of deformation (5-20%) at which intense grain growth occurs in recrystallization annealing. As will be seen from Fig. 29, in annealing cold deformed specimens of molybdenum, no substantial grain growth is observed up to 1300°C; annealing at 1400°C results in appreciable grain growth for a degree of deformation of 10%; above 1400° increase in grain growth occurs for all degrees of deformation; the critical degree of deformation, at which grain growth is particularly intense is displaced to 7.5%. The recrystallization diagram of high-purity rhenium after cold deformation by rolling of polycrystalline specimens is given in [174]. The presence of interstitial impurities in the rhenium only varies the quantitative proportions in recrystallization; qualitatively, however, the character of the recrystallization diagram is not altered, as is also the case for other refractory metals. The temperature of the commencement of recrystallization of high-purity rhenium (oxygen 0.002%, hydrogen 0.005%, and nitrogen 0.006%) is in the range 100-1500°C, depending on the degree of previous deformation [174]; for cast rhenium of commercial purity (99.8%), it is in the range 1200-1750°C [146].

Intense grain growth on annealing cold deformed tantalum is observed at a temperature of 1800°C (Fig. 30). The critical degree of deformation is in the limits of 7-10%. For niobium, the rapid increase in grain size commences at 1300°C, for vanadium at 1100°C, and for rhenium at 1600°C. In the thermomechanical treatment of components made of the above-mentioned metals, therefore, it is undesirable to exceed these critical temperatures. In the treatment of refractory metals by pressure (cold and hot), low degrees of reduction in the last stages are inadmissible since, on subsequent annealing, this leads to a sharp increase in grain size and an impairment of the mechanical properties. It is well known that for strongly deformed metals of commercial purity, there is a relationship between the temperature of commencement of recrystallization and the melting point [22], i.e., $T_{recr}$ is 0.3-0.4 $T_{mp}$. This relationship does not hold for polymorphic metals. Processing of numerous experimental data [151] has shown that the ratio of recrystallization temperature to the temperature of polymorphic transition does not exceed 0.9, and $T_{recr}$ does not depend on $T_{mp}$.

The recrystallization diagram of a polymorphic metal must be regarded as a combination of the recrystallization diagram of each modification [152]. Figure 31 shows the recrystallization diagrams of the polymorphic metals titanium, zirconium, and hafnium [151, 166, 167]. The low-temperature modification is always fine-grained, and the high-temperature modification is coarse-grained. This is also true of alloys based on polymorphic metals [24, 153]. For producing finely crystalline components, the thermomechanical treatment of semifinished products of polymorphic metals and alloys, therefore, should be carried out at temperatures at which the low-temperature modifications exist. Research on recrystallization processes made by a number of investigators on metals of different degrees of purity has shown that the temperature of the commencement of recrystallization depends strongly on the degree of purity with respect to interstitial impurities. For example it has been shown [154] that the temperature of the commencement of recrystallization of high-purity aluminum (99.9999%) is -50°C, while for

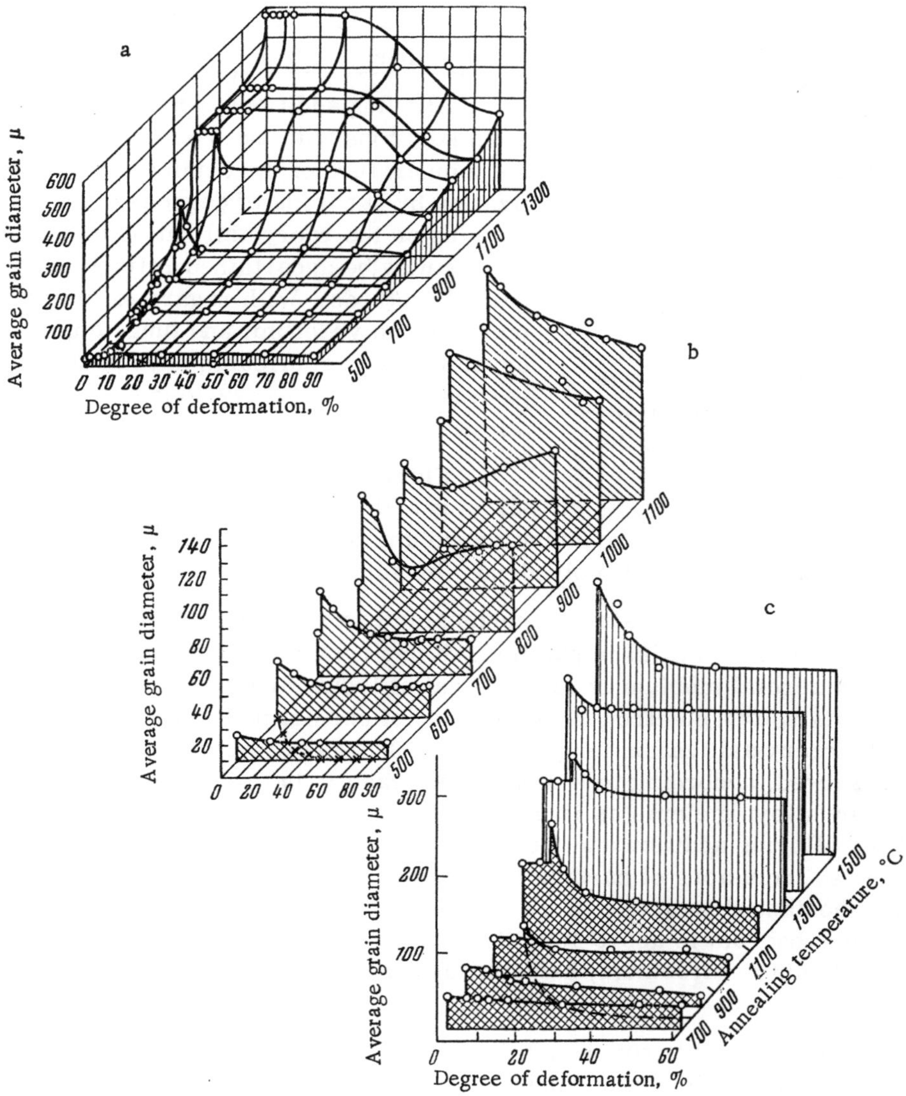


Fig. 31. Recrystallization diagrams of titanium (a), zirconium (b), and hafnium (c).

commercial aluminum (99.7%) it is +250°C; the temperature of the commencement of recrystallization for deformed single-crystal molybdenum produced by electron-beam zone refining is 650°C as against 935°C [155] for arc-melted molybdenum; for deformed single-crystal tungsten it is 850°C as against 1250°C for arc-melted tungsten. In single crystals, $T_{recr}$ depends strongly on orientation (influence of the initial crystallographic orientation on the temperature of commencement of recrystallization of deformed single-crystal refractory metals will be dealt with in Chapter V). The recrystallization temperature of niobium, vanadium, and tantalum is reduced by 200-300°C after electron-beam remelting of these metals. The cause of such a strong influence of interstitial impurities on the increase in recrystallization temperature has not yet been clarified. It is possible that this is due to the existence in the structure of the metals of disperse nonmetallic compounds, impeding direct contact of the grains and the development of diffusion processes.

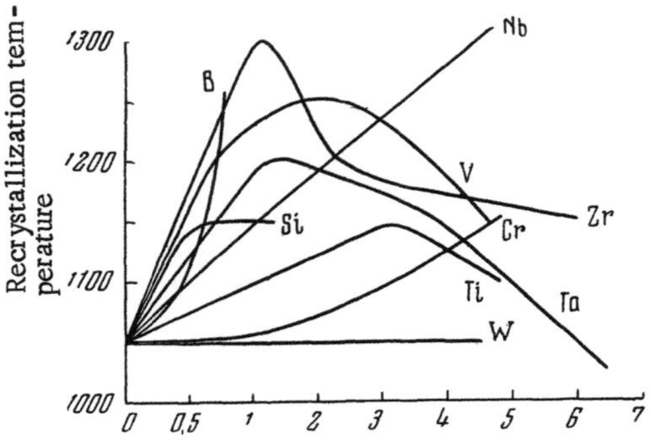


Fig. 32. Recrystallization temperature of commercially pure molybdenum versus content of different alloying additions.

## 2. Influence of Alloying Additions on the Recrystallization Temperature

Alloying additions, which form substitution solid solutions with the basis metal, may increase the temperature of the commencement of recrystallization by hundreds of degrees. The influence of alloying on the recrystallization temperature of metals may be ascertained reliably only by experimental methods. With regard to titanium, chromium, niobium, molybdenum, and tungsten, effective recrystallization retardants have been found, an important factor in the development of heat-resistant alloys.

Figure 32 presents data on the influence of alloying additions on the recrystallization of molybdenum [151]. Additions of boron, zirconium, vanadium, and niobium in amounts of 0.5-2% raise the recrystallization temperature of molybdenum by some hundreds of degrees.

The recrystallization temperature depends also to a considerable extent on the purity of the metal with regard to interstitial impurities. It has been found that the recrystallization temperature of commercial niobium is 1100°C, that of iodide niobium is 940°C, and that of electron-beam niobium is 900°C; electron-beam melting of cermet molybdenum or vacuum arc remelting lowers the recrystallization temperature by 300-400°C. At the same time, a definite relationship is observed, according to which the increase in the recrystallization temperature of the metal base by alloying elements is greater, the greater is the difference between the atomic radii of the alloying addition and the metal base. It is possible by uniform distribution of the finely disperse inclusions of carbides and oxides of the thermodynamically active elements to effect a substantial increase in the recrystallization temperature of refractory metals, for example, for molybdenum to above 1500°C. The maximum effect is produced in molybdenum by boron, silicon, titanium, zirconium, hafnium, and in niobium by tungsten, zirconium, and molybdenum.

The pronounced effect of boron, which is only slightly soluble in molybdenum, is evidently due to the formation of refractory molybdenum borides impeding the recrystallization of the basis metal. Some additions increase the recrystallization of the basis metal even in low contents. For niobium, such additions are vanadium, zirconium, and titanium. Evidently, the increase in the crystallization temperature of niobium by small additions of vanadium, zirconium, and titanium is due to the high thermodynamic activity of these elements which, with the impurities occurring in niobium, form refractory nonmetallic compounds inhibiting recrystallization. Rhenium has been found to be the most effective recrystallization inhibitor of body-centered cubic refractory metals and alloys. The recrystallization temperature of tungsten and molybdenum is increased to 400°C on the introduction of rhenium, due to the increase in the forces of the interatomic bonds in the solid solutions formed [153].

Considerable interest attaches to research on recrystallization processes in metals and alloys of high purity with regard to interstitial impurities. The influence of interstitial impurities (carbon, oxygen, and nitrogen) on the recrystallization temperature has been investigated [156]. For 70% of total deformation, the recrystallization of tantalum containing both carbon (0.0995%) and oxygen (0.056%) impurities commenced at a temperature of about 1000°C. Tantalum containing nitrogen impurity (0.0225%) did not recrystallize up to 1100°C.

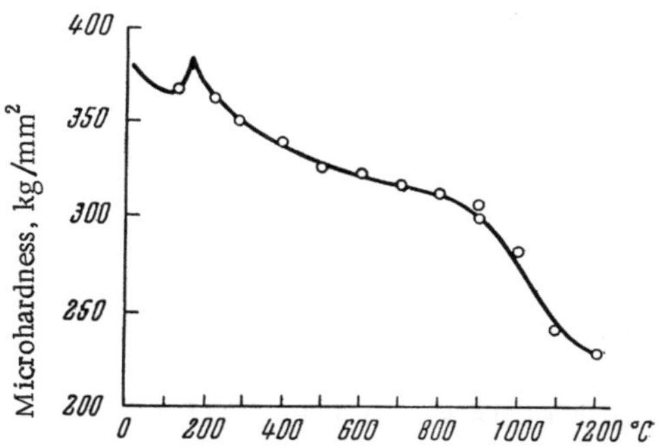


Fig. 33. Variation in microhardness of irradiated molybdenum during annealing.

## Effect of Irradiation on the Properties of Refractory Metals

In connection with the application of refractory metals for the constructional material of reactors using fast neutrons, the study of the change in properties of refractory metals exposed to the action of irradiation is very important. Generally speaking, it may be observed that irradiation leads to an increase in strength and a reduction in plasticity of metals. The structural variations and the associated changes in properties are reversible to a considerable extent, i.e., the original properties of the material may be restored by subsequent high-temperature annealing.

The influence of neutron irradiation on molybdenum was studied in [104]. After irradiation of molybdenum by an integral neutron flux of $10^{20}$ nvt (n is the number of neutrons per unit volume in the incident flux, v is the velocity of the neutrons, and t is the irradiation time of the material), accumulation of point defects in the structure, an increase in strength, and an increase in the ductile-brittle transition temperature were found. In addition to the radiation damage, interaction of the neutrons with the atoms of the irradiated material was also observed. As a result of this interaction, as pointed out in [104], gaseous helium was formed in the crystal lattice. Irradiation resulted in the separation of carbides. After annealing the irradiated molybdenum at 1200°C, microhardness was practically completely restored to the former value (about 225 kg/mm$^2$). A characteristic feature of the curve of the variation of the microhardness of irradiated molybdenum during anneal was the existence of a maximum at a temperature of 150°C (Fig. 33) [104].

Electron microscope examination, however, failed to reveal any changes in structure to explain the hardness maximum. Without dwelling on the changes in the dislocation structure of molybdenum by the action of a neutron flux followed by annealing, we should mention the very important part played by interstitial impurities in the distribution and behavior of radiation damage.

Considerable interest attaches to the assumption made in [159] that the removal from molybdenum (or any other metal having a body-centered cubic lattice) of interstitial impurities reduces the negative influence of irradiation on its plastic properties.

The influence of neutron irradiation on the transition temperature to the brittle state in molybdenum and tungsten has been studied [157]. The molybdenum and tungsten were irradiated by a flux of $5 \cdot 10^{19}$ neutron/cm$^2$. The increase in the ductile-brittle transition temperature for tungsten after irradiation was very slight (from +118 to +126°C), while for molybdenum the transition temperature increased from −136°C to −73°C. The authors of [157] draw the conclusion that the magnitude of the increase in the transition temperature after irradiation depends largely on the initial transition temperature; the lower is the initial transition temperature, the more does it increase as the result of irradiation of the metal.

The influence of neutron irradiation on the properties of niobium has been studied in [158]. It was found that by the action of irradiation, the tensile strength was increased (from 50 to 54 kg/mm$^2$ as also was the yield point (from 42 to 53 kg/mm$^2$), while the elongation of the irradiated was considerably reduced (from 20 to 7%). The initial properties of the niobium were restored after annealing for 1 h at 600°C. The increase in strength was explained by "condensation" of vacancies (formed during irradiation) on dislocation lines.

A similar effect of irradiation on the mechanical properties of tantalum are mentioned in [159].

Most recent data on the effect of irradiation on the properties of metals and alloys are given in the book by S. T. Konobeevskii [168].

## LITERATURE CITED


1. N. V. Ageev, Nature of the Chemical Bond in Metal Alloys [in Russian], Izd. Akad. Nauk SSSR (1947).
2. N. F. Mott and H. Jones, The Theory of the Properties of Metals and Alloys, Oxford (1936).
3. S. T. Konobeevskii, The Nature of Bonds in Metals. Transaction of Conference on the Theory of Metallic Alloys [in Russian], Izd. MGU (May, 1952).
4. S. T. Konobeevskii, Izv. Sektora Fiz.-khim. Analiza, 16(1):19, (1943).
5. S. T. Konobeevskii, Uchenye Zapiski MGU, No. 74, p. 13 (1944).
6. E. S. Makarov, Structure of Solid Phases Having a Variable Number of Atoms in the Unit Cell [in Russian], Izd. Akad. Nauk SSSR (1947).
7. N. F. Mott, Progress in Physics, 25:218 (1962).
8. F. Seitz, Modern Theory of Solids, McGraw-Hill, Inc, New York (1940).
9. S. V. Vonsovskii and Yu. A. Izyumov, Uspekhi Fiz. Nauk, 77:377 (1962).
10. E. M. Savitskii, G. S. Burkhanov, and Ch. V. Konetskii, Izv. Akad. Nauk SSSR, Metallurgiya i Gornoe Delo, No. 6, p. 12 (1963).
11. W. Hume-Rothery, Atomic Theory for Students of Metallurgy, Institute of Metals, London (1946).
12. L. Pauling, Phys. Rev., 54:899 (1938).
13. W. Hume-Rothery, H. M. Irving, and R.P.S. Williams, Proc. Roy. Soc., A, 208:31 (1951).
14. R. Kiessling, "Bonding in metal," Metals Rev., No. 2, p. 77 (1957).
15. R. S. Weiss and L. S. de Marco, Rev. Mod. Phys., 30:59 (1958).
16. M. I. Korsunskii and Ya. E. Genkin, Izv. Akad. Nauk SSSR, Seriya Fizicheskaya, 28(5):832 (1964).
17. Transactions of Conference "Nature of Metallic Phases and Its Character of the Chemical Bond in Them" [in Russian], Neorganicheskie Materialy, No. 11 (1965).
18. Collection: Ultrapure Metals [in Russian], Izd. "Metallurgiya" (1966).
19. G. Zaiman, Principles of Solid State Theory [Russian translation], Izd. "Mir" (1966).
20. Collection: Electronic Structure of Transition Metals and the Chemistry of Their Alloys [in Russian], Izd. "Metallurgiya" (1966).
21. Ya. I. Frenkel', Introduction to the Theory of Metals [in Russian], Moscow, Tekhteoretizdat (1950).
22. A. A. Bochvar, Physical Metallurgy [in Russian], Metallurgizdat (1956).
23. J. J. Gilman, The Physics and Chemistry of Ceramics, New York (1963).
24. E. M. Savitskii, Metallovedenie i Termoobrabotka, No. 7, p. 3 (1965).
25. E. M. Savitskii, Dokl. Akad. Nauk SSSR, 73(75):5 (1950).
26. W. H. Zachariasen, Acta Crystallogr., No. 5, p. 5 (1952).
27. E. Gebchardt, Segnezzi, Reaktorenwerkstoffen (1964).
28. V. I. Kutaitsev, Alloys of Thorium, Uranium, and Plutonium [in Russian], Atomizdat (1962).
29. W. H. Zachariasen, J. Inorg. and Nucl. Chem., 15(1-2):185 (1960).
30. Express Information, Metallovedenie i Termoobrabotka, VINITI (Ref. 34) (1964).
31. K. A. Gschneidner, Alloys of the Rare-Earth Metals [in Russian], Izd. "Mir" (1965).
32. K. Mendelson, Low-Temperature Physics [Russian translation], IL, p. 159 (1963).

33. E. M. Savitskii and Ch. V. Konetskii, Dokl. Akad. Nauk SSSR, 131(5):1137 (1960).
34. G. S. Zhdanov, Solid State Physics [in Russian], Izd. MGU (1962).
35. V. M. Amonenko, V. E. Ivanov, G. F. Tikhinskii, V. A. Finkel', and I. V. Shpagin, Fizika Metallov i Metallovedenie, 12(6):865 (1961).
36. A. E. Vol, Structure and Properties of Binary Metallic Systems [in Russian], Fizmatgiz (1959).
37. E. G. Ponyatovskii and A. I. Zakharov, Kristallografiya, 6(3):461 (1962).
38. M. A. Filyand and E. I. Semenova, Properties of the Rare Elements (Handbook) [in Russian], Izd. "Metallurgiya" (1964).
39. E. M. Savitskii, The Influence of Temperature on the Mechanical Properties of Metals and Alloys. Stanford University Press (1961).
40. A. G. Knapton, J. Savile, and R. Siddall, Less-Common Metals, 2(5):357 (1960).
41. E. M. Savitskii, V. F. Terekhova, I. V. Burov, I. A. Markova, and O. P. Naumkin, Alloys of the Rare-Earth Metals [in Russian], Izd. Akad. Nauk SSSR (1962).
42. A. D. Makvillen and M. K. Makvillen, Titanium [in Russian], Metallurgizdat (1958).
43. C. A. Hampel, Rare Metals Handbook, Reinhold, New York (1961).
44. D. E. Thomas and E. T. Hayes, The Metallurgy of Hafnium, Washington (1960).
45. W. Rostoker, The Metallurgy of Vanadium, Wiley, New York (1958).
46. G. L. Miller, Tantalum and Niobium, London (1959).
47. R. I. Jaffee, D. J. Maykuth, and R. W. Douglass, Refractory Metals and Alloys, New York p. 383 (1961).
48. D. R. Stuhl and G. C. Sinke, Thermodynamic Properties of the Elements, ASM (1956).
49. A. H. Sully, Metallurgy of the Rarer Metals. Chromium, Academic Press, New York (1954).
50. K. Agte and I. Vatsek, Tungsten and Molybdenum [in Russian], Izd. "Énergiya" (1964).
51. A. N. Krestovnikov and V. N. Vigdorovich, Chemical Thermodynamics [in Russian], Metallurgizdat (1962).
52. C. R. Tottle, Nucl. Engng., 3(26):212 (1958).
53. G. L. Miller, Tantalum and Niobium, London (1959).
54. A. Eucken and K. Ditrich, Z. Phys. Chem., 125(3):211 (1927).
55. B. G. Lifshits, Physical Properties of Metals and Alloys [in Russian], Mashgiz (1956).
56. K. A. Gschneidner, Physical Properties and Interrelationships of Metallic and Semimetallic Materials, New York (1964).
57. N. Mott and K. Stevens, Phil. Mag., No. 2, p. 1364 (1957).
58. D. M. Sivertsen and M. E. Nikol'son, Structure and Properties of Solid Solutions [in Russian], Izd. "Metallurgiya" (1964).
59. A. A. Smirnov, Theory of the Electrical Resistance of Alloys [in Russian], Kiev, Izd. Akad. Nauk Ukrainian SSR (1960).
60. E. A. Lynton, Superconductivity, Wiley, New York (1962).
61. Ch. Raub, Z. Metallkunde, 55(4):195 (1964).
62. I. Bardeen, L. N. Cooper, and I. R. Schieffer, Phys. Rev., 108:1175 (1958).
63. V. L. Ginsburg and L. D. Landau, Zh. Éksp. Teor. Fiz., 20(12) (1950).
64. E. M. Savitskii and V. V. Baron, Izv. Akad. Nauk SSSR, Metallurgiya i Gornoe Delo, No. 5, p. 3 (1963).
65. E. M. Savitskii, V. V. Baron, V. R. Karasik, S. Sh. Akhmedov, V. Ya. Pakhomov, and M. I. Bychkova, Pribory i Tekhnika Éksperimenta, No. 1, p. 182 (1963).
66. Collection: Superconducting Materials, Russian translation from the English, edited by E. M. Savitskii, Izd. "Mir" (1965).
67. E. M. Savitskii, in: Metallurgy and Metal Physics of Superconductors [in Russian], Izd. "Nauka," p. 3 (1965).
68. A. A. Rudnitskii, Thermoelectric Properties of the Noble Metals and Their Alloys [in Russian], Izd. Akad. Nauk SSSR (1956).

69. B. M. Tsarev, Radiotekhnika i Elektronika, 2(6):675 (1957).
70. C. J. Smithells, Tungsten: A Treatise on Its Metallurgy, Properties, and Applications (3rd ed.) Chemical Publishing Co., New York (1952).
71. W. Espe, Werkstoffkunde der Hochvakuumtechnik, Berlin (1959).
72. Collection: Thermoemission Conversion of Energy [in Russian], Atomizdat, p. 1. (1964).
73. G. N. Shuppe, Electronic Emission of Metallic Crystals [in Russian], Tashkent, Izd. Sredne-aziatski. Gos. Univ. (1959).
74. C. G. Linde and G. K. Karpetcher, J. Electrochem. Soc., 108:1079 (1961).
75. H. Webster, J. Metals, 72 (1963).
76. H. Webster, J. Appl. Phys., 32:1802 (1961).
77. W. R. Savage, Bull. Am. Phys. Soc., 8(2):198 (1963).
78. M. Drechsler and R. Vannelow, Z. Kristallogr., 107:161 (1956).
79. J. Bardeen, Phys. Rev., 49:653 (1936).
80. V. P. Shrednik, Fiz. Tverd. Tela, 3(6):1750 (1961).
81. V. S. Fomenko, Emission Properties of Elements and Chemical Compounds [in Russian], Kiev, Izd. Akad. Nauk Ukr. SSR (1962).
82. E. M. Savitskii, I. V. Burov, O. P. Naumkin, and V. F. Terekhova, Izv. Akad. Nauk SSSR, Neorg. Materialy, No. 6 (1966).
83. E. M. Savitskii, I. V. Burov, et al., Zh. Tekh. Fiz., No. 7, p. 1310 (1966).
84. K. Hering and M. Nichols, Thermionic Emission [Russian translation], IL (1950).
85. L. N. Dobretsov, Electronic and Ionic Emission, Tekhteoretizdat, Moscow (1952).
86. G. R. Fond, Phys. Rev., 21:343 (1923).
87. G. R. Fond, Phys. Rev., 31:260 (1928).
88. L. Richardson, M. Gottlieb, and R. Solweg, in: Thermionic Emission Energy Conversion [in Russian], Vol. 1, Atomizdat, p. 144 (1964).
89. D. E. Fornwall, B. R. Gourley, and A. V. Manzione, Sympos. Electrochem. Soc. Meeting, Washington (October 1964) (CNLM-5942).
90. D. M. Scruggs, The Electrochemical Soc., New York (1963).
91. M. Slivka, in: Thermionic Emission Energy Conversion [in Russian], Vol. 1, Atomizdat, p. 129 (1964).
92. J. H. Port, Metalle für die Raumfahrt, Metallwerk Plansee A. G., Reutte/Tyrol, p. 613 (1965).
93. A. Weinberg and L. Yang, in: Thermionic Emission Energy Conversion [in Russian], Vol. 1, Atomizdat, p. 74 (1964).
94. A. B. McIntosh and K. Q. Bagley, J. Inst. Metals, 84:251 (1955-56).
95. A. B. McIntosh, Problemy Sovremennoi Metallurgii, No. 4, p. 78 (1957).
96. D. Dimpel, Metall, No. 3, p. 211 (1962).
97. W. D. Wilkinson, J. Inst. Metals, 83:881 (1954).
98. E. M. Sominskaya and A. A. Nikitina, in: Rhenium [in Russian], Izd. "Nauka," p. 90 (1964).
99. A. V. Gresse, Naturforschung, 8:533 (1953).
100. V. K. Grishin, M. G. Glazunov, A. G. Arakelov, A. V. Vol'dent, and T. S. Makedonskaya, Properties of Lithium [in Russian], Metallurgizdat (1963).
101. G. Boll, Problemy Sovremennoi Metallurgii, No. 5, p. 82 (1956).
102. G. Macguty and S. Simmons, Problemy Sovremennoi Metallurgii, No. 4, p. 107 (1961).
103. N. Orangeburg, Am. Metal Market, 65(107):14 (1963).
104. M. E. Waldron, J. Adam, and B. L. Eyre, Metalle für die Raumfahrt, Metallwerk Plansee A. G., Reutte/Tyrol, p. 859 (1965).
105. B. B. Gulyaev, O. N. Magnitskii, and A. A. Demidova, Casting of Refractory Metals [in Russian], Izd. "Mashinostroenie" (1964).
106. G. M. Pew, Problemy Sovremennoi Metallurgii, No. 1, p. 74 (1956).

107. T. Sims, M. Craighead, and R. I. Jaffee, Problemy Sovremennoi Metallurgii, No. 5, p. 139 (1960).
108. E. M. Savitskii and G. E. Chuprikov, Izv. Akad. Nauk SSSR, OTN, Metallurgii i Toplivo, No. 6, p. 167 (1956).
109. A. T. Churchman, Trans. AIME, 218:262 (1960).
110. G. A. Hitch, R. A. Jeffery, and E. Smith, Problemy Sovremennoi Metallurgii, No. 5, p. 139 (1960).
111. Ito, Rept. Tohoku Univ., No. 12, p. 52 (1923).
112. A. V. Mincher and V. F. Sheely, Problemy Sovremennoi Metallurgii, No. 4, p. 123 (1960).
113. R. T. Begley and W. N. Platte, Wright Air Development Center, Techn. Report WADC-TR-57-344, Part 4 (April 1960), p. 131.
114. Columbium and Tantalum, Frank T. Sisco and Edward Epremian (eds.), Wiley, New York (1963).
115. R. W. Fountain and C. R. McKinsea, in: Columbium and Tantalum, Wiley, New York, p. 198 (1963).
116. B. L. Mordyke, Problemy Sovremennoi Metallurgii, No. 4, p. 149 (1960).
117. J. H. Bechtold, Acta Metallurgica, 3:249 (1955).
118. W. Pugh, Trans. ASM, 48:677 (1956).
119. F. F. Schmidt, W. D. Klopp, W. M. Albrecht, F. C. Holden, H. X. Ogden, and R. I. Jaffee, Wright Air Development Division Techn. Report WADD-TR-59-13 (March 1960), p. 152.
120. M. Schussler and J. S. Braunhouse, Trans. AIME, 218:893 (1960).
121. E. M. Savitskii and V. F. Terekhova, in: Chemistry of the Rare Elements [in Russian], No. 2, Izd. Akad. Nauk SSSR (1955).
122. J. L. Ratcliff and H. R. Ogden, Defense Metals Inform. Center Mem., No. 157 (1962).
123. R. H. Gassner, Some Problems of Refractory Metals and Alloys [Russian translation], IL (1963).
124. M. Semchishen, in: Molybdenum [Russian translation], IL, p. 189 (1962).
125. R. I. Jaffee, in: Molybdenum [Russian translation], IL, p. 227 (1962).
126. N. N. Morgunova, Refractory Alloys Based on Molybdenum, VINITI (1959).
127. J. E. Dorn, Some Fundamental Experiments on High-Temperature Creep, Proc. Nat. Phys. Lab. Symposium on Creep and Fracture of Metals at High Temperatures, H.M.S.O. (June 1954).
128. N. P. Allen and W. E. Carrington, J. Inst. Metals, 82:525 (1953-54).
129. T. J. Heal, in: Niobium and Tantalum, Academic Press, New York, p. 88 (1960).
130. C. R. Tottle, Nuclear Engng., 3:212, 226 (1958).
131. J. A. Gurklis, L. D. McCraw, and C. L. Faust, Defense Metals Inform. Center Mem., No. 98, p. 16 (1961).
132. S. W. Pugh, Trans. ASME, 48:677 (1956).
133. C. T. Sims and R. I. Jaffee, J. Metals, 8:913 (1956).
134. S. W. Pugh, J. Metals, 10:35 (1958).
135. S. W. Pugh, Trans. ASME, 47:984 (1955).
136. High Temperature Materials, Vol. 2, G. M. Ault, W. F. Barclay, and H. R. Munger (eds.), Interscience Publishers, New York, p. 81 (1963).
137. C. R. Tottle, J. Inst. Metals, 85:375 (1956-57).
138. R. F. Rolsten, Trans. AIME, 215:472 (1959).
139. E. M. Savitskii, V. V. Baron, and K. N. Ivanova, Dokl. Akad. Nauk SSSR, 126(4):771 (1959).
140. I. A. Tsyganova, Reaction of Tantalum with Refractory Metals and the Properties of Some Alloys. Author's Abstract of Candidates Dissertation [in Russian], Moscow, IMET (1965).
141. G. Tammann, Physical Metallurgy, ONTI NKTP, Moscow (1935).
142. A. A. Bochvar, Principles of the Heat Treatment of Alloys [in Russian], Moscow (1940).

143. E. M. Savitskii, V. V. Baron, and K. N. Ivanova, Dokl. Akad. Nauk SSSR, 113(5):1070 (1957).
144. E. M. Savitskii, M. A. Tylkina, A. S. Stroev, and A. I. Pekarev, Dokl. Akad. Nauk SSSR, 140(6):1301 (1961).
145. E. M. Savitskii, V. F. Terekhova, and A. V. Kholopov, Dokl. Akad. Nauk SSSR, 109(4):794 (1956).
146. E. M. Savitskii, M. A. Tylkina, and K. B. Povarova, Dokl. Akad. Nauk SSSR, 119:274 (1958).
147. E. M. Savitskii, V. V. Baron, and K. N. Ivanova, Dokl. Akad. Nauk SSSR, 126(4):771 (1959).
148. E. M. Savitskii, V. V. Baron, and Yu. V. Emifov, Dokl. Akad. Nauk SSSR, 145(3):612 (1962).
149. E. M. Savitskii, M. A. Tylkina, and I. A. Tsyganova, Dokl. Akad. Nauk SSSR, 118(4):720 (1958).
150. E. M. Savitskii, M. A. Tylkina, and A. N. Turanskaya, in: Titanium and Its Alloys [in Russian], Izd. Akad. Nauk SSSR, p. 33 (1958).
151. E. M. Savitskii, Rare Metals and Alloys [in Russian], Izd. Doma Tekhniki (1959).
152. E. M. Savitskii, M. A. Tylkina, and K. B. Povarova, Rhenium Alloys [in Russian], Izd. "Nauka" (1965).
153. E. M. Savitskii and A. A. Tylkina, in: Research on Refractory Alloys, Vol. 4 [in Russian], Izd. Akad. Nauk SSSR, p. 218 (1959).
154. F. Montoriol, R. Reich, R. Albert, and G. Chandron, Compt. Rend., 238(7):815 (1954).
155. E. M. Savitskii, G. E. Chuprikov, and A. S. Grokhochinskii, Byull. "Obmen Opytom v Radioelektronnoi Promyshlennosti," No. 10, p. 1 (1964).
156. E. Gebhardt, H. D. Segherri, and W. Durrschnabel, Metallwerk Plansee A. G., Reutte/Tyrol, p. 291 (1959).
157. M. J. Makin and E. Gillis, Problemy Sovremennoi Metallurgii, No. 4, p. 58 (1958).
158. M. J. Makin and F. J. Minter, Acta Metallurgica, 7:361 (1959).
159. D. O. Liser, in: Effect of Nuclear Radiation on the Structure and Properties of Metals and Alloys [in Russian] Metallurgizdat, p. 156 (1957).
160. E. M. Savitskii and A. M. Zakharov, Izv. Vyssh. Uch. Zaved. Chernaya Metallurgiya, No. 1, pp. 104-119 (1965).
161. R. Kieffer and H. Braun, Vanadin-Niob-Tantal, Berlin (1963).
162. Nucl. Week, No. 35, p. 5 (1964).
163. L. K. Hansen and N. S. Rasor, Evaluation of Metal Emitters for Thermionic Converters, Metallurgy of Semiconductor Materials.
164. Thermionic Efficiencies Increased, Nucleonics, 20(5) (1962).
165. P. N. Flagella and C. O. Tarr, AIME Technical Meeting, French Lick, Indiana (1965).
166. M. A. Tylkina, K. V. Kirilenko, and E. M. Savitskii, Dokl. Akad. Nauk SSSR, 127(2):310 (1959).
167. E. M. Savitskii and V. F. Terekhova, Trudy Inst. Metallurgii im. A. A. Baikova, No. 3, p. 181 (1958).
168. S. T. Konobeevskii, Effect of Irradiation on Materials. Introduction to the Radiation Study of Materials [in Russian], Atomizdat (1968).
169. J. M. Ziman, Advances in Physics, No. 13, p. 89 (1964).
170. Collection: Quantum Physics and the Solid State [in Russian], Izd. "Znanie" (1966).
171. F. Heiniger, Elektronische spezifische Wärme und Antiferromagnetismus in Chromlegierungen, Geneva (1966).
172. F. I. Morin and I. P. Maita, Phys. Rev., 129(3):1115 (1963).
173. B. Chalmers, Physical Metallurgy, Wiley, New York (1959).

CHAPTER III

# PHYSICOCHEMICAL INTERACTION OF REFRACTORY METALS WITH ELEMENTS OF THE PERIODIC SYSTEM

Refractory metals have an optimum combination of a number of valuable physical properties which makes these metals irreplaceable in the new technology. The properties of but a few pure metals have already been discussed, but alloys of two or more refractory metals have also found wide application in technology. However, even at the commencement of this century, Academician N. S. Kurnakov discovered the laws governing the variation in many physical properties in binary equilibrium systems ("Kurnakov laws"). For example, he explained that the formation of solid solutions of metals was accompanied by an increase in hardness and electrical resistance compared with their values for the original components (there may be exceptions in ternary systems); he found that hardness depended on the composition of the mechanical mixture of the components, and he showed that metallic compounds had much higher hardness and electrical resistance values than the metals forming them. The teaching of N. S. Kurnakov and his school played an important part in the production of new metallic materials with given properties, and is still the most fruitful theoretical line of advance in this field [1-5].

It is impossible to conduct a search for new alloys without a knowledge of phase or constitutional diagrams which clearly represent the phase constitution of alloys under equilibrium conditions as a function of composition and temperature; the diagrams serve as basic, theoretical foundation for the choice of the composition of alloys. More than half of the known phase diagrams in world literature [8] which formed the basis of the development of commercial alloys were constructed by workers of Kurnakov's school [6, 7, 9]. Phase diagrams are constructed experimentally, using various methods of physicochemical analysis (thermal, microstructure, x-ray analysis, and the variation in electrophysical and mechanical properties as a function of the composition).

The refractoriness and high thermodynamic activity at elevated temperatures of most of the metals under consideration call for the application of special methods in the preparation of the alloys and the study of their properties in the construction of phase diagrams. Currently, only a few attempts to calculate theoretically the simplest (binary and ternary) phase diagrams from thermodynamic data (determination of the solubility lines and surfaces, liquidus lines, and surfaces) are known [11-15]. Considering the large volume and laboriousness of work on the construction of phase diagrams, the problem of enlisting the aid of computer facilities and the mechanization and automation of the measurements is becoming one of increasing urgency.

The type of phase diagram and the characteristics of the interaction of the components are primarily determined by the outer electron structure of the latter. As pointed out by N. V. Ageev [16], the phase diagram is the expression of the interatomic interaction of the components; the structure of the phase diagram is determined by the chemical properties of its components, i.e., by the tendency of the atoms of one component or the other to part with or receive valence electrons. Electrons situated in the deeper shells, however, have a definite influence on the behavior of metals in alloying; in the example of the rare-earth metals, this is shown in connection with the f-shell [10].

Phase diagrams describe the structure of alloys in the solid and liquid states in every possible combination of the components. This is still not enough for the development of new alloys; it is essential to know how structure is connected with the physical and chemical properties of alloys. Therefore, closely bound up with phase diagrams and of no less importance are composition–property diagrams, characterizing the variation in mechanical, electrical, technological, and other properties of alloys as a function of composition and of the type of phase diagram. Actually, phase diagrams and composition–property diagrams contain almost all the information on the physicochemical research of alloys necessary for the development of new compositions. Unfortunately, in many metallographical investigations, the construction of composition–property diagrams is underestimated.

## Present State of the Theory of Metal Alloys

As already stated, there is as yet no quantitative metallic alloy theory. Therefore, the choice of materials for the new technology has to be made experimentally, taking into account already revealed laws in the variation in mechanical, electrical, and physicochemical properties of alloys as a function of their composition, structure, melting point, systems of active forces, rate of deformation (strain), and sensitivity to the action of aggressive media, penetrating radiations, acceleration, vibration, heat cycling, and presence of a vacuum [17].

The results of numerous investigations of the interaction of different elements of the periodic system, as well as certain theoretical considerations, has made it possible to establish a number of semiempirical criteria, for determining the tendency of components to form liquid and solid solutions and chemical compounds. In Russia, the maximum contribution to the development of these concepts has been made by Kurnakov [1], Ageev [3], and Konobeevskii [5], and abroad by Hume-Rothery [18, 19], Mott [20], Kubaschewski [15], Gschneidner [21], Seitz, Darken, and Gurry [14], and others.

According to modern views, the solubility of elements in one another is determined by three factors: crystal-chemical (similarity or difference in the crystal lattices of the solvent or dissolved element), dimensional (difference in the atomic radii of the components), and chemical (value of electronegativity, as reflecting the position of the elements in the periodic system, more exactly the energy of the bond between two elements). For the formation of continuous solubility or a wide range of solid solutions between metals, there must be a similar structure of the outer electron shells of the atoms of the interacting components, isomorphism of their crystal lattices, and only a slight difference in their atomic diameters (not more than 8-15%).

When there is a considerable difference in these criteria, the metals do not dissolve well in one another in the liquid and solid states and reveal a considerable tendency to form with each other chemical compounds with the existence of different forms of mixed bond between the atoms ranging from the metallic to the purely ionic bond. As shown by S. T. Konobeevskii, the boundary of a solid solution corresponds to minimum thermodynamic potential and critical electronic concentration. When atoms of a different valence dissolve in a metal, there will be a change in the filling of the energy zones. The result will be that the limit solubility will vary appreciably.

When an alloying component dissolves, there is elastic deformation of the crystal lattice of the solid solution, the distortion of the crystal lattice occurring around a dissolved atom being greater, the greater is the difference in the atomic diameters of solvent and alloying component. The solubility limit corresponds to the composition for which the force of elastic deformation of the crystal lattice no longer increases, and a second phase is precipitated, leading to a reduction in the stresses. Nepton [23] cites the Lowson equation expressing the dependence of the deformation energy in the alloying of a solid solution on the shear modulus and molar volumes of the pure components.

An empirical rule, connecting the solubility limit in the solid state with the dimensional factor was first formulated by Hume-Rothery. The criterion for determining the dimensions of an atom was the shortest distance between the atoms of the crystal lattice. According to the opinion of Spedding and Daan [24], this method of determining the geometrical dimensions of the atom is quite arbitrary, since complex interactions of the atoms in the solid state are reflected in the dimensions of the dissolved atom.

Recently, in publications on the theory of alloys, considerable attention has also been paid to the electronegativity factor, which characterizes the affinity for an electron and is a function of the electronic structure of the atom [14]

$$x = 0.31\left(\frac{n+1}{r}\right) + 0.50,$$

where n is the number of valence electrons and r is the atomic radius; the electronegativity x is related to the work function $\varphi$ by the linear equation $x = 0.44\varphi - 0.15$ [25]. The greater is the difference between the electronegativity of two elements, the greater is the probability of the formation of a stable ionic compound.

As pointed out by Gschneidner [21], the tendency to form compounds is an electrochemical factor, limiting the range of formation of solid solutions. The mutual solubility is always low if the difference in the value of the electronegativity of the elements exceeds ±0.4 unit.

Another factor which limits the range of the existence of solid solutions is the valence factor. According to Hume-Rothery and Raynor [26], the bond energy ought to decrease if the

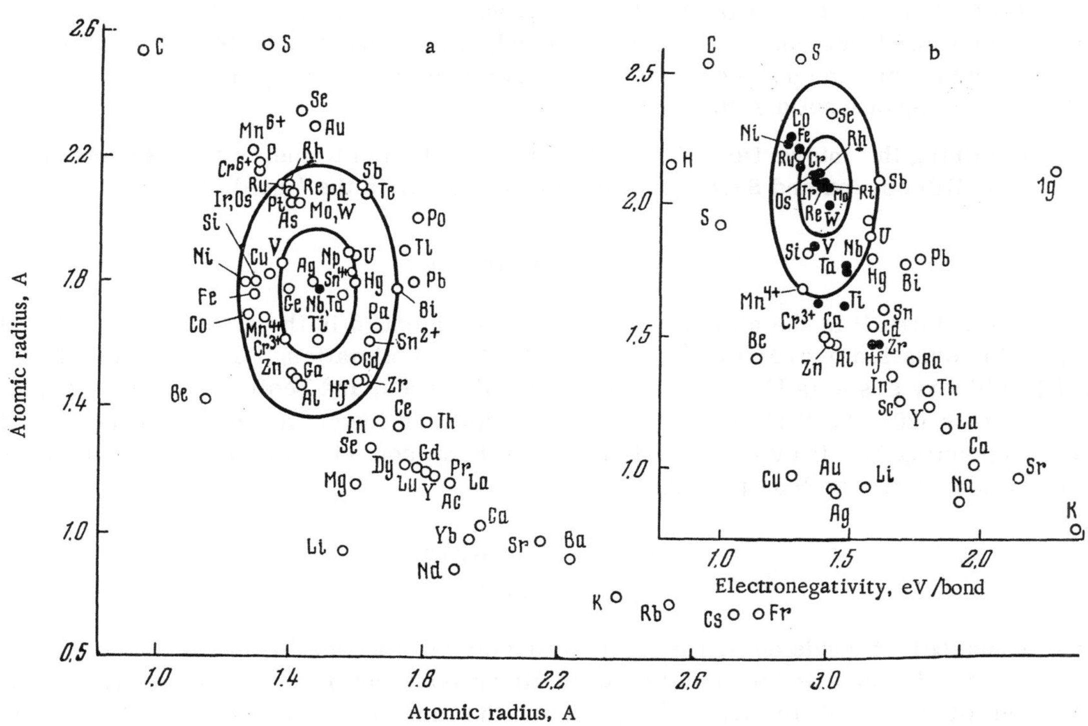


Fig. 34. Solubility graphs for niobium (a) and rhenium (b).

number of valence electrons in the dissolving element is less than the number of valence electrons in the solvent. The solubility of an alloying element in the metal solvent is greater, the higher is its valence.

Darken and Gurry [14] proposed graphs for expressing variation of the value of maximum solubility in the solid state of different alloying elements with the atomic diameters of solvent and dissolved addition, and with the value of the electronegativity of these elements. In these graphs, the atomic radii of the elements are plotted along the axis of abscissas, the electronegativity along the axis of ordinates. To define the solubility limits, accessory ellipses are constructed, an inner ellipse (major axis ±0.2 electronegativity unit, minor axis ±8% of the difference in the atomic radii), and an outer ellipse (±0.4 electronegativity, ±15% difference in the atomic radii). The graphs are constructed for definite metal solvents and may be termed solubility graphs. In the boundary of the small ellipse are metals forming unlimited solid solutions with the given metal solvent. Between the small and large ellipses are metals with limited solubility in the basis metal. Beyond the boundary of the large ellipse, the dimensional and valence factors are unfavorable for the formation of solid solutions. Thus, graphs have been constructed for yttrium [27], niobium (Fig. 34a) [28], rhenium (Fig. 34b) [29], and ohter metals. These graphs show, for example, that niobium ought to form a continuous series of solid solutions with metals of groups Va and VIa. The rare-earth metals, far removed from the majority of refractory metals, are not very soluble in the latter in the liquid and solid states. In the case of polymorphic metals, solubility graphs must be constructed for a definite modification with corresponding recalculation of the atomic radius. The solubility of the same metal in the different modifications of a polymorphic metal may be quite different. There are evidently some laws connecting the position of the element in the periodic system with its ability to stabilize one or the other modification of a polymorphic metal.

It should be noted that in the interaction of different elements, there are a number of exceptions to the semiempirical solubility rules described above. Among the most refractory metals, such exceptions are the tungsten–chromium, vanadium–tantalum, and other systems, in which the formation of intermediate phases is found, despite the isomorphism of the crystal lattices of the components and a dimensional factor and electronegativity factor favorable for the formation of continuous solid solutions.

For determining the interaction of the elements, in addition to the criteria enumerated in the foregoing, the Hildebrand and Scott criterion for solubility in the liquid state [30]

$$\frac{1}{2}(V_1+V_2)(\delta_1-\delta_2)^2<2RT$$

is very important. Here V is the atomic volume of the element, $\delta$ is the solubility parameter determined by the equation $\delta = (\Delta E/V)^{1/2}$ ($\Delta E$ is the heat of sublimation). On the basis of an analyses of 529 binary systems Mott showed that the Hildebrand and Scott criterion was true for 59% of the cases [20]. Mott introduced into the Hildebrand inequality the difference in electronegativity, assuming that if there was a tendency to form compounds, the absence of solubility in such a case would be less probable:

$$K=\frac{\frac{1}{2}(V_1+V_2)(\delta_1-\delta_2)^2-2RT}{23\,060\,(en_1-en_2)^2},$$

where K is the number of bonds determining the character of the interaction of the components; for $K < 2$ the metals dissolve in the liquid state in all proportions, for $K > 6$ a separation range is assumed, and for $2 < K < 6$, interaction depends on the dimensional factor. For the latter case, a statistical curve is constructed (Fig. 35). In an analyses of the same 529 systems, the

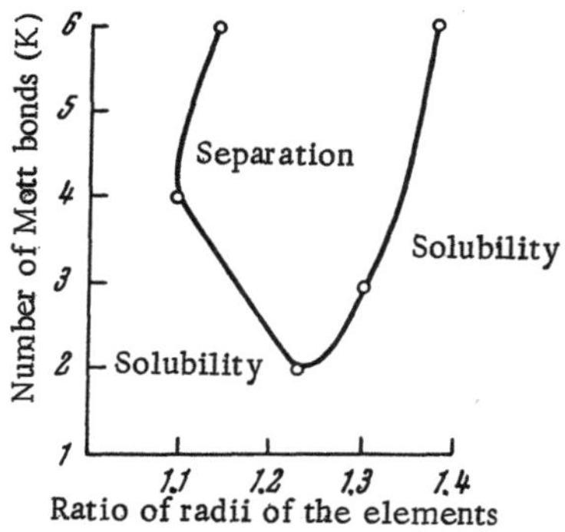


Fig. 35. Statistical curve of separation of liquid phases.

Mott criterion is found to be true for 80% of the cases. Teatum, Gschneidner and Waber [31] have compiled tables of Hildebrand and Mott criteria, electronegativity and other quantities necessary for predicting the interaction of two elements for all possible binary systems.

According to Tammann [32], the solubility of a fusible metal in a refractory metal is higher than that of the refractory metal in the fusible metal. It is also known that in eutectic systems, consisting of refractory and fusible components, the eutectic approaches the fusible component in composition and temperature. In a binary system, in which there is unlimited solubility of the components in the liquid state, in addition to the "lever rule," determining the phase content of the alloy, the eutectic or peritectic point and quantity of eutectic may be found by the Tammann geometric method (construction of the Tammann triangle). The Cordes empirical equation is also quite useful, since it enables the composition of the eutectic point to be calculated from the values of the melting points of the initial components and eutectic

$$\frac{T_B - T_E}{T_B} : \frac{T_A - T_E}{T_A} = x_A : x_B,$$

where $T_A$, $T_B$, and $T_E$ are the melting points of the components A, B, and the eutectic (on the Kelvin scale); $X_A$ and $X_B$ are the concentrations of the components in the eutectic (at.%).

This equation, however, is rather approximate and does not by any means take into account all the factors determining the composition corresponding to the eutectic point. A statistical analysis of metallic systems of eutectic type led the authors of [33] to the conclusion that there was a linear correlation connection between the calculated and experimentally determined eutectic compositions. The correlation coefficient was calculated for metallic systems ( k = 0.86) and an equation was found for determining the mean probable value of the actual composition of a eutectic:

$$\left[\frac{\bar{x}_A}{x_B}\right] = 0.024 + 0.008\left[\frac{x_A}{x_B}\right]_{calc},$$

where $[x_A/x_B]$ is the mean probable value of the actual composition of the eutectic; $[x_A/x_B]_{calc}$ is the composition of the eutectic calculated from Cordes equation.

Some laws governing the structure of simple eutectic phase diagrams of metallic systems are considered by V. M. Vozdvizhenskii [34].

For determining the tendency of metals to form intermediate metallic phases, Kubaschewski introduced the parameter of the ratio of the heats of sublimation of the components $L_A/L_B$, since the heat of sublimation is a quantitative characteristic of the bond energy of the pure metal. Kubaschewski proposed the following formula for the expected compound,

$$-\Delta H_f = \frac{N_A L_A\,[C^*(A)_{alloy}\,C^*(A)_{metal}]}{C^*(A)_{metal}} + \frac{N_B L_B\,[C^*(B)_{alloy} - C^*(B)_{metal}]}{C^*(B)_{metal}},$$

where N are the molar fractions, L is the heat of sublimation, and C* is the mean coordination number.

Kubaschewski calculated from this formula the heats of sublimation for a number of chemical compounds, and constructed curves in which $\Delta H_f = 0$ for different combinations of $L_A/L_B$ and $R_A/R_B$. From these curves, starting from the values of the heats of sublimation and atomic radii, it is possible in a number of cases to ascertain the possibility of the formation of compounds between two metals.

Metallic compounds and phases based on them are formed in many metallic systems. They have individual crystal structures (as a rule, complex) and differ sharply in chemical and physical properties from the pure metals and also from solid solutions. Metallic compounds are characterized by the existence of a mixed (metallic, covalent, and ionic) bond. However, in contrast to compounds of ionic and covalent types, chemical compounds of metals are distinguished by the predominance of the metallic bond between oppositely charged atoms and possess metallic properties (metallic luster, electrical conductivity, positive temperature coefficient of electrical resistance, thermal conductivity, and, under certain conditions, a capacity for plastic deformation).

Chemical compounds are mainly formed between those components which differ most strongly from each other in their electronic structure and chemical and physical properties, i.e., are situated remote from each other in the periodic system, have very different atomic volumes and different crystal structures. At the present time, about ten thousand two-component, three-component, and more complex metallic compounds are known. In the crystallization of a melt, the metallic compounds either solidify congruently or are formed as the result of corresponding reactions between the solid and liquid phases or in the solid state. According to N. S. Kurnakov's classification, metallic compounds are divided into two large classes, daltonides and berthollides.

A definite chemical compound (a daltonide, in Kurnakov's classification) has its own type of space lattice, ordered arrangement of the atoms, simple stoichiometric composition, and singular points on composition–property diagrams. Metallic compounds of variable composition (berthollides, in Kurnakov's classification) possess a partly ordered arrangement of the atoms; their degree of ordering varies with composition and temperature. They have their own crystal structure but have no singular points on the composition–property diagrams. The metallic bond is more definitely pronounced in them than in daltonides. N. V. Ageev proposed that berthollides should be regarded as solid solutions on the basis of "imaginary" chemical compounds, without independent existence and detectable only in solid solution. In a number of systems, a continuous transition may be found from daltonides to berthollides in the solid state. Metallic compounds have a complex and characteristic crystal structure: $\beta$-phases have a cesium chloride type of structure, $\gamma$-phases have a complex cubic structure with 52 atoms in the unit cell, and $\varepsilon$-phases belong to the hexagonal system. Metallic compounds on melting behave like the independent components and may form between themselves mixtures and limited or even continuous solid solutions, in which the constituent metals isomorphously replace each other in the lattice of the compound. Many metallic compounds dissociate completely or partly when heated, the stability of the compounds is in direct proportion to the value of the free energy, heat of formation, and melting point.

There is a more or less generally accepted classification of metallic compounds according to the type of chemical interaction of the components and similarity of crystal structure: Kurnakov compounds, compounds with a valence ratio of the atoms, electronic compounds, Laves phases, nickel arsenide phases, and compounds of metals with nonmetals (hydrides, borides, carbides, nitrides, silicides, metal oxides).

*Kurnakov compounds* are found in many systems having large solubility ranges in the solid state. There are now many reasons to assume that in all systems with continuous solubility of the components, Kurnakov compounds ought as a rule to be formed under certain conditions.

Kurnakov compounds ought also to include many representatives of metallic compounds of metals of groups Va and VIa with metals of groups VIIa or VIII, called sigma-phases. These phases have a hardness of the order of 1200 $kg/mm^2$ and high electrical resistance.

Metallic compounds of valence type are formed between strongly electropositive and electronegative elements, far removed from each other in the periodic system. Characteristics of these compounds are high values of the heat of formation, high solidification points, low electrical conductivity, and high brittleness at ordinary temperatures.

Electronic compounds according to Hume-Rothery are characterized by a constant ratio of the number of outer electrons to the number of atoms: 3/2, 21/13, and 7/4. They do not obey the valence rule. Electronic compounds are formed in the interaction of elements of subgroup Ib or certain transition elements with elements of groups II-V (subgroup a). In the phase diagrams, homogeneity regions of various extent correspond to electronic compounds. Four types of crystal structure are known for electronic compounds: face-centered cubic lattice of $\beta$-brass type, complex cubic lattice of $\beta$-manganese type, complex cubic lattice of $\gamma$-brass type, and hexagonal lattice of the type of the compound $CuSn_3$. An explanation of the structure of electronic compounds from the quantum mechanical position has been given by Mott and Jones [20, 35], and S. T. Konobeevskii [36].

Laves phases are usually understood to be metallic compounds having for the larger part the composition $AB_2$ and crystallizing in one of the characteristic close-packed cubic or hexagonal structures of the type of $MgCu_2$, $MgNi_2$, and $MgZn_2$. Typical of Laves phases is the ratio of the atomic radii of the components in the limits 1.15-1.68 and the predominance of the metallic bond. A peculiarity of some type $AB_2$ phases is the presence of polymorphous transformations; this refers, for example, to compounds of rhenium with the rare-earth metals [37, 38].

Nickel arsenide phases include metallic compounds crystallizing in a hexagonal lattice of NiAs type with a coordination number of eight. They were discovered by N. V. Ageev and E. S. Makarov. According to Makarov [39], a system of such phases may represent a homologous series of compounds having a variable number of atoms in the unit cell and with a set of different types of chemical bond from ionic to metallic.

There are numerous compounds of metals with nonmetals (oxygen, hydrogen, carbon, nitrogen, boron, silicon), which ought also to be included in metallic compounds, since the metallic bond predominates in their structure and they have typical metallic properties: electronic conductivity, high thermal conductivity, metallic luster, etc. Such compounds are usually divided into two groups. If the difference in the atomic radii of the metalloid (A) and metal (B) is relatively small ($r_A/r_B > 0.59$), phases are formed with special, often complex, types of ordered crystal lattices, representing ordinary metallic compounds; if, however, $r_A/r_B \leq 0.59$, then as shown by Hagg, the larger atoms of the metal form their own cubic or hexagonal lattice, and the smaller atoms of the nonmetal are situated in the spaces between the atoms of the metal, so-called interstitial phases being formed, which are very common in metallic compounds containing nonmetals. Interstitial phases differ from the ordered arrangement of the atoms by the simple stoichiometric proportions of the components and also by the outer, distinguishing signs of metallic bodies.

A mixed interatomic bond is characteristic of all compounds of metals with nonmetals; in oxides, it is of the ionic type, in hydrides evidently covalent-metallic; the metallic bond predominates in borides, carbides, nitrides, and silicides, but there is an ionic bond of different strengths.

The variety of the types of interatomic bond and crystal structures gives rise in metallic compounds to the presence of special physicochemical, electrical, emission, magnetic, optical, mechanical, and other properties. Thus, for example, the electrical properties may vary from superconductivity in liquid rhenium to semiconductor properties at the ordinary temperature. Molybdenum disilicide, as distinct from molybdenum, is stable in air up to 1700°C; the hexaborides of the rare-earth metals are noted for their high thermoemission properties; etc.

## Characteristics of the Methods of Investigating Refractory Metal Alloys

### 1. Production of the Alloys

The reliability of experimental data obtained on the structure and properties of metallic materials (metals, alloys, and compounds) is largely determined by the purity and methods of preparation of the starting materials. Therefore, the quality of the melting of refractory metals and alloys, prepared for investigation, requires special attention.

One of the commonest methods of melting refractory metals is the method of arc melting in a vacuum or in an inert atmosphere. In recent years, in addition to melting and casting refractory metals in a vacuum, argon, or helium in electric high-frequency or arc furnaces, electron-beam vacuum melting, ensuring the production of refractory metals and alloys having a high degree of purity, has begun to be used. The laboratory apparatus for electron-beam melting is intended for the production of small ingots in the form of "tablets" weighing 50-80 g; the metal is melted on a water-cooled copper hearth in a vacuum of $10^{-5}$ to $10^{-6}$ mm Hg. Different modifications of this apparatus exist. The apparatus is provided with various devices for increasing the output or for producing ingots of different shapes. Figure 36 shows a water-cooled copper eight-hole turntable. Such a device ensures the melting of eight disk-shaped ingots without opening the furnace. Annular ingots may be melted in an electron-beam turntable apparatus.

Since cooling in a water-cooled copper mold is quite rapid (of the order of 600°C per minute) the specimens of alloys produced are quench-hardened and have a large amount of internal stresses. Such specimens must be annealed; this is done in a vacuum or purified inert gases. More reliable results of physicochemical research on alloys are obtained with specimens not as cast, but deformed and subjected to long-continued homogenization.

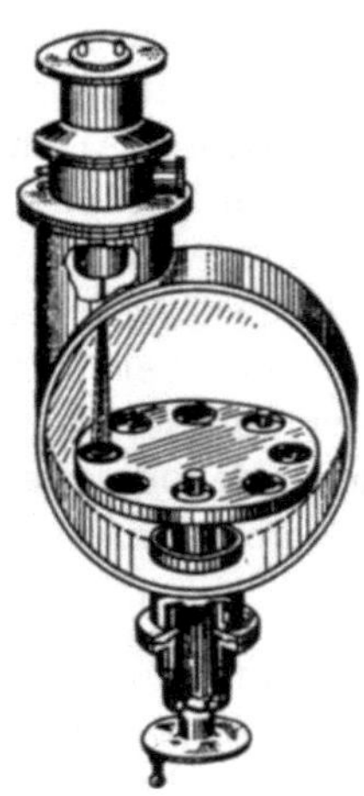

Fig. 36. Melting chamber of laboratory rotating hearth furnace.

A great advantage in the production of alloys in vacuum crucible induction furnaces and resistance furnaces is thorough mixing of the charge during the melt down. The applicability of the above-mentioned melting methods to chemically active refractory metals is difficult, however, owing to the absence of a sufficiently refractory crucible material. The possibility of induction melting in crucibles of cerium sulfide or titanium carbide and in water-cooled copper or silver crucibles is being investigated [40].

The method of electron-beam vacuum zone refining has acquired considerable importance in laboratory investigations [41]. Single crystals of nearly all refractory metals and a number of their alloys have been obtained by this method. The melting of a narrow zone of metal is produced by electronic bombardment of the specimen acting as anode. The liquid metal in the molten zone is held together by the forces of surface tension. The evaporation of volatile impurities in the vacuum and the zone-refining effect ensures profound purification of the refractory metal. Grown single crystals of refractory metals and alloys, owing to the insignificant impurity content (of the order of ten thousandths of a percent) and the absence of nonmetallic inclusions along the grain boundaries, have exceptionally high plasticity, permitting their deformation at room temperature to micron cross sections without intermediate annealing. A method has been developed for the production of single crystals of refractory metals and alloys of given crystallographic orientation by means of "seeds."

## 2. Methods of Impurity Determination

Impurities do not exert the same quantitative influence on metals. Therefore, in a purity determination, it is important to know not only the total impurity content but also what impurities are left in the metals after purification, melting, and treatment. Interstitial impurities, constituents of the air and carbon, exercise a great influence on the properties of refractory metals.

However, the methods of analysis of metals and alloys with the necessary accuracy still remains a bottleneck, and without an exact knowledge of the content of interstitial impurities it is impossible to ascertain the laws governing the variation in properties of alloys as a function of their alloying proportions, and to control the physical properties of alloys. Several methods are used for determining the gas content of metals, vacuum melting, activation analysis, methods of isotope dilution and infrared absorption, and mass-spectrometric analysis. Physical methods of impurity analysis, which are capable of ensuring a sensitivity of impurity determination of $10^{-4}$ to $10^{-6}$ %, have not yet been employed extensively, owing to the short supply and complexity of the apparatus. Scientific literature has very little information on work devoted to the development of these methods [42].

Vacuum melting is the method which is most widely used [43-46]. The analyzed metal is most usually melted in a graphite crucible in a vacuum of $10^{-5}$ mm Hg, the hydrogen and nitrogen being separated in elementary form, and the oxygen reacting with the carbon of the crucible forming carbon monoxide. Complete separation of nitrogen and oxygen from the specimen is ensured by carbonization of the latter. To create the optimum conditions for this reaction, the specimen of the analyzed metal is immersed in a bath of some metal or other, previously melted in the crucible. This metal should be a good solvent for carbon and should have a low vapor tension. Platinum, iron, and nickel are often used as metal for this bath. The metal bath is melted by induction heating. Analysis of the gases separated during melting is often done by the chromatography method, permitting the determination of the hydrogen, nitrogen, methane, and carbon monoxide contents.

The principal advantages of the vacuum-melting method are the simple technique of making the analysis and the possibility of automation of the measurements. The disadvantages of the method include the impossibility of determining the oxygen contained in the metal in the form of a high refractory oxide. The sensitivity of the vacuum-melting method ($10^{-3}$ to $10^{-4}$ %) is already failing to satisfy modern requirements of science and production.

The development of physical methods for the development of the analysis of "homeopathic" impurities in refractory metals, in particular interstitial impurities (nitrogen, oxygen, hydrogen, and carbon), to an accuracy of from $10^{-5}$ to $10^{-7}$ % is one of the most important problems of the day. The mass-spectrometric method is one of the most widely used of these methods. By its means, it is possible to analyze the gases liberated in the melting and heat treatment of refractory metals and alloys [47, 48]. Local x-ray spectral analysis of impurities (metallic and nonmetallic) is being used on an increasing scale [49-51]. The French firm "Comeca" and a number of American and English firms have developed the series production of such apparatus, ensuring quantitative analysis of almost all the elements of the periodic system; the analysis may be carried out over an area of 1 $\mu$ in diameter. In the USSR, instruments of the microprobe type are made by GOI; their resolving power is 3 $\mu$. The value of this method for physicochemical investigations of metal systems cannot be overestimated.

The comparatively simple method of estimating the purity of metals by measuring the residual resistance in helium has received some popularity. This method, however, is not without some shortcomings, which will be discussed later (in Chapter V).

In the study of the reaction of metals with each other and with interstitial impurities, intermetallide phase analysis is quite an effective method, since by means of the electrochemical method it enables the intermediate metallic phase to be separated from the alloy, and its chemical composition and crystal lattice to be determined [52]. Mineralogical analysis offers considerable interest in research on the interaction of refractory metals with interstitial impurities, but at present it has been developed and used more for the study of steels [53] than refractory metals.

In studying the distribution of nonmetallic inclusions and the nature of the fracture of refractory metals and alloys, the simple and at the same time quite effective method of fractography, consisting of the metallographic examination of fractures, is extensively used.

The determination of the carbon content is very important in the study of refractory metals, especially those of group VIa. Combustion is usually employed for this method. For determining the content of excess carbon (above the limit of solubility in the solid state) in refractory metals, metallographic analysis (as is usually employed for the determination of the amount of carbide phase in steels) may be successfully used. Unfortunately, the possibilities of the metallographic determination of carbon content have so far been little explored with regard to refractory metals.

## 3. Microstructural and X-Ray Structural Analysis

The structural details of pure metals include grain size, shape of the grains, treatment and annealing substructure, and texture (of crystals, pores, and impurities). In turn, the following forms of substructure are known: subgrain boundaries, dislocation interlocking, stacking defects, microstresses, and point defects.

Metallographic analysis of homogenized or quenched specimens in the construction of phase diagrams enables the variation of microstructure to be studied as a function of temperature and chemical composition of alloys.

It is possible to determine with fair accuracy by microscopic analysis the extent of homogeneous and heterogeneous regions, the presence of metallic phases in the system, and the size and mutual disposition of structural components. Here we must emphasize the need for reducing alloys to the equilibrium condition, which is not always done in the construction of phase diagrams, especially in non-Russian work. Homogenization of specimens should be carried out until with further increasing in annealing period there is no longer any further change in microstructure, lattice constants, and the measured property (hardness, electrical resistance, etc.). Sometimes, the period required for attaining equilibrium may be very lengthy. Thus for example, according to data of Kubaschewski et al. [15], in the chromium–molybdenum system, true equilibrium at 800-600°C may be reached in from 1 to 250 years. The most reliable results may be obtained by measuring complex physical properties of the specimen, the measurement being preferable made simultaneously on the same specimens. In homogenization and other heat treatment of refractory metals, it is very important to ensure purity of the vacuum or inert gases. Experience shows that even with the use of getters, there is always some oxidation of the specimens; it is preferable that it should be slight. For preventing carburization of the material when annealed in a vacuum produced by oil-vapor pumps, it is preferable to use getter-ion or sorption pumps.

A special apparatus has been developed in the Refractory and Rare Metal Alloys Laboratory of the A. A. Baikov Institute of Metallurgy for the metallographic analysis of high-temperature phase conditions of metallic systems and the simultaneous measurement of complex physical properties. With this apparatus, it is possible to carry out not only the quenching of specimens from any temperatures in a vacuum, but vacuum examination and photography of the microstructures of prepared polished sections of metals and alloys heated to a high temperature, as well as electrical resistance and hardness measurements. The specimens are heated by passing current through them. For quenching the specimens from high temperatures, the electrical contacts are unclamped and the specimen falls into a small bath of vacuum oil contained in the vacuum chamber.

It should be noted that lengthy (lasting several days) vacuum high-temperature (2000-3000°C) annealing of chemically active refractory metals and alloys followed by their rapid annealing in liquid or gaseous helium presents a very difficult problem. Without its solution, however, data on the limit mutual solubility of refractory metals and their interaction with nonmetals at high temperatures remain incomplete.

After suitable heat treatment, the investigated specimens of metals and alloys are subjected to microstructure analysis with the aid of optical metallurgical microscopes. The quality of the preparation of the metallographic section for this purpose is determined by the choice of the methods used and the polishing and etching conditions, which depend on the physicochemical properties of the metal or alloy.

In the preparation of metallographic polished sections of tungsten and molybdenum, a number of special methods are used, differing from the methods of preparing polished sections of other metals, owing to the high hardness of tungsten and molybdenum, as well as to their high chemical activity.

Cutting of the specimens is usually carried out by means of thin corundum or diamond wheels only 0.3 mm in thickness with intense water cooling. The cut-out specimens are usually fixed in plastic for subsequent polishing of the surface and etching. Polystyrene is the most convenient plastic to use. The specimens prepared for polishing are pressed into the polystyrene softened by warming. The low temperature used in mounting the specimen precludes any possibility of change in its composition. There are also no sources of possible influence on the structure of the specimens during their polishing and etching. This mounting of the specimens for polishing and etching is also applicable for other refractory metals and their alloys. When it becomes necessary to examine the structure of small specimens of refractory metals such as wires, Wood's metal is poured around them; very thin wires (of micron cross sections) may be placed in small Pyrex tubes which are then ground and polished together with the specimen [54].

In mechanical polishing, the surface layer of the metal is subjected to work hardening, which considerably distorts its structure and properties. Electrolytic and chemical polishing occur as the result of the solution of the metal surface in the corresponding chemical reagent, and therefore this method of preparing metallographic sections introduces no appreciable changes in the structure of the metal. The application of these methods is particularly important for metals which are susceptible to work hardening during twinning. This applies chiefly to hexagonal metals, for example rhenium, titanium, zirconium, and hafnium. The work-hardened layer of refractory metals subjected to preliminary mechanical polishing may be completely removed by electrolytic polishing. Chemical polishing is rarely used in the practice of metallographic examinations since chemical polishing is difficult to control and adapt to the treatment of different metals. Solutions for chemical polishing contain a substance which reacts with the metal, and a substance which controls the rate of this reaction. For chemical polishing of niobium and tantalum a mixture of sulfuric, nitric, and hydrofluoric acids is used

[55]. The rate of the reaction of the polishing solution with the metal is regulated by varying the temperature and the hydrofluoric acid concentration.

Electrolytic polishing appears to be the most convenient since it ensures the preparation of high-quality sections in a short time. Electrolytic polishing removes from the specimen the work-hardened layer formed during mechanical treatment and enables the true structure of the metal to be revealed. In the case of tungsten, molybdenum, and rhenium, electrolytic polishing is carried out in a 2-5% aqueous solution of NaOH with a current density of 2-3 $A/cm^2$ and a voltage of about 6 V. Electrolytic etching is carried out with a lower current density for several seconds. For revealing the high-quality structure of tungsten and molybdenum, alternate electrolytic etching and polishing has been found useful. For the electrolytic etching of tungsten, the use of an etchant consisting of 25 cc of a normal solution of NaOH and 20 cc of $H_2O_2$ is recommended [54].

For the chemical etching of tungsten, a 3% boiling $H_2O_2$ solution may be used. The etchant 10% NaOH + 30% $K_3[Fe(CN)_6]$ at room temperature rapidly reveals the grain boundaries in a few seconds. An ammonical solution of $K_3[Fe(CN)_6]$ has quite an intense action on tungsten. For etching tungsten, the following reagents may also be recommended [57]: 1) 10 ml $HNO_3$ + 10 ml HF + 10 ml $H_2O$; 2) 10 ml $HNO_3$ + 10 ml HF + 10 ml $CH_3COOH$. The etching time with these reagents is 30-90 sec at room temperature. A hot mixture of hydrofluoric and nitric acids has a much more rapid action. A rapidly acting reagent is Murakami's: 5 cc 10% $K_3[Fe(CN_6)]$ + 0.5 g KOH. Murakami's reagent may be used either for chemical or electrochemical etching.

For etching molybdenum, apart from a 2-10% aqueous NaOH solution, an ammoniacal copper sulfate solution may be used [54]. This etchant reveals grain boundaries very clearly, practically without etching the actual surface of the grains.

Existing methods of electrolytic polishing of niobium, tantalum, vanadium, and titanium are described in [55].

For niobium, an electrolyte of the following composition is recommended: 175 ml 40% HF + 175 ml 70% $HNO_3$ + 650 ml $H_2O$, current density 20.4-34 $A/dm^2$ with a voltage of 12-20 V; the cathode is platinum, the electrolyte temperature about 50°C; for etching, the current density is reduced. An electrolyte of the following composition is also recommended: 170 ml 70% $HNO_3$ + 50 ml 40% HF = 5 ml citric acid + 510 ml methanol, current density 495-605 $A/dm^2$. For the chemical etching of niobium and its alloys with tungsten, molybdenum, zirco-

TABLE 25. Etchants for Refractory Metals

| Metal | Etchant | Action |
|---|---|---|
| Tungsten | Boiling mixture: 3% $H_2O_2$ + 97% $H_2O$; 2 parts 25% $CuSO_4 \cdot 5H_2O$ + 1 part $NH_4OH$ for 15 sec at 40°C | Reveals subboundaries and grain boundaries<br>Etch figures in single crystals |
| Molybdenum | Boiling mixture: 1.5% $H_2O_2$ + 98.5% $H_2O$; 1 part 30% $K_3Fe(CN)_6$ + 1 part 10% NaOH for 10-15 sec. | Reveals subboundaries and grain boundaries<br>Etch figures corresponding to edge or screw dislocations |
| | Electrolytic etching by immersion in a 10% aqueous solution of NaOH at 0.75 V. | Reveals subboundaries and grain boundaries |
| | Electrolytic etching in a 0.5% aqueous oxalic acid solution at 3 V. | Etch figures corresponding to edge and screw dislocations |
| Tantalum | Hot mixture (50-60°C): 1 part $NH_4OH$ (20% aqueous solution) + 1 part HF (30% aqueous solution) | Reveals individual dislocations and subboundaries and grain boundaries |
| Niobium | 1 part HF (48% aqueous solution) + 1 part $H_2SO_4$ (conc.) + 1 part $H_2O$ + a few drops of $H_2O_2$. | Reveals subboundaries and grain boundaries |

nium, and other elements, a mixture of concentrated HF and $HNO_3$ acids is used in different proportions, depending on the composition of the alloy. Etching time is a few seconds.

For electrolytic polishing and etching of vanadium, an electrolyte of the following composition may be used [55]: 60 ml perchloric acid + 350 ml butyl cellosolve + 590 ml methyl alcohol, current density about 16 $A/dm^2$, room temperature; stainless steel cathode. In ordinary chemical etching of vanadium, a mixture of hydrofluoric and nitric acids is used, the proportion and concentration of the acids being varied according to the chemical composition of the vanadium alloys.

The following electrolytes are used for the electrolytic polishing of tantalum:

1) 170 ml 70% $HNO_3$ + 50 ml 40% HF + 30 g ammonium fluoride + 510 ml methanol; current density 355 $A/dm^2$;

2) 2-7% HF + 32-38% $H_2SO_4$, the rest water; current density 6-7 $A/dm^2$.

A mixture of hydrofluoric and nitric acids is used for the chemical etching of tantalum.

Metals of the platinum group are resistant to the action of nitric and hydrofluoric acids. For etching platinum and alloys based on it, a saturated solution of bromine and ethyl alcohol is used. Iridium and its alloys may be etched electrolytically in a 10% solution of HCl using alternating current.

The microstructure of a single-phase polycrystalline specimen subjected to preliminary deformation, followed by recrystallization annealing, consists of grains of polygonal form. Such a grain form is due to the tendency of the polycrystal to reduce its surface energy by reduction in the surface of the grain boundaries. In the process of holding the metal at sufficiently high temperatures, the curvature of the grain surface gradually diminishes, tending to become a straight line. In coarsely crystalline specimens, it is also possible to observe a network of subboundaries, which etch much more weakly. On heating, holding, and subsequent slow cooling of the specimen, the subboundaries are revealed much more clearly, this evidently being due to the segregation of impurities. Under high magnifications, the subboundaries are revealed in the form of etch pits, the sites of emergence of dislocations on the crystal surface.

Metallographic analysis is an effective method in the investigation of the processes of plastic deformation and recrystallization of metals and alloys (revealing slip traces, deformation and recrystallization twins, character of distribution of nonmetallic inclusions during deformation).

Optical, "white light" microscopy, is still the most effective method of studying the structure of metals and alloys. In recent years, however, extensive use has been made of color and vacuum metallography, microscopy in polarized light, infrared and ultraviolet rays, electron, iron, neutron, and proton microscopy with a magnification of one million times, making it possible to observe even individual atoms and vacancies in the crystal lattice of refractory metals [59].

In the optical microscope (limit of resolution 2000 Å), the following substructure details may be observed: shape and size of subgrains, shape and character of distribution of etch pits, and also twins and deformation bands. In the reflecting electron microscope details down to 100 Å are resolved.

For fine structure investigations in the USSR, USA, and other countries, work is continuing on the improvement of the design of various types of microscopes. A unique electron microscope with an electron energy of 557 KeV has been developed by N. M. Popov [60, 61]; the high resolving power of this microscope and the high accelerating voltage have enabled it to be used in the examination, by transmission, of single crystals of tungsten films of a thickness of the order of 1000Å [62].

Fig. 37. Crystal lattice of molybdenum ("shadow" method).

According to [63], a new ionic field microscope has been developed for the examination of refractory metals and alloys; the microscope gives maximum magnification (2 million times), and enables the crystal lattice of a solid substance to be observed directly; by means of this powerful microscope, it is possible to reveal and photograph the smallest crystal defects. A review [64] examines the construction of microscopes operating on the principle of thermionic or secondary electron (or ion) emission from the examined object.

For the construction of phase diagrams involving refractory metals and also for crystal-chemical investigations, the high-temperature x-ray analysis method is of great value. Various designs of high-temperature cameras have been proposed in this connection [65-67].

A very effective and graphic method of examining the structure of metals has been found to be what is termed the "shadow" method [68], based on the interaction of a directional proton beam with the ions of the crystal lattice of the metal. This method has been used for observing the crystal lattice of molybdenum (Fig. 37).

The scientific value of the experimental data obtained on the structure and properties of metals and alloys definitely depends on the variety and quality of the scientific equipment used. Industrial production of the necessary instruments is not organized in all cases, and therefore self-designed instruments and apparatus are often used in investigations. As example, some of the equipment used in the Refractory and Rare Metal Alloys Laboratory of the Institute of Metals will now be described.

## 4. Thermal Analysis

The determination of the temperatures of the commencement and end of melting of refractory metals and alloys often presents considerable difficulties, mainly due to the lack of

Fig. 38. Vacuum apparatus for solidus temperature determination

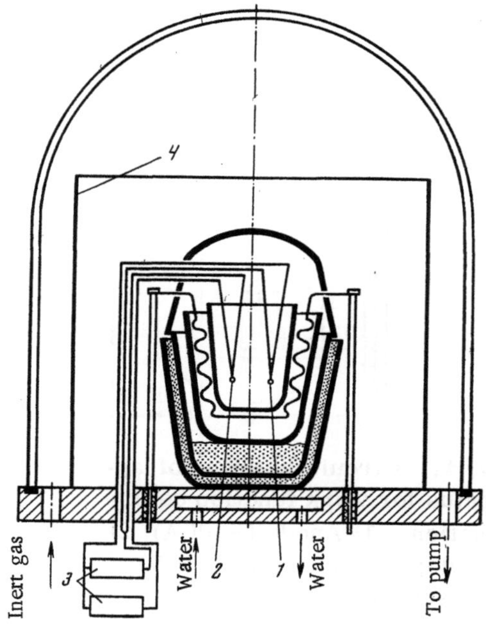


Fig. 39. Diagram of apparatus for differential thermal analysis of microspecimens. 1) Working thermocouple with specimen; 2) differential thermocouple; 3) galvanometer; 4) tantalum screen.

thermocouples and refractory inert ceramics for temperatures of 3000°C and above. The color pyrometry method is unfortunately unsuitable for arc melting conditions [69].

The melting points of the majority of refractory systems are usually determined by means of a special vacuum apparatus, a general view of which is shown in Fig. 38. In our laboratory, the solidus line for all alloys of the system of the two most refractory metals, tungsten and rhenium, has been determined with this apparatus [70]. The determination of the temperature of the commencement of melting is carried out by the "drop" method. The commencement of melting corresponds to the appearance of the first drop of melt in a fine orifice previously drilled in the specimen. The accuracy of the method is ±25°C. As a tentative determination of the liquidus temperature in the apparatus, a thin strip specimen may be rapidly melted. The same apparatus may be used to measure electrical resistance in a vacuum and in an atmosphere of inert gases at different temperatures.

For the determination of the liquidus line in the construction of phase diagrams, no more accurate method has so far been developed than the conventional method of thermal analysis by means of thermocouples. Substantial improvements have been made in recent years in the method of thermal analysis since the time of N. S. Kurnakov. The thermal analysis of steels is carried out not only in air but in a vacuum and in inert gases. Methods have been developed for the differential thermal analysis using microspecimens, which practically exclude errors due to the thermal inertia of the crucible ceramic and of the heating furnace, and are very suitable for the investigation of rare and expensive metals. And perhaps the most important factor, in replacement of the Le Chatelier platinum–rhodium thermocouple, new thermocouples have been devised from alloys of the noble metals, particularly tungsten–rhenium thermocouples, enabling the "ceiling" of thermal analysis to be raised from 1300 to 2000°C and to use it for refractory systems.

Figure 39 shows the junction of a tungsten–rhenium thermocouple with a specimen of metal weighing 0.2 g clamped between the electrodes. The thermocouple with specimen is placed under a glass bell with a ground-in joint connected to a vacuum. The heating elements and screens are molybdenum. Recording of the cooling or heating curves is by means of a Kurnakov pyrometer or sensitive potentiometer. Due to the existence of the laboratory-made apparatus for microthermal analysis in a vacuum or an inert gas, it is now possible to study a number of phase diagrams involving refractory metals in microspecimens (weighing 0.2 g and less).

## 5. Determination of Physicomechanical Properties

The investigation of the physicomechanical properties of chemically active refractory and rare metals in the heated condition must also be conducted in a vacuum or an inert gas. A universal apparatus developed in our laboratory [71, 72] has made it possible to determine different mechanical properties by the method of the axial stressing of microspecimens at

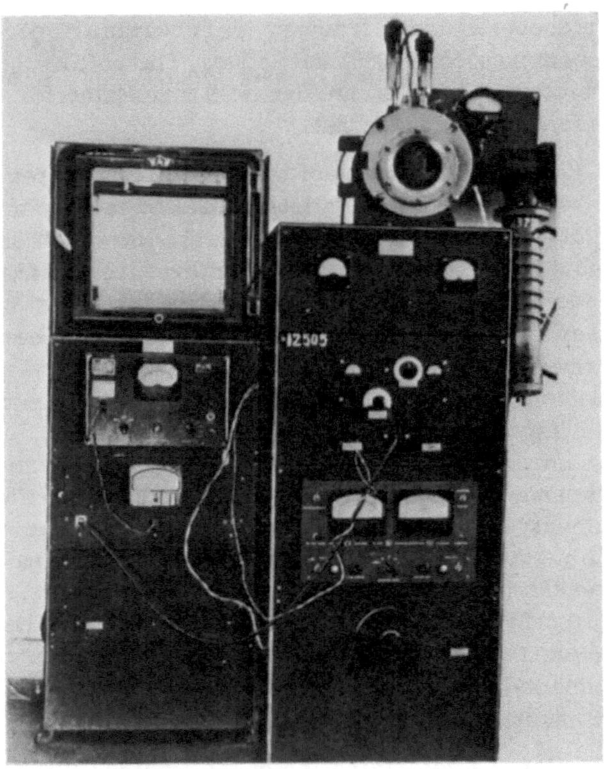

Fig. 40. Vacuum capacitance dilatometer constructed by A. V. Panov.

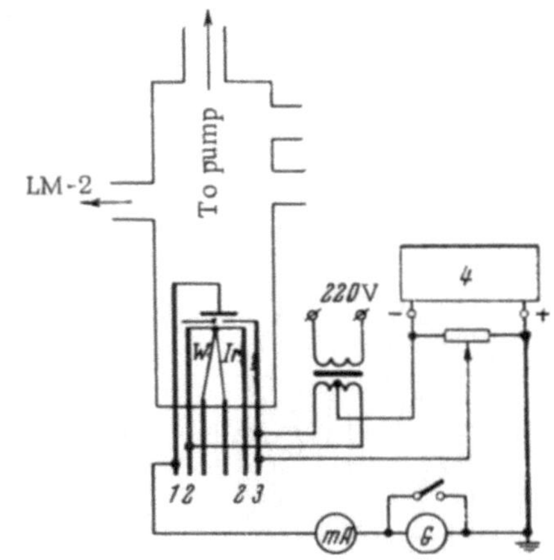


Fig. 41. Circuit diagram of apparatus for work function determination. 1) Anode (+400 A); 2) cathode; 3) diaphragm (+50V); 4) dc source.

different temperatures in a vacuum or in inert gases. The methods of high-temperature mechanical testing has been discussed in [4, 71-73].

For the construction of phase diagrams, various forms of dilatometers are used. In the Refractory and Rare Metal Alloys Laboratory of the A. A. Baikov Institute of Metallurgy, dilatometer measurements in inert media are made with the use of a high-sensitivity volume dilatometer constructed by A. V. Panov (Fig. 40) [74]. The pickup is a variable capacitor; the variation in dimensions of the specimen with temperature produce a variation in capacitance of the capacitor; the dilatometer curve is recorded by means of an electronic potentiometer.

Other constructions of vacuum dilatometers have been developed [75, 76].

E. Savitskii and A. I. Dashkovskii [77] have shown that the internal friction method may be used successfully for the physicochemical analysis of metal systems, including those of refractory metals; a series of investigations have been made on the internal friction of uranium, zirconium, niobium, cerium, and lanthanum, and their alloys, at different temperatures in a vacuum [77, 78].

For investigating the superconductivity of alloys based on niobium, vanadium, and other metals, an apparatus developed in our laboratory is used for determining the critical temperature of transition to the superconducting state and the critical current which destroys this state.

An important scientific problem is that of discovering the laws governing the variation in the emissive properties of metals, alloys, and compounds as a function of the variation of their composition, crystal structure, and the type of chemical interatomic bond. Figure 41 shows the circuit diagram of an apparatus developed for determining the work function of refractory metals and alloys in a vacuum [80]. The work function is determined by the conventional method according to the density of the emission current using the well-known Richardson equation. The

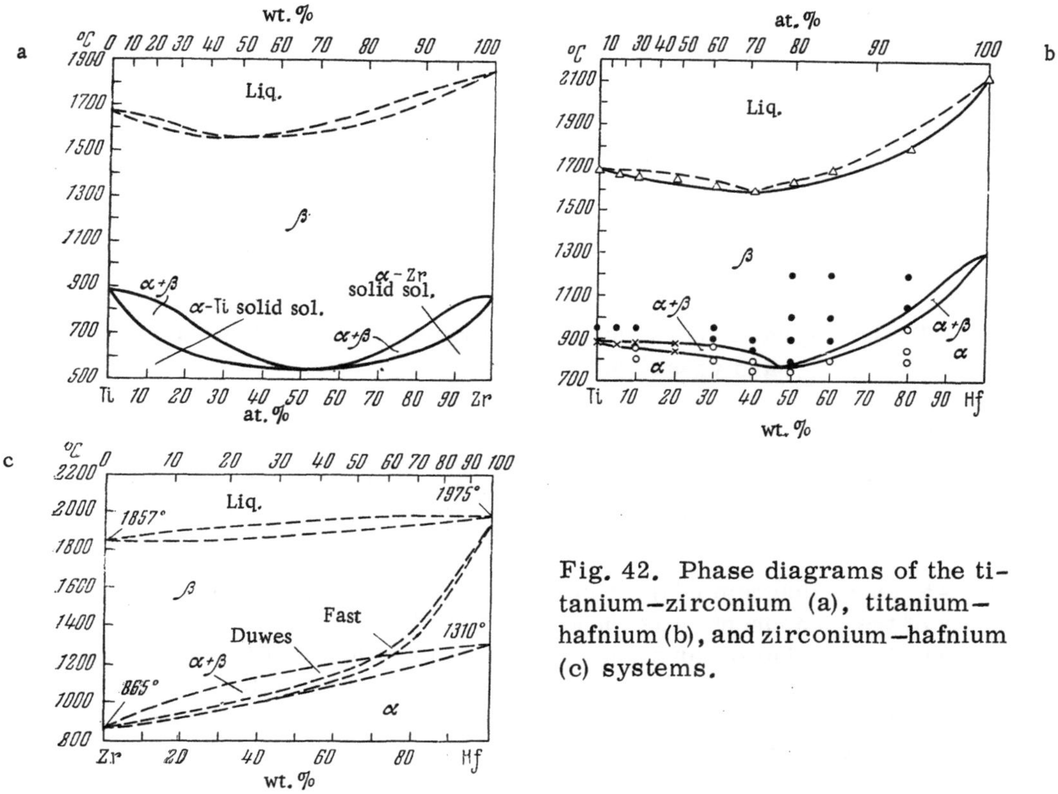


Fig. 42. Phase diagrams of the titanium–zirconium (a), titanium–hafnium (b), and zirconium–hafnium (c) systems.

most important problem was to devise a vacuum system for ensuring measurements at a residual pressure of not more than $10^{-8}$ mm Hg. Nitrogen traps are also provided in the vacuum system. Preliminary evacuation to $10^{-5}$–$10^{-6}$ mm Hg is produced by ordinary and oil-diffusion pumps. The use of an electric discharge pump enables the pressure in the system to be reduced to $10^{-8}$ mm Hg. A system has been developed for producing a preliminary vacuum without the use of oil-diffusion pumps, which is extremely important in conducting research connected with the study of the surface properties of metals and alloys.

## Interaction of Refractory Metals of Groups IV-VIII (In the Limits of Each Group)

### 1. Group IVa Metals

The isomorphism of the crystal lattices of both modifications of group IVa metals, the favorable dimensional factor, and the slight difference in electropositivity result in the complete solubility of these metals in each other. Additions of zirconium to titanium or titanium to zirconium result in a reduction in the temperature of polymorphic transformation of these metals. The $\alpha + \beta$ region in the diagram has a minimum [81]. A similar character of interaction is also found for the titanium–hafnium [82] and hafnium–zirconium systems [8]. The phase diagrams of the titanium–hafnium, titanium–zirconium, and zirconium–hafnium systems are shown in Fig. 42.

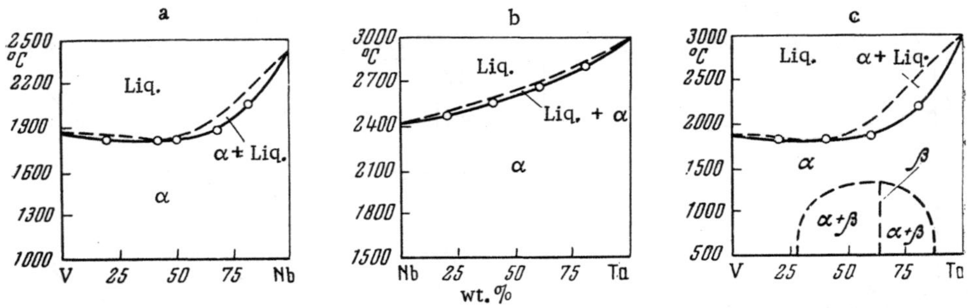


Fig. 43. Phase diagrams of the vanadium–niobium (a), niobium–tantalum (b), and vanadium–tantalum (c) systems.

## 2. Group Va Metals

Their position in one subgroup of the periodic system and the isomorphism of the crystal structures predetermine the high mutual solubility of the group Va metals – vanadium, niobium, and tantalum.

The complete phase diagram of the niobium–tantalum system was studied in [83]. Continuous solubility of the components was found. The solidus line rises in a flat slope from the melting point of niobium to that of tantalum. No traces were found of transformation in the solid state. In the opinion of the authors of [83], tantalum–niobium alloys are characterized by a very small crystallization interval; a slight increase in the solidus point of the alloys resulted in the occurrence of a considerable quantity of liquid phase.

Components in the vanadium–niobium and vanadium–tantalum systems interact similarly. The formation of an intermediate phase $TaV_2$ in the vanadium–tantalum system has been mentioned in a number of publications, but there is no unified opinion as to the nature of this phase. According to the results of V. N. Eremenko and collaborators [84], in the vanadium–tantalum system, decomposition of the solid solution occurs below 1420°C with the formation of $\sigma$-phase, the range of existence of which widens with reduction in temperature, being in the limits of 30-43 at.% Ta at 1000°C. In the alloy with 32 at.% Ta, the $\sigma$-phase has the lattice parameters a = 6.15 kX, c = 8.85 kX. Rostoker [85] also considers the compound $V_2Ta$ has a hexagonal structure and has no homogeneity region [86]. According to Elliot, this phase is isomorphous with $MgCu_2$ [87]. According to data of M. Yu. Teslyuk [276], $V_2Ta$ is a Laves phase with the structure of $MgCu_2$. The temperature range of crystallization in alloys of these systems is somewhat wider than that of tantalum–niobium alloys (Fig. 43c).

Rostoker and Jamamoto [85] report on the formation of $\sigma$-phase in the niobium–vanadium system. According to data of [88 and 89], no intermediate phases were found in this system, although it is included in systems having a minimum on the liquidus curve, for which ordering or the decomposition of solid solutions at low temperatures is characteristic.

## 3. Group VIa Metals

The molybdenum–tungsten system in cermet and vacuum-melted specimens has been studied by many authors. The lattice parameter of tungsten–molybdenum alloys, representing a continuous series of solid solutions, is a linear function of their chemical composition [91].

The density of the alloys also varies according to a linear law. Figure 44a shows the variation of the temperature of the commencement of melting with composition [91]. The liquidus line for the tungsten–molybdenum system, like that for other refractory systems, has not been determined exactly. The curve of the resistivity of the alloy has a maximum for

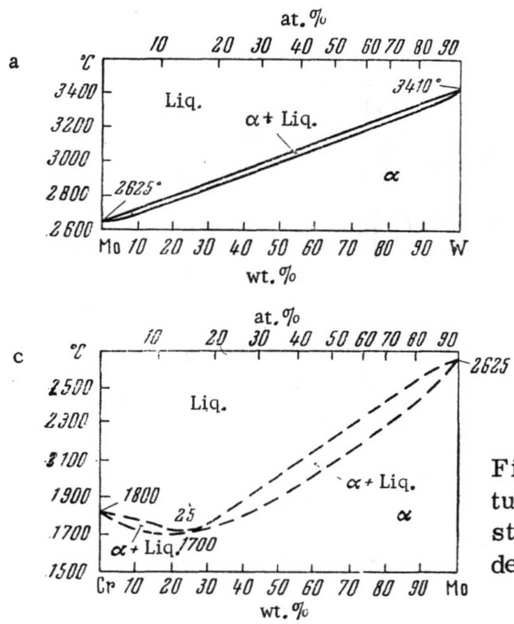


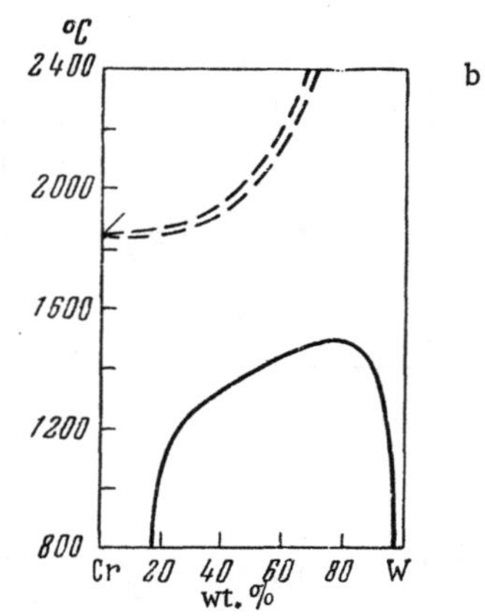


Fig. 44. Phase diagrams of the tungsten–molybdenum (a), tungsten–chromium (b), and molybdenum–chromium (c) systems.

a 50% Mo content, to which corresponds a minimum of the temperature coefficient of resistivity [92].

In the tungsten–chromium system, complete intersolubility is not found throughout the entire temperature range, despite the considerable difference in atomic diameters and electronegativity. At temperatures below 1500°C, according to [93–95], there exists a considerable range of limited solubility in the solid state (Fig. 44b). The tungsten content in the chromium solid solution at 1000°C is about 20 wt.%, while the solubility of chromium in tungsten at this temperature is about 4 wt.%. The existence of a two-phase region is in agreement with the data of [96]. Continuous solubility in the liquid and solid states is found in the molybdenum–chromium system (Fig. 44c) [97].

### 4. Group VIII Metals

The noble metals having an isomorphic crystal lattice form systems of unlimited solubility. Unlimited solubility is found in the platinum–rhodium [98, 99], platinum–palladium [100], platinum–iridium [101], palladium–rhodium [102, 103], iridium–rhodium, and ruthenium–osmium [106] systems.

In the platinum–iridium, platinum–rhodium, and palladium–rhodium systems, decomposition of the solid solution occurs with the formation of a two-phase region. In the palladium–iridium system, despite isomorphism of the crystal lattices and favorable dimensional factors, a peritectic reaction occurs [104].

In the interaction of noble metals having different crystal lattices, a system of peritectic type having two solid solutions is formed. There are no intermediate phases in these systems. Such systems include the palladium–ruthenium system [107].

## Interaction between Metals of Groups Va and VIa and Their Interaction with Metals of Other Groups

Metals of groups Va and VIa belong to the basic, central group of refractory metals: We shall therefore discuss first of all their interaction among themselves and then with metals of other groups.

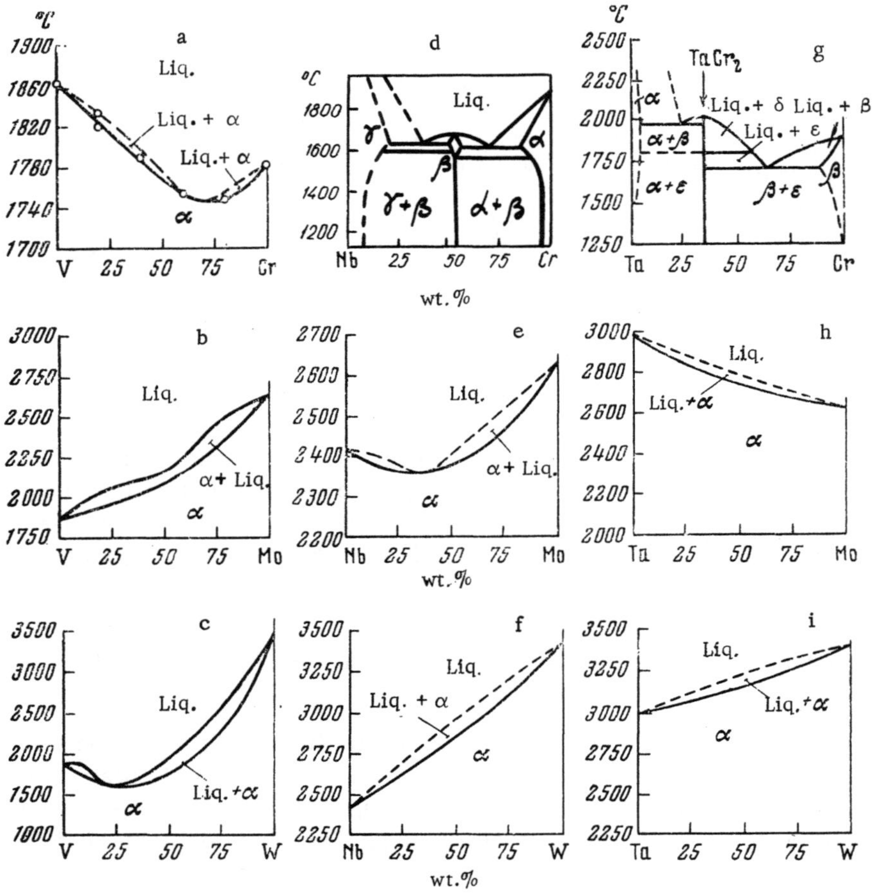

Fig. 45. Phase diagrams of the vanadium–chromium (a), vanadium–molybdenum (b), vanadium–tungsten (c), niobium–chromium (d), niobium–molybdenum (e), niobium–tungsten (f), tantalum–chromium (g), tantalum–molybdenum (h), and tantalum–tungsten (i) systems.

## 1. Interaction between Group Va and VIa Metals

The similarity in crystal lattices, and the slight difference in electrical resistance and atomic diameters signify complete intersolubility or the formation of a wide range of substitution solid solutions between metals of groups Va and VIa.

The phase diagram of the vanadium–chromium system (Fig. 45a) was investigated by E. M. Savitskii et al. [108]. The system represents a continuous series of solid solutions.

Pipitz and Kieffer [109] and Savitskii et al. [110] found that molybdenum and vanadium form a continuous series of solid solutions (Fig. 45b). These investigations do not confirm the assertion made by Rostoker and Jamamoto [85] concerning the decomposition of the solid solution in alloys containing 10-60 wt.% Mo.

The vanadium–tungsten system was investigated in [111] by the methods of thermal, microscopic, and x-ray analyses. The solidus temperature was determined by the drop method. A continuous series of solid solutions was found. At 4.5 at.% W, a minimum was observed on the liquidus curve (1635°C). According to [112], vanadium and tungsten also form a continuous series of solid solutions and the phase diagram of these alloys (Fig. 45c) is characterized by the presence of a minimum on the liquidus curve.

Niobium and molybdenum form a continuous series of solid solutions (Fig. 45e) [113, 114]. The fusibility diagram has a minimum corresponding to 80-70% Nb at 2345°C [114]. The form of the solidus curve on the fusibility diagram of the molybdenum–niobium system was confirmed in [115] with the sole difference that the minimum temperature corresponded to 2300°C for 70-65% Nb.

The tungsten–niobium system was investigated in [116-118]. The results of the investigations showed complete intersolubility of these metals. Investigation of the fusibility diagram of the niobium–tungsten system in [119] showed that the melting point of alloys with a tungsten content of 25-100% was in good agreement with the data of [118]. The solidus temperature of alloys containing up to 25 wt.% W was found to be lower. The inflection on the fusibility curve noted by the authors of [118] was also not found. Figure 45f shows the fusibility diagram of the niobium–tungsten system according to the results of investigations [119]. According to data of [112], the hardness and electrical resistance curves of niobium–tungsten alloys have a definite maximum at a content of 50-60% Nb, which appears to confirm the existence of a continuous series of solid solutions.

Limited solubility in the solid state is found in the niobium–chromium [120] and tantalum–chromium [121] systems. The compounds $NbCr_2$ and $TaCr_2$ having an isomorphic crystal structure (Laves phases) are formed in these systems. According to [120], the limit solubility of niobium in chromium at 1640°C is about 17 wt.%, and the limit solubility of chromium in niobium is about 16% at 1660°C. The compound $NbCr_2$ has the polymorphic transformation $\varepsilon \rightleftarrows \beta$ at 1585°C. The ε-phase has a cubic lattice of $MgCu_2$ type (lattice parameter 6.972 kX).

According to the data of [121], the compound $TaCr_2$ melts consistently at 2020°C. A polymorphic transformation of this compound at 1805°C has been found. The system has two eutectics Cr + $Cr_2Ta$ (13% Ta) at 1700°C and TaCr + Ta (46% Ta) at 1980°C. The solubility of tantalum in chromium at the eutectic temperature is 1.5 at.%, and that of chromium in tantalum at 1800°C is 28 at.%. Figures 45d and g show the niobium–chromium and tantalum–chromium phase diagrams.

In [116, 122, 123] continuous solubility was found in the tantalum–molybdenum (Fig. 45h) and tantalum–tungsten (Fig. 45i) systems.

## 2. Interaction of Refractory Metals with Group Ia and IIa Metals

The interaction of refractory metals with metals of groups Ia and IIa is of interest in connection with the high thermodynamic activity of the latter with regard to interstitial impurities, especially oxygen. The heats of formation of the oxides of these metals (from the elements) are characterized by the following data [124]:

| | | | |
|---|---|---|---|
| Oxide . . . . . . . . . . . . . . . . . . | $Li_2O$ | $Na_2O$ | $K_2O$ |
| $-\Delta H_{298}$, kcal/mole . . . . . . . . . . | 142.3 ± 2 | 102.9 | 86.0 ± 2 |

Theoretically, these elements may be used as deoxidizers for refractory metals. The low melting and boiling points of the alkali and alkaline earth metals as yet present considerable technical difficulties for their use in this direction. More detailed studies of the interaction of refractory metals with the alkali and alkaline earth metals are also very important in connection with the problem of the compatibility of refractory metals and alloys with the plasma, vapors, and melts of alkali and alkaline earth metals.

Experimental data are available to show the absence of considerable interaction of most of the metals of groups Va and VIa with metals of group Ia, even on long-continued action of the latter (provided that the alkali metals are in a state of high purity). It has been found, for instance, that the solubility of niobium, tantalum, and molybdenum in liquid lithium at 1000°C is only $10^{-4}\%$ [125]. Vanadium and chromium, however, are unstable in lithium at this temperature. Refractory metals evidently do not form alloys with metals of group Ib (copper, silver, and gold). In industry, however, extensive use is made of the impregnation of cermet tungsten by copper (contacts) and by silver (cooling of some components of rocket engines by vaporization of silver) [126].

There are no published data regarding the character of the interaction of metals of groups Va and VIa with alkaline earth metals. The alloying of these metals presents very considerable difficulties on account of the relatively low sublimation temperature of metals of group II. To suppress intense vaporization during melting, inert gas pressures are applied which demands the construction of special high-strength reactors. There are reports of unsuccessful attempts to alloy tungsten and molybdenum with magnesium [127].

A phase diagram of the niobium–zinc system was constructed comparatively recently, showing the existence of a number of compounds of moderate refractoriness, which were stable in an oxidizing atmosphere [128]. This gave rise to the possibility of using zinc for producing protective coatings on niobium alloys.

Less difficulty is found in alloying metals of the platinum group with group Ia and IIa metals, although no systematic researches of the corresponding systems have been made. It has been reported that molten sodium acts appreciably on platinum at temperatures above 450°C. It is also assumed that the metals form the compound NaPt under the action of pressure [8]. Platinum reacts with molten lithium at temperatures a little above the melting point of lithium (186°C) [8].

Of the metals of group IIa, the reaction of magnesium with platinum has been studied. It is assumed that the compound $Mg_2Pt$ exists in this system. These data, however, require refinement [8]. According to [127] osmium alloys slightly with magnesium; the osmium–magnesium system forms a eutectic type of phase diagram with a eutectic point close to the ordinate for pure magnesium.

Available experimental data indicate the presence of some region of solubility of magnesium in group IVa metals. According to [129], the limit solubility of magnesium in $\alpha$-titanium is about 0.64%. It has been found that zirconium dissolves a certain amount of magnesium in the solid state [8]. Hafnium, however, has been found to be stable in liquid magnesium at a temperature of 1000°C [126].

## 3. Interaction of Refractory Metals with Rare-Earth Metals (Group IIIa)

The examination of the character of the physicochemical interaction between these groups of metals in the present scope is of interest not merely because the rare-earth elements include metals having a high melting point, but also in connection with the high thermodynamic activity of the rare-earth metals, especially with regard to interstitial impurities, which may be successfully used in the refinement and modification of alloys of refractory metals.

As was to be expected in accordance with available data on the solubility factors (see Fig. 34), the study of the interaction of metals of groups IVa, Va, and VIa with rare-earth metals revealed a wide range of separation (absence of solubility) in the liquid and solid states. This circumstance, together with the high affinity of the rare-earth metals for oxygen (Table 26), considerably facilitates the process of refining refractory metals by rare-earth metals, which act like a deoxidizing flux. In the investigation of the interaction of rare-earth metals with group IVa metals, regions of immiscibility in the liquid state have been found in the titanium–lanthanum, titanium–cerium, and titanium–neodymium systems [131]. Absence of alloying is also assumed in the case of titanium with praseodymium and dysprosium [132].

TABLE 26. Heat of Formation of Oxides of the Rare-Earth Metals [130]

| Element | Oxide | Heat of formation kcal/mole | Element | Oxide | Heat of formation kcal/mole |
|---|---|---|---|---|---|
| Lanthanum . . | $La_2O_3$ | 428.6 | Dysprosium . . . | $Dy_2O_3$ | 445.8 |
| Cerium. . . . . | $Ce_2O_3$ | 435.0 | Holmium . . . | $Ho_2O_3$ | 449.5 |
| Praseodymium | $Pr_2O_3$ | 436.8 | Erbium . . . | $Er_2O_3$ | 453.6 |
| Neodymium . | $Nd_2O_3$ | 432.1 | Thullium . . . | $Tu_2O_3$ | 451.8 |
| Samarium . . . | $Sm_2O_3$ | 433.9 | Ytterbium . . . | $Yb_2O_3$ | 433.7 |
| Gadolinium . . | $Gd_2O_3$ | 433.9 | Luthecium . . . | $Lu_2O_3$ | 452.8 |
| Terbium . . . . | $Tb_2O_3$ | 436.8 | | | |

According to [133 and 134], the titanium–gadolinium system is of eutectic type. Most rare-earth metals form systems of eutectic type with zirconium and hafnium [132]. As a rule, addition of rare-earth metals slightly increase the temperature of the polymorphic transformation of titanium, zirconium, and hafnium. In the interaction of rare earth metals with group IVa metals (titanium, zirconium, and hafnium), no metallic compounds are formed. The solubility of the rare-earth metals in both modifications of titanium, zirconium, and hafnium is low, although it is much higher than in metals of groups Va–VIII of the periodic system. According to [131], the solubility of lanthanum in $\alpha$-titanium at 600°C is 1.5 wt.% (0.4 at.%), that of cerium is 1.2 wt.% (0.4 at.%), and that of neodymium is about 1.8 wt.% (0.6 at.%). Taylor gives a much lower solubility of cerium in titanium [135]. The maximum solubility of yttrium in $\alpha$-titanium at 600°C is about 0.8–0.9 wt.% (0.5 at.%), and that in $\beta$-titanium about 1.5 wt.% (0.9 at.%). According to Lundin [132], the maximum solubility of yttrium in $\beta$-titanium attains 2 wt.%); a wider solubility range in the solid state is found in systems of the rare-earth metals with zirconium.

Gschneidner [21] points out that the intersolubility of the components in systems of zirconium with the rare-earth metals increases with increase in the atomic number of the rare-earth metal. Our experiments and an analysis of published data permit the conclusion that such a relationship, due to lanthanide compression effect, also exists for systems of titanium and hafnium with the rare-earth metals [131, 136].

The different crystal lattices of the rare-earth metals and group Va and VIa metals, the considerable difference in the atomic diameter of these metals and also in the electronegativity is the cause of their very limited solubility in one another. Most of the work on elucidating the character of the interaction of refractory metals with the rare-earth metals has been carried out in the Laboratory of Refractory Metal and Rare Earth Metal Alloys of the Institute of Metallurgy [27]. Reviews of American investigations are given in [21, 137, 138].

E. M. Savitskii, V. V. Baron, and Yu. V. Efimov have investigated the phase diagrams of vanadium with lanthanum and cerium [139, 140]. They found wide ranges of separation in the solid and liquid states. Maximum solubility in the solid state of lanthanum in vanadium was 0.4% and that of cerium was 0.1%. The solubility of vanadium in $\alpha$- and $\beta$-lanthanum was less than 0.5 wt.%. With increase in temperature, the solubility increased somewhat. No chemical compounds were found in the systems.

The vanadium–yttrium system also shows a wide range of separation in the liquid and solid states [141]. On the yttrium side at a temperature of 1455°C, a eutectic containing 7 wt.% vanadium is formed. The phase diagrams of vanadium with cerium, praseodymium, gadolinium, dysprosium, erbium, and luthecium are similar to the vanadium–lanthanum and vanadium–yttrium systems.

In the niobium–cerium and niobium–lanthanum systems, investigated by E. M. Savitskii, V. F. Terekhova, and I. V. Burov, there are wide ranges of separation in the liquid state. The solubility of both lanthanum and cerium in niobium does not exceed 0.05 wt.% [142].

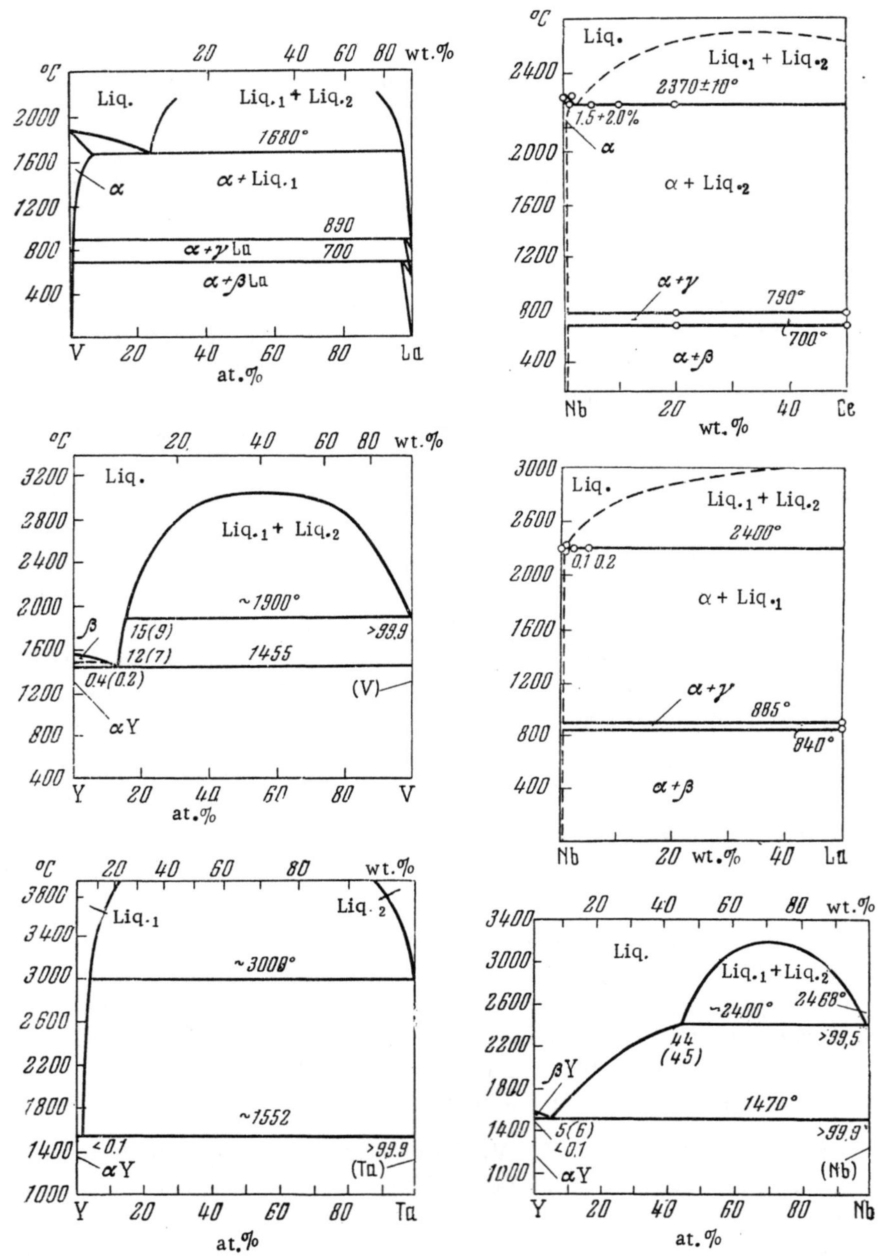


**Fig. 46. Interaction of group Va metals with rare-earth metals.**

The interaction of tantalum with the rare-earth metals bears the same character. Of the refractory metals, tantalum is the most inert to the rare-earth metals. The intersolubility of these metals does not exceed 0.1 at.%. No chemical compounds are formed. Lanthanum, cerium, neodymium, praseodymium, and other rare-earth metals having a melting point of up to 1000°C, after being melted in tantalum crucibles, contain tantalum in an amount of less than 0.05 wt.%. Tantalum is the best material for crucibles used for melting rare-earth metals and their alloys [24].

Some phase diagrams of group Va metals with rare-earth metals are shown in Fig. 46.

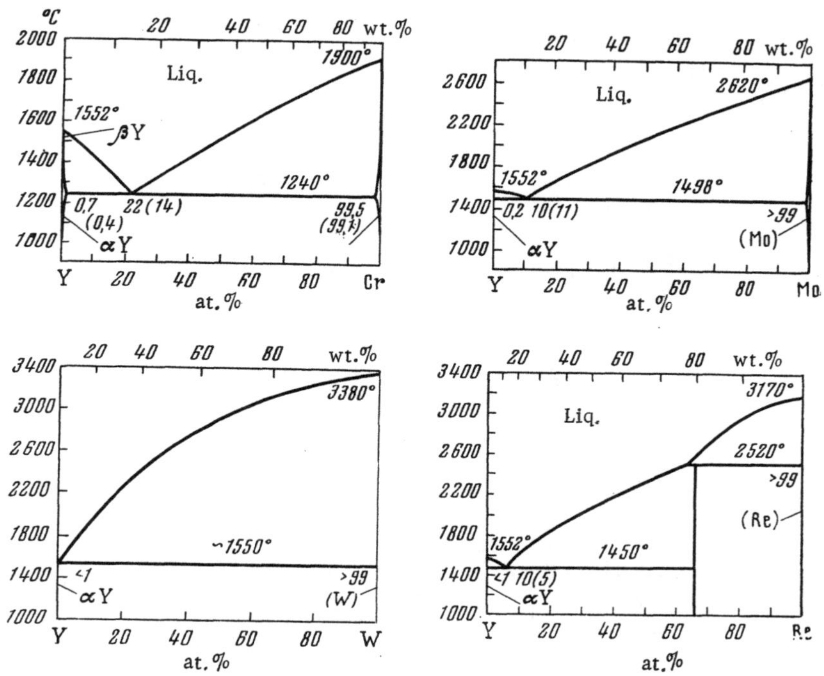


Fig. 47. Interaction of yttrium with chromium, molybdenum, tungsten, and rhenium.

Wide regions of separation on the solid and liquid states are found in the phase diagrams of lanthanum, cerium, yttrium, and other rare-earth metals with chromium [141, 143, 144], molybdenum [141, 145], and tungsten [141].

According to Lundin, the phase diagram of tungsten with yttrium is a eutectic type of diagram. According to data, requiring refinement, in the tungsten–cerium system, the existence of two chemical compounds has been found in the tungsten–cerium system. One of these compounds has the formula $W_2Ce$ [146].

The chromium–yttrium phase diagram was studied in [143]. At room temperature chromium dissolved 0.4 wt.% Y. The solubility increased with increase in temperature, attaining 1 wt.% at 1700°C. No chemical compounds were found in the system. A eutectic containing 13 wt.% Cr was formed on the yttrium side at 1330°C.

The character of the interaction of the components is also the same in the molybdenum–yttrium system [141]. The interaction of rare-earth metals (lanthanum, cerium, praseodymium, neodymium) with molybdenum was studied in [147, 148]. It was found that at room temperature, the solubility of lanthanum and cerium in molybdenum did not exceed 0.15%, that of praseodymium 0.07% [147]. No intermediate phases are formed in these systems. Alloying of molybdenum with rare-earth metals in the solid solution limits lowers the ductility–brittleness transition temperature and increases the technological ductility of molybdenum. The occurrence of a second phase produces rapid embrittlement of the molybdenum [147, 148].

The phase diagrams of chromium, molybdenum, tungsten, and rhenium are shown in Fig. 47.

A knowledge of the interaction of the rare-earth metals with rhenium, a group VIIa element, is very important. Lundin and Klodt constructed the rhenium–yttrium phase diagram for the entire concentration range (see Fig. 47) [141]. It is of interest that the compound YRe

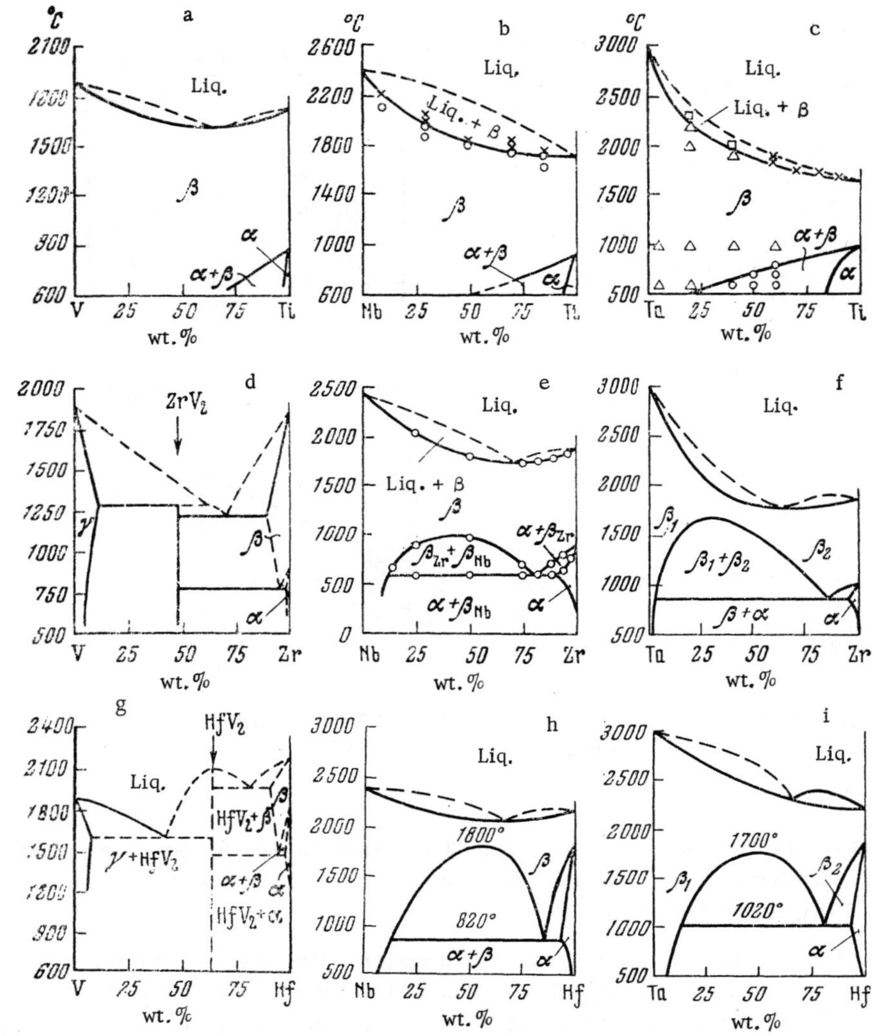

Fig. 48. Phase diagrams of the vanadium–titanium (a), niobium–titanium (b), tantalum–titanium (c), vanadium–zirconium (d), niobium–zirconium (e), tantalum–zirconium (f), vanadium–hafnium (g), niobium–hafnium (h), and tantalum–hafnium (i) systems.

with a hexagonal structure of $MgZn_2$ type [141, 149, 150], formed by peritectic reaction at 2520°C, was found. The solubility of yttrium in rhenium at the peritectic temperature is less than 1 at.%.

E. M. Savitskii and O. Kh. Khamidov have investigated the phase diagrams of rhenium with scandium, gadolinium, and erbium [37, 38]. The rhenium–scandium system is of peritectic type with two metallic compounds $Sc_5Re_{24}$($\chi$-phase) and $ScRe_2$ (Laves phase). There is a narrow region of existence of solid solutions (maximum solubility of scandium in rhenium is 5-6 at.%, that of rhenium in scandium is less than 1 at.%). According to the results of E. M. Savitskii and O. Kh. Khamidov, the rhenium–gadolinium and the rhenium–erbium systems also belong to the peritectic type with the existence of compounds to $GdRe_2$ and $ErRe_2$, respectively. Slight intersolubility of components in the solid state has been found.

The results of work on the study of the interaction refractory metals of groups Va, VIa, and VIIa with the rare-earth metals show that separation in the liquid state and the formation of a eutectic is characteristic of these systems.

The interaction of refractory noble metals of group VII with the rare-earth metals was investigated by V. Compton and B. Matthias [150, 151] and also by E. Savitskii and O. Khamidov [152]. They found the formation in these systems of compounds $AB_2$ having a structural type of cubic or hexagonal Laves phases. Compounds of the rare-earth metals with ruthenium, osmium, platinum, and iridium of Laves phase type possess superconductivity and ferromagnetism at liquid helium temperatures. Complete phase diagrams of group VIII refractory metals with rare-earth metals have not yet been constructed (they offer considerable interest for the development of cathode materials for thermionic emission converters). Taking into account the considerable difference in electronegativity and atomic diameters of these elements, it may be assumed that the phase diagrams will be complex with the presence of several chemical compounds and low solubility of the elements in each other (as in the case of the interaction of iron, cobalt, and nickel with the rare-earth metals).

## 4. Interaction of Group Va and VIa Metals with Group IVa Metals

Group IVa refractory metals (titanium, zirconium, and hafnium) are well known to be polymorphic. Their low-temperature modification has a hexagonal, close-packed lattice, and the high-temperature modification has a body-centered cubic lattice. Group Va and VIa metals form a continuous series of solid solutions with the isomorphic high-temperature $\beta$-modification of titanium and limited solid solutions with $\alpha$-titanium (hexagonal close-packed lattice). An exception is the tungsten–titanium system, in which there is limited solubility of tungsten in both modifications [153]. In the chromium–titanium system, the compound $TiCr_2$ with melting point 1350°C is formed [154]. In the remaining systems with the participation of group IVa and VIa metals, there are no intermediate phases [153-158].

The interaction of group Va metals with group IVa metals is shown in Fig. 48.

Niobium (group Va) forms a continuous series of solid solutions with the high-temperature modification of zirconium (body-centered cubic lattice), while with the $\alpha$-modification (hexagonal close-packed lattice) it forms limited substitution solid solutions, as in the interaction with titanium. Figure 48e, from data of [159], shows the phase diagram of the niobium–zirconium system.

In the tantalum–zirconium system [160], limited solubility is found for both zirconium modifications. However, in later work, continuous solubility of tantalum with the $\beta$-modification of zirconium has been found [161]. A number of solid solutions of $\beta$-zirconium and tantalum decompose according to a monotectoid reaction at 800°C. The maximum solubility of tantalum in $\alpha$-zirconium is less than 2% and is reduced to 0.5% at 735°C. The limit solubility of zirconium in tantalum at room temperature is less than 2%. The tantalum–zirconium phase diagram, according to [161] is shown in Fig. 48f.

The character of the physicochemical interactions in the tantalum–hafnium system [162] and the niobium–hafnium system [163] are similar to the tantalum–zirconium and niobium–zirconium systems.

Group VIa metals (chromium, molybdenum, and tungsten), which are more electronegative elements with respect to zirconium than group Va metals, form limited solid solutions with both modifications of zirconium. Figure 49 shows the complete zirconium–molybdenum and zirconium–tungsten phase diagrams [164]. The incongruently melting compounds $ZrMo_2$ (melting point 1880°C) and $ZrW_2$ (melting point 2175°C) have been found. The solubility of molybdenum in $\alpha$-zirconium at all temperatures does not exceed 0.18%, the solubility of tungsten in $\alpha$-zirconium is less than 0.5% at all temperatures. The solubility of molybdenum and tungsten in $\beta$-zirconium at the eutectic temperature is 22 and 8%, respectively. The compounds

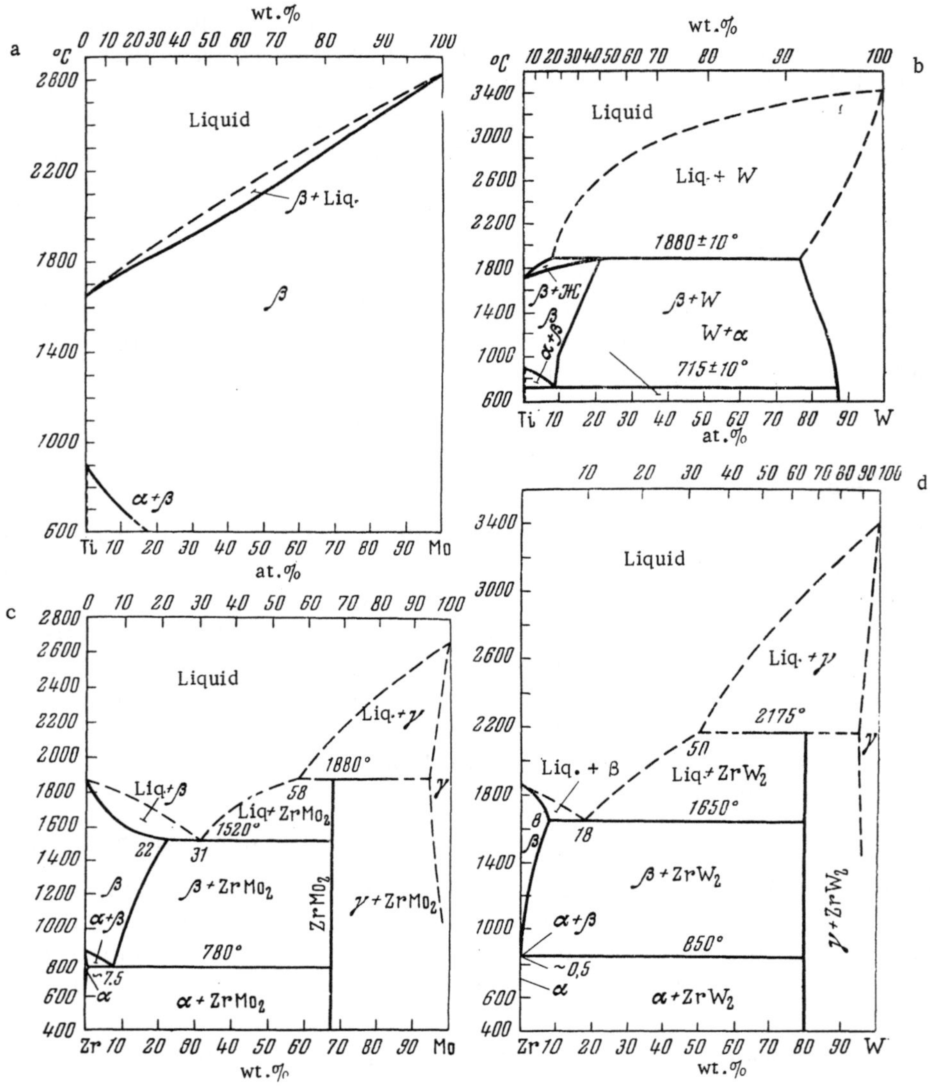


Fig. 49. Phase diagrams of the molybdenum–titanium (a), tungsten–titanium (b), molybdenum–zirconium (c), and tungsten–zirconium (d) systems.

$ZrMo_2$ and $ZrW_2$ are formed in the respective systems according to the peritectic reactions: liquid melt + solid solution on a molybdenum (tungsten) base → $ZrMo_2$ ($ZrW_2$). The solubility of zirconium in molybdenum and tungsten is less than 10 wt.%.

The tungsten–hafnium phase diagram was studied in [165, 166]. The results were very similar. In this system also the formation of a compound $HfW_2$ having a structure isomorphic with $ZrMo_2$ and $ZrW_2$ (cubic lattice C15 of $MgCu_2$ type) was found. The compound is formed by peritectic reaction at a temperature of 2540°C: melt + solid solution on a tungsten base → $HfW_2$ [165] (according to later work [166], the temperature of the peritectic reaction is rather higher, 2650°C). The solubility of hafnium in tungsten at the peritectic temperature [165, 166] is 6-15%, the solubility of tungsten in β-hafnium at the eutectic temperature is about 11-14%.

The molybdenum–hafnium phase diagram is similar to that of the tungsten–hafnium system [167].

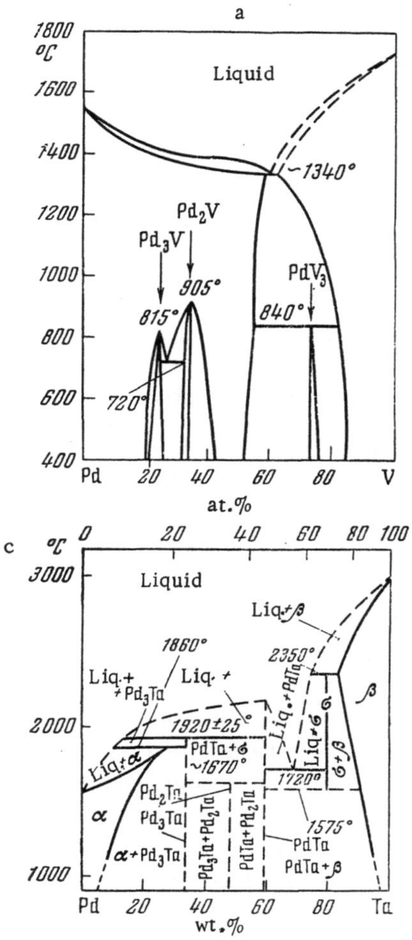


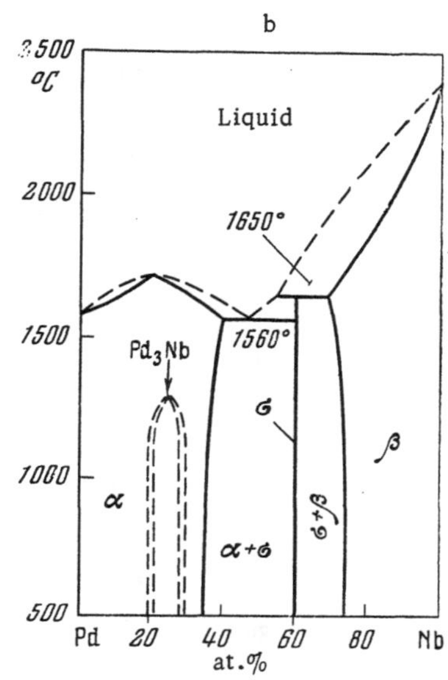


Fig. 50. Phase diagrams of the palladium–vanadium (a), palladium–niobium (b), and palladium–tantalum (c) systems.

## 5. Interaction of Group Va and VIa Metals with Group VIII Metals

In systems with the participation of palladium and group Va metals, a wide range of solid solutions has been found, as well as some metallic compounds. The solubility of group Va metals in palladium decreases as their atomic number increases. Thus, the maximum solubility of vanadium in palladium is 58 at.% [168], niobium 40 at.% [169], and tantalum 18 at.% [170].

Group Va metals having a higher atomic radius than palladium form $\sigma$-phase with palladium. In the palladium–niobium system, the $\sigma$-phase corresponds to 60 at.% Nb, and in the palladium–tantalum system it corresponds to 70 at.% Ta.

Vanadium, with a smaller atomic radius than palladium forms with the latter a compound of $\beta$-tungsten type ($PdV_3$), resembling a $\sigma$-phase in structure [168]. It was found in [171] that the $\sigma$-phase had a tetragonal symmetry; the unit cell contained 30 atoms. The palladium–vanadium systems also includes the compounds $Pd_2V$ and $Pd_3V$ (isomorphic with the corresponding compound $Pd_3Ta$ in the palladium–tantalum system).

In systems of group Va metals with palladium, in addition to $\sigma$-phase, the formation of a further series of compounds is observed. In the palladium–vanadium, palladium–niobium, and palladium–tantalum systems four compounds are found [105]. For example in the palladium–tantalum system, in addition to the $\sigma$-phase, there are compounds PdTa (body-centered cubic lattice with the parameters a = 3.28 kX, c = 6.00 kX, c/a = 1.84), $Pd_3Ta$) tetragonal crystal structure of $TiAl_3$ type, a = 3.87 kX, c = 7.94 kX, c/a = 2.05), and $Pd_2Ta$ (crystal structure not interpreted) [170]. The phase diagrams of these systems are shown in Fig. 50.

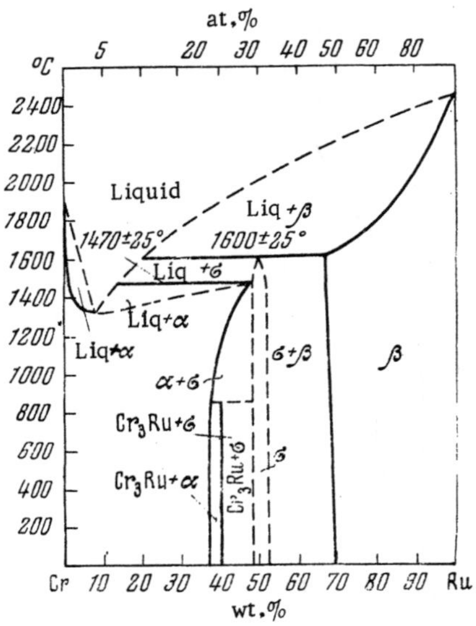


Fig. 51. Phase diagram of the chromium–ruthenium system [273].

Investigation of alloys of the niobium–platinum system was limited to a determination of the liquidus line on the platinum side. The liquidus line passes through a maximum close to the compound $NbPt_3$ [23]. It may be assumed that this system will be very like the niobium–palladium system.

For alloys of tantalum with iridium, the existence of two intermediate phases has been found: $\sigma$-phase with a homogeneity range from 75 to 85 at.% Ta [172] and the compound $TaIr_3$ with an ordered structure of $AuCu_3$ type [173]. The $\sigma$-phase is formed by peritectic reaction. The compound $TaIr_3$ melts congruently.

In alloys of the tantalum–platinum system, the existence of $\sigma$-phase and also the compound $TaPt_3$ has been found [83]. The phase diagram has not been fully investigated, but it is evidently similar to the tantalum–iridium diagram [23].

In systems of group Va metals and the hexagonal metals ruthenium and osmium, high solubility of these metals in each other has been found, despite the different types of crystal lattices. In the niobium–ruthenium system, the existence of an intermediate phase having a tetragonal structure is assumed [23]. A similar character of the interaction of the components is also found in the tantalum–ruthenium system. In niobium–osmium alloys, the existence of $\sigma$-phase in the concentration range from 30 to 54 at.% Os and $\chi$-phase ($\alpha$-Mn structural type) in the range from 55 to 65 at.% Os has been found. $\sigma$- and $\chi$-phases have also been found in alloys of the tantalum–osmium system [23]; the $\sigma$-phase in this system is stable from 1200°C to the melting point, the $\chi$-phase is stable throughout the entire temperature range to the melting point. The solubility of tantalum in osmium at 2200°C is 22 at.%, and the solubility of osmium and tantalum at the same temperature is 14 at.% [174, 175].

Alloys of chromium with palladium were investigated in [174, 175]. The components form a eutectic, in which the high-palladium solid solution crystallizes along with the chromium. The solubility of palladium in chromium is insignificant. The compound PdCr, having a tetragonal face-centered lattice is formed at a temperature below 570°C.

Alloys of chromium and platinum were investigated in [176-178]. A region of existence of solid solutions on the platinum side was found. The solubility limit is found to be 62% Cr. Commencing with 23 at.% an ordered phase is formed. The temperature of formation of the superstructure is unknown. The formation of the compound $PtCr_3$ with a crystal structure of $\beta$-tungsten type was discovered; there is a eutectic on both sides of the compound.

In the iridium–chromium system [178], as distinct from platinum–chromium alloys, the formation of two metallic phases is observed. The solubility limit of chromium in iridium is 27 at.%. In the solid-solution range, an ordered $AuCu_3$ type of phase occurs. Between 31.5 and 58 at.% Cr is a region of $\varepsilon$-phase homogeneity with hexagonal closest packing. The second metallic phase is $IrCr_3$. The range of existence of solid solutions based on this compound lies within the limits of 72.5-78.5 at.% Cr. $IrCr_3$, like the compound $PtCr_3$, crystallizes in a $\beta$-tungsten type of lattice.

In the rhodium–chromium system [178], an $\varepsilon$-phase with hexagonal closest packing is also found; the $\varepsilon$-phase has a wide homogeneity range (from 24.4 to 60 at.% Cr). No compound isomorphic with $PtC_3$ and $IrCr_3$ has been found in this system.

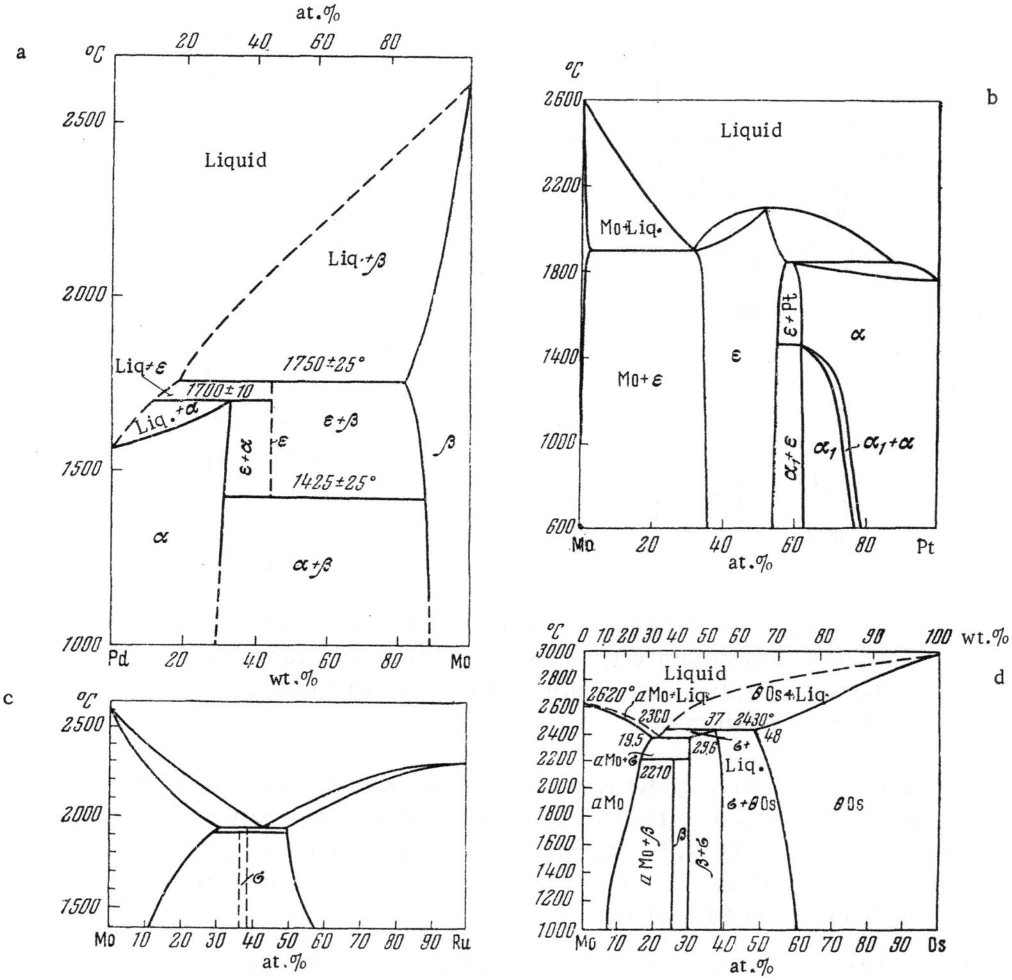


Fig. 52. Phase diagrams of the molybdenum–palladium (a), molybdenum–platinum (b), molybdenum–ruthenium (c), and molybdenum–osmium (d) systems.

An interesting feature of the interaction of ruthenium with chromium is the fact that the hexagonal ruthenium lattice dissolves a large amount of chromium with a body-centered cubic lattice (Fig. 51). The metallic compound $Cr_3Ru$ and $\sigma$-phase are found in this system.

The diagram of the chromium–osmium system is published in [8]. The existence of two metallic compounds: CrOs with a $\beta$-tungsten type of structure, and $Cr_2Os$ ($\sigma$-phase). The maximum limit solubility of osmium in chromium is 28 wt.%, and of chromium in osmium it is 22 wt.%.

The molybdenum–palladium [179] and molybdenum–platinum systems [180] (Fig. 52) are characterized by a wide range of solid solutions on the palladium and platinum side.

In alloys of molybdenum with rhodium and iridium considerable intersolubility of the components is found [181, 182]. An intermediate ε-phase having a wide homogeneity range is formed. In the molybdenum–rhodium system this range extends from 45 to 82 at.% Rh [181]. The ε-phase has a hexagonal close-packed lattice, the parameters of which diminish with increase in rhodium content, the c/a ratio having a minimum at 75 at.% Rh.

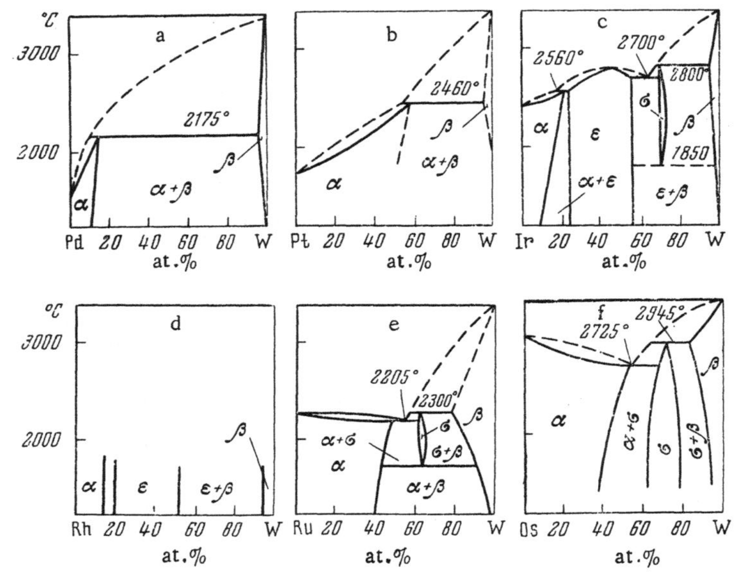


Fig. 53. Phase diagrams of the tungsten–palladium (a), tungsten–platinum (b), tungsten–iridium (c), tungsten–rhodium (d), tungsten–ruthenium (e), and tungsten–osmium (f) systems.

The molybdenum–osmium system, shown in Fig. 52d, was investigated in [183]. The maximum solubility of Os in Mo is 19.5 at.% at the eutectic temperature of 2380°C, and that of molybdenum in osmium is 48 at.%. Intermediate σ- and β-phases, having the crystalline lattice of β-tungsten ($Mo_3Os$) are formed in the system. The homogeneity range of the σ-phase at 1000°C is from 30 to 39 at.% Os.

Only one σ-phase has been found in the molybdenum–ruthenium system. The system is characterized by a wide intersolubility of the components. The maximum solubility of molybdenum in ruthenium is 50 at.% at 1600°C, and that of ruthenium in molybdenum is 30 at.% [18].

Tungsten does not form metallic compounds with platinum and palladium. The phase diagram of these metals is of peritectic type with limited solubility of the metals in each other. At the same time, a wide range of solid solutions is found on the palladium and platinum sides. Figure 53a shows the tungsten–palladium phase diagram according to the data of [185] and Fig. 53b shows a modification of the tungsten–platinum diagram according to the combined results of various investigators.

The complete tungsten–iridium diagram is given in [187]. The maximum solubility of tungsten in iridium is 22 at.%; the solubility diminishes with decrease in temperature and at 1000°C is about 10% W; the solubility of iridium in tungsten is insignificant. The authors of [187] discovered the existence in the system of σ-phase in addition to the previously discovered ε-phase [188]. The σ-phase exists at high temperatures in a narrow concentration range (69-71 at.% W).

The tungsten–rhodium phase diagram has so far not been studied completely. The existence of limited solid solutions on the tungsten and rhodium side and a metallic compound of ε-phase type (hexagonal close-packed lattice), having a wide homogeneity range, has been found [188]. The maximum solubility of tungsten in rhodium is 12 at.%, while that of rhodium in tungsten is insignificant.

The phase diagrams of tungsten with hexagonal ruthenium and osmium are similar in many respects. The tungsten–ruthenium system was studied in [190]. At a temperature of 2300°C, $\sigma$-phase is formed in the system according to a peritectic reaction. The maximum solubility of tungsten in ruthenium is 48 at.% and that of ruthenium in tungsten is 18 at.%. A strong temperature dependence of the solubility is found.

In the tungsten–osmium system [191], $\sigma$-phase is also formed by peritectic reaction at a temperature of 2945°C. However, the $\sigma$-phase does not undergo decomposition, as in the tungsten–ruthenium system; the concentration range of existence of the $\sigma$-phase is much wider. Maximum solubility of tungsten in osmium is 48 at.%, and that of osmium in tungsten is 15 at.%.

The formation of $\sigma$- and $\varepsilon$-phases is characteristic of systems of tungsten with group VIII refractory metals. In the rhodium–tungsten system, by analogy with the iridium–tungsten system, the formation of a high-temperature $\sigma$-phase may be expected.

Despite the difference in the crystal structures of tungsten and group VIII metals, considerable solubility of tungsten in group VIII metals is found. The solubility in hexagonal ruthenium and osmium is particularly high.

The interaction of group Va and VIa metals with iron is discussed in [286].

Summing up the character of the physicochemical interaction of group Va and VIa refractory metals with group VIII metals, some general features may be noted:

1) Considerable solubility of group Va-VIa metals in group VIII metals, especially in hexagonal ruthenium and osmium (this latter circumstance has not yet been fully elucidated);

2) Limited solubility of group VIII metals in group Va and VIa metals;

3) Formation in a number of systems of some metallic compounds, including $\chi$-phases having a hexagonal close-packed lattice, $\sigma$-phases, or phases having a $\beta$-tungsten type of structure.

In [192], the opinion is expressed that the formation of $\sigma$-phase depends on the following conditions: The atomic diameters of the alloying elements should not exceed 8% and one of the two metals should have a bcc lattice, while the other should have a face-centered lattice (at least one of the allotropic modifications of this element).

The work of Savitskii and collaborators, Raub, and other investigators has shown that the $\sigma$-phase is formed under conditions in which one element has a bcc lattice and the other a hexagonal close-packed lattice.

In [192], the view is expressed that the $\sigma$-phase is an electronic compound with a ratio of valence electrons and atoms equal to 1.7. These problems are discussed in [274].

## Interaction of Rhenium with Other Refractory Metals

Rhenium occupies a special place among refractory metals by reason of its exceptional influence on the plastic properties of tungsten and molybdenum. In what follows, some general relationships of the interaction of rhenium with refractory metals will be examined. The problems of the interaction of rhenium with refractory metals have been discussed in the recently published monograph on rhenium [193].

The character of the physicochemical interaction of rhenium with group IVa, Va, VIa, and VIII metals is of interest. An extremely detailed study of the interaction of rhenium with various elements of the periodic system has been made in the USSR in the Refractory and Rare Metal Alloys Laboratory of the A. A. Baikov Institute of Metallurgy. The results of these investigations have been published in numerous articles and papers and in the monograph [193].

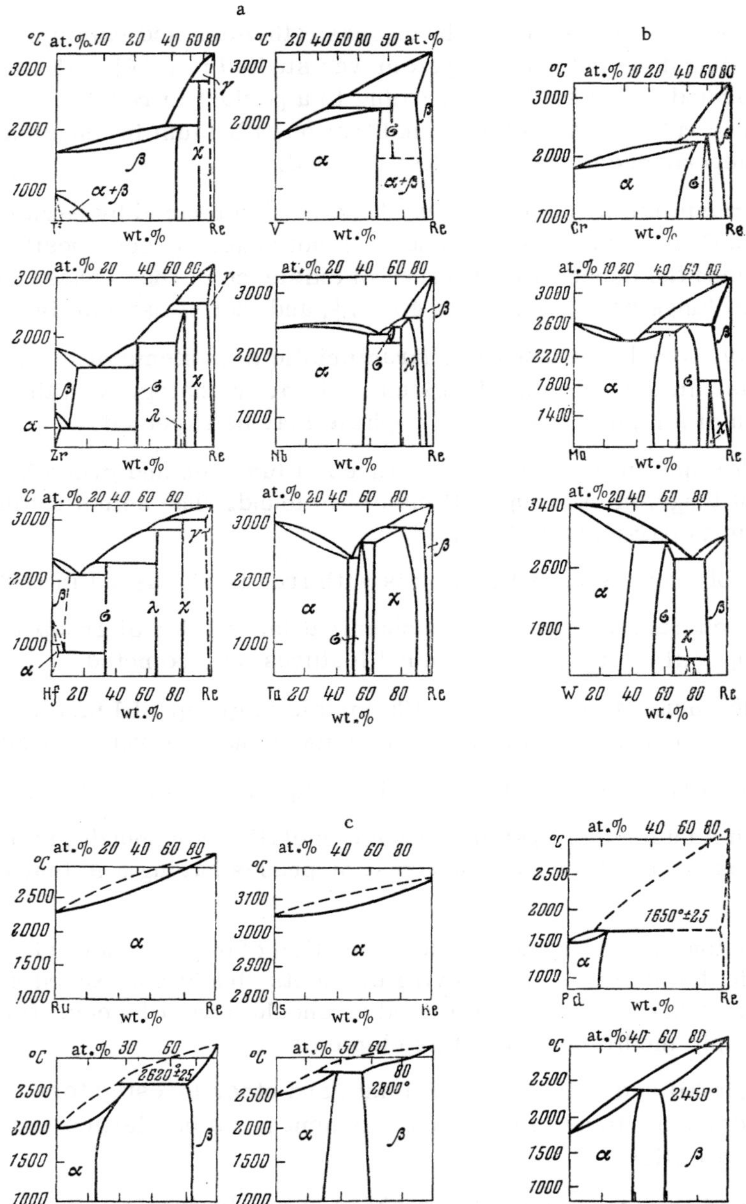


Fig. 54. Phase diagrams of systems of rhenium with transition metals of groups IVa and Va (a), group VIa (b), and group VIII (c).

The investigations made in western countries are described to a considerable extent in [194, 195]. A characteristic feature of the interaction of rhenium with group IVa-VIa metals is the formation of $\sigma$- and $\chi$-phases ($\alpha$-Mn structure type) in the corresponding binary systems. In systems with group IV metals (zirconium, hafnium), Laves phases having a $MgZn_2$ type of structure are also formed; all these compounds melt incongruently. It has been reported [186] that under conditions of the combined reduction of tungsten and rhenium hexafluorides by hydrogen at 500°C in alloys containing 13-45 wt.% Re, a phase having a $\beta$-tungsten structure is formed.

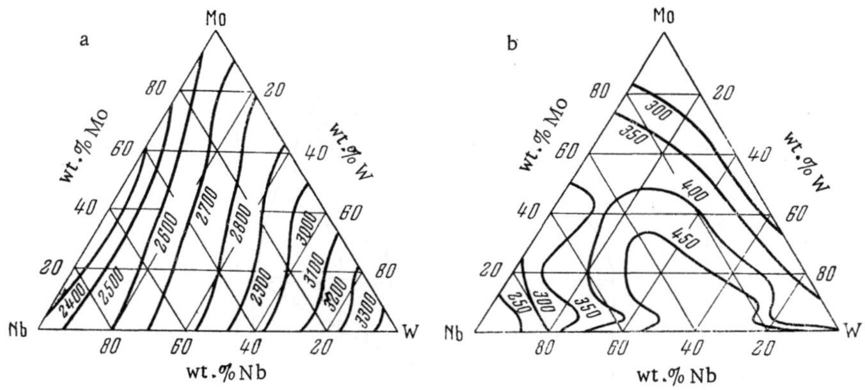


Fig. 55. Projections of solidus isotherms on the concentration triangle of the niobium–molybdenum–tungsten system (a) and isohardness lines of alloys of the same system (b).

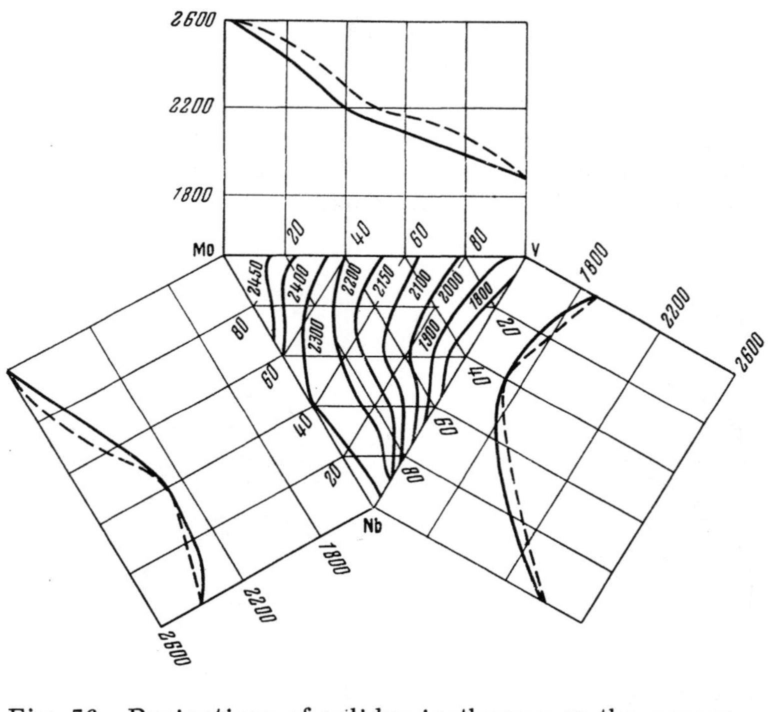


Fig. 56. Projections of solidus isotherms on the concentration triangle of the niobium–molybdenum–vanadium system.

An interesting feature of the hexagonal metal rhenium is its high solubility in group Va and VIa metals having a body-centered cubic lattice; for example, the solubility of rhenium in vanadium is 65 at.%; up to 32 wt.% of rhenium is dissolved in tungsten at 1100°C [193]. Such high solubility for metals having different crystal structures, together with the high solubility of a number of body-centered cubic metals in hexagonal ruthenium and osmium is a unique phenomenon among metal systems. These facts have not as yet found a sufficiently convincing theoretical explanation. The solubility of rhenium in isomorphic $\alpha$-modifications of titanium, zirconium, and hafnium is very low (0.5-0.4 at.%), while its solubility in high-temperature modifications of these metals having a body-centered cubic lattice is much greater, attaining 50 at.% in the system with titanium [29].

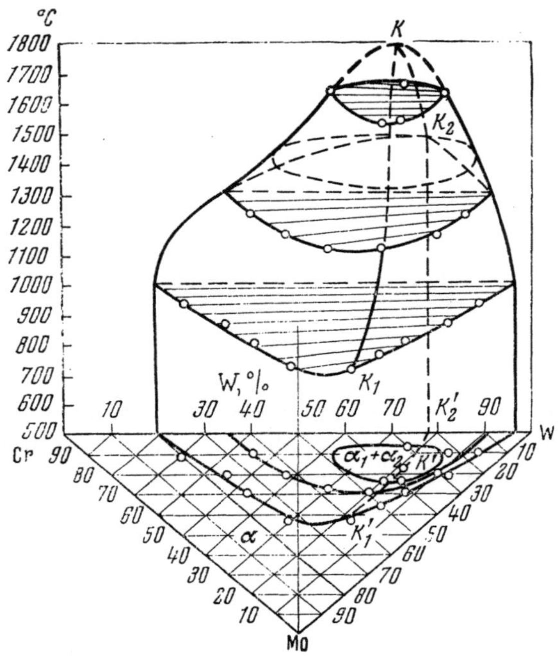


Fig. 57. Three-dimensional diagram of separation of ternary solid solutions of the tungsten–molybdenum–chromium systems.

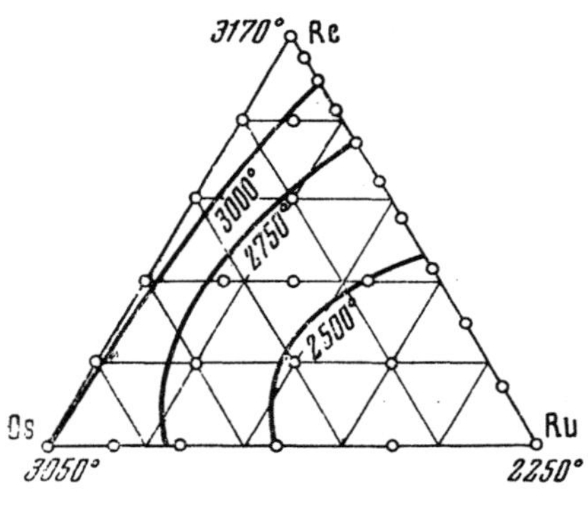


Fig. 58. Projection of solidus isotherms on the concentration diagram of the ruthenium–osmium–rhenium system.

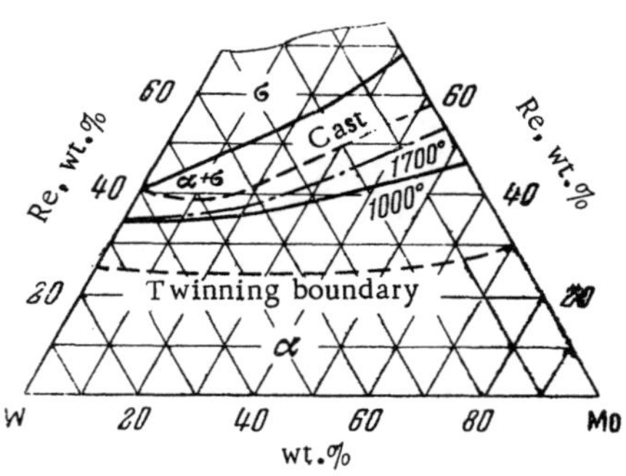


Fig. 59. Isothermal cross section of the tungsten–molybdenum–rhenium system at 1000°C.

Rhenium interacts with the noble refractory metals (group VIII) with the formation of either a continuous series of solid solutions or limited solid solutions. No intermediate phases are formed in these systems. In systems of rhenium with isomorphic ruthenium and osmium, a continuous series of substitution solid solutions is found. The rhodium–rhenium, iridium–rhenium, and palladium–rhenium phase diagrams are diagrams of peritectic type with an appreciable region of existence of solid solutions of these metals in each other.

The fundamental phase diagrams of rhenium with the transition metals are shown in Fig. 54 [29, 193].

The influence of electronegativity and the geometric factor on the solubility of different metals in rhenium was shown in Fig. 34b. Elements which form continuous series of solid solutions with rhenium are arranged within the small ellipse (cobalt, ruthenium, osmium); elements forming a wide region of solid solutions (not less than 10 at.%) are arranged in the interval between the large and small ellipses (vanadium, niobium, chromium, molybdenum, tungsten, etc.). All the other elements situated outside the limits of the large ellipse form very limited solutions or are practically insoluble in rhenium.

## Some Ternary and Multicomponent Systems of Refractory Metals

Most of the industrially important alloys belong to complex systems consisting of several components. There is, therefore, a definite interest in the examination of some ternary and multicomponent systems of refractory metals, on the basis of which a number of the more promising alloys are being developed.

The niobium–molybdenum–tungsten diagram was first investigated in [119]. A continuous series of solid solutions was found in the ternary system. Figure 55a shows a projection of the solidus isotherms on the concentration triangle of the niobium–molybdenum–tungsten system; the alloys of the tungsten corner have the highest temperatures. As will be seen from Fig. 55b, showing the lines of equal hardness of the niobium–molybdenum–tungsten system, alloys in the central part of the diagram and those adjacent the niobium–tungsten side of the triangle have the maximum hardness.

Alloys of the niobium–molybdenum–vanadium system were studied in [115]. A continuous series of solid solutions was found in this system. Projections of the sodium isotherms on the concentration triangle of the niobium–molybdenum–vanadium system are shown in Fig. 56.

The tungsten–molybdenum–chromium ternary system was investigated by N. V. Grum-Grzhimailo and D. I. Prokof'ev [196]. A region of decomposition of the ternary solid solution was found, consisting of a mixture of the ternary solid solutions ($\alpha_1 + \alpha_2$) having a body-centered cubic lattice (Fig. 57).

A continuous series of solid solutions was found in the tungsten–molybdenum–tantalum, tungsten–molybdenum–vanadium, and vanadium–niobium–tantalum systems [197-199].

The ruthenium–osmium–rhenium phase diagram is the first example of unlimited solubility in the liquid and solid states of three hexagonal metals (Fig. 58) [200].

In the tungsten–molybdenum–rhenium system a ternary $\alpha$-solid solution with a body-centered cubic lattice is formed [201]. Regions of existence of $\alpha$-phase and a two-phase ($\alpha + \beta$) region have been found. Figure 59 shows an isothermal cross section of the ternary system at 1000°C. Deformation twins are formed in alloys situated in the region of the ternary $\alpha$-solid solution [201].

The isothermal cross section of the molybdenum–niobium–chromium system at 1500°C has been studied [202]. A wide range of ternary solid solutions is found. On the chromium–niobium side, a two-phase region of $\alpha$-solid solution and the compound $NbCr_2$ is formed. Figure 60 shows the isothermal cross section of the molybdenum–niobium–chromium system according to the data of [203]. The authors of this article found a $\delta$-phase homogeneity region (based on the compound $NbCr_2$), situated in the limits of 62-68 at.% Cr and extending in the ternary system to 11 at.% Mo at 63 at.% Nb.

The tungsten–niobium–zirconium system has been studied by E. M. Savitskii and A. M. Zakharov [204]. The combined solubility of tungsten and zirconium falls from 16-18% at 2000°C

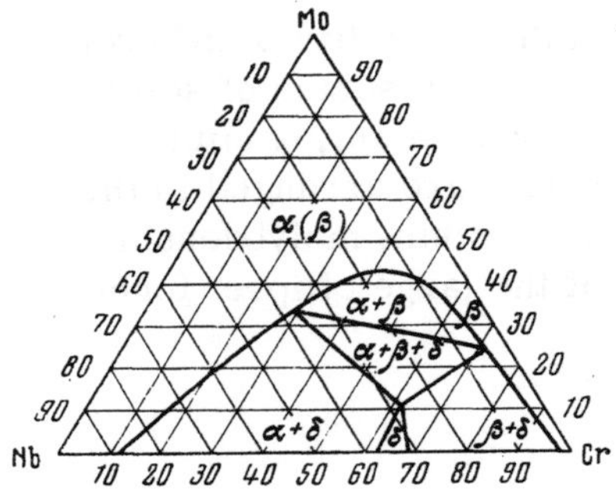


Fig. 60. Isothermal cross section of the molybdenum–niobium–chromium system at 1500°C.

to 8-9% at 1100°C. Figure 61 shows polythermal cross sections and an isothermal cross section of this system.

The niobium–molybdenum–zirconium system was also investigated by E. M. Savitskii and A. M. Zakharov [205]. It was shown that the molybdenum–niobium–zirconium system was analogous to the tungsten–niobium–zirconium system. Both systems are characterized by transition of the monovariant peritectic transformations L + $\alpha \rightarrow W_2Zr$ and L + $\alpha \rightarrow Mo_2Zr$ to the eutectic transformations L $\rightarrow \beta + W_2Zr$ and L $\rightarrow Mo_2Zr + \beta$, where L is liquid, and $\alpha$ and $\beta$ are solid solutions based on tungsten (or molybdenum) and $\beta$-zirconium.

The tungsten–molybdenum–zirconium system was studied by A. M. Zakharov and E. M. Savitskii [206, 207]. Figure 62 shows the polythermal cross sections and isothermal cross section at 1000°C, as given in [206]. This system is characterized by transition of the peritectic transformation L + $\alpha \rightarrow W_2Zr$ and the eutectic transformation L $\rightarrow \beta + W_2Zr$, occurring in the tungsten–zirconium binary system at 2260 and 1640°C (respectively), to the peritectic L + $\alpha \rightarrow Mo_2Zr$ and eutectic L $\rightarrow \beta + Mo_2Zr$ transformations, occurring in the molybdenum–zirconium binary system at

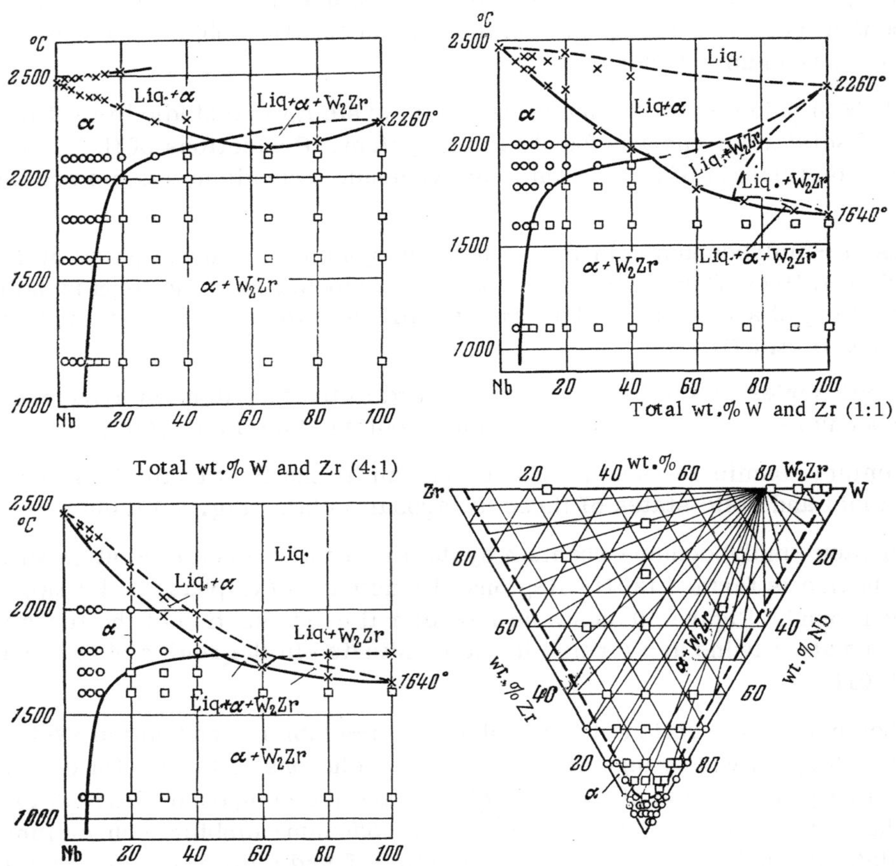


Fig. 61. Polythermal cross sections and isothermal cross section (at 1600°C) of the tungsten–niobium–zirconium system.

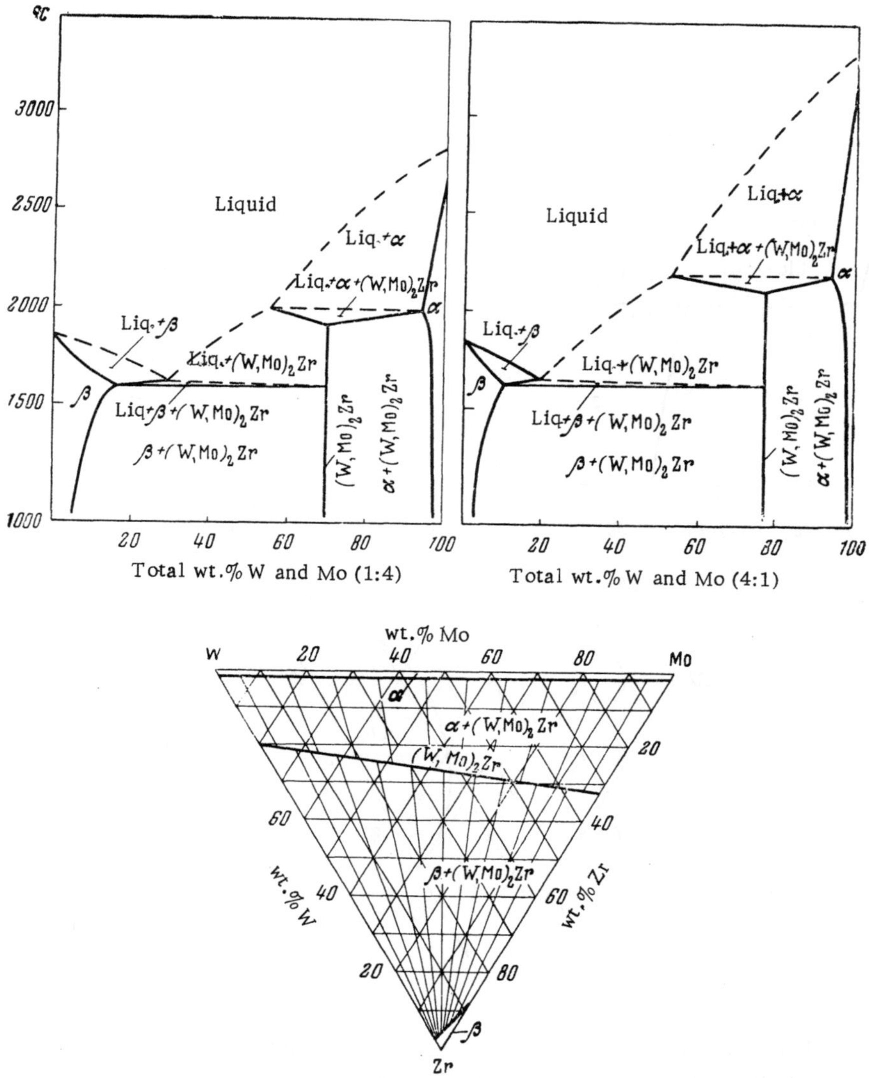


Fig. 62. Polythermal cross sections and an isothermal cross section (at 1000°C) of the tungsten–molybdenum–zirconium system.

1890 and 1570°C. The isomorphic compounds $W_2Zr$ and $Mo_2Zr$, having a crystal lattice of $MgCu_2$ type, form a continuous series of solid solutions [206].

A. M. Zakharov and E. M. Savitskii studied the tungsten–molybdenum–titanium system [207]. Figure 63 shows polythermal cross sections and an isothermal cross section (at 1000°C). Tungsten and titanium have considerable combined solubility in molybdenum in the solid state; this solubility diminishes somewhat on reduction in temperature. Addition of molybdenum to alloys of the tungsten–titanium binary system increases the limited solubility of the components in the solid state to complete solubility. At 1000°C continuous series of solid solutions are formed between tungsten and $\beta$-titanium with a molybdenum content equal to 25 wt.%.

Alloys of the vanadium–molybdenum–titanium system crystallize as a continuous series of solid solutions with a body-centered cubic lattice [275]. The system has a two-phase region of ternary $\alpha$ and $\beta$ solid solutions with hexagonal close-packed and body-centered lattices.

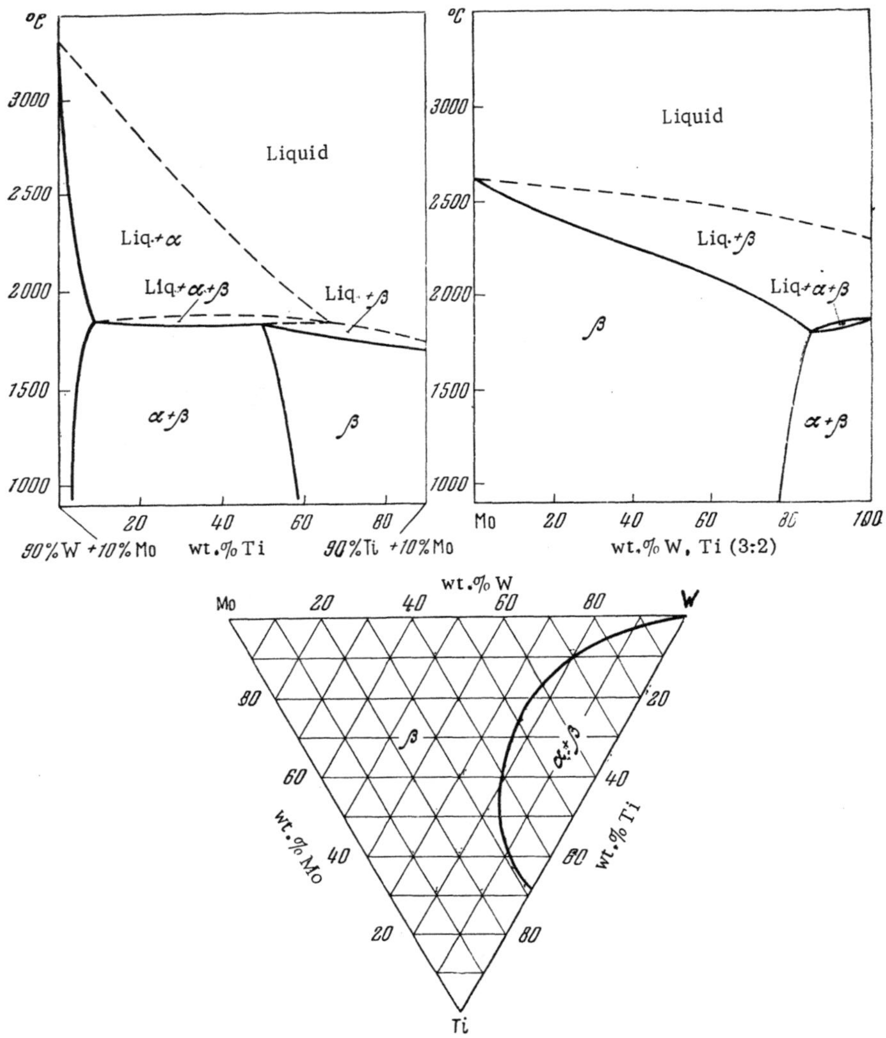


Fig. 63. Polythermal cross sections and isothermal cross section (at 1000°C) of the tungsten–molybdenum–titanium system.

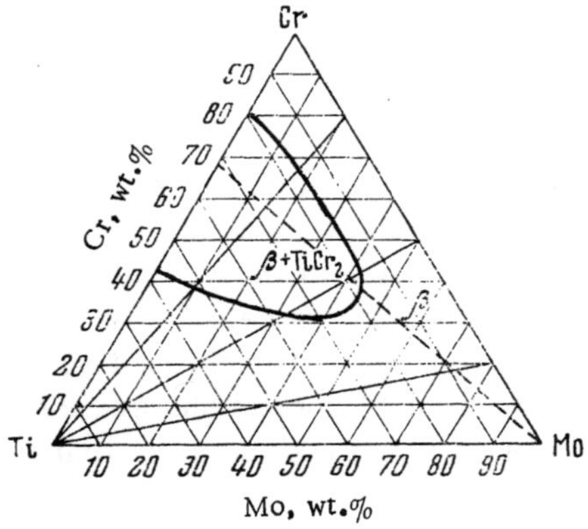


Fig. 64. Isothermal cross section of the molybdenum–titanium–chromium system at 1200°C.

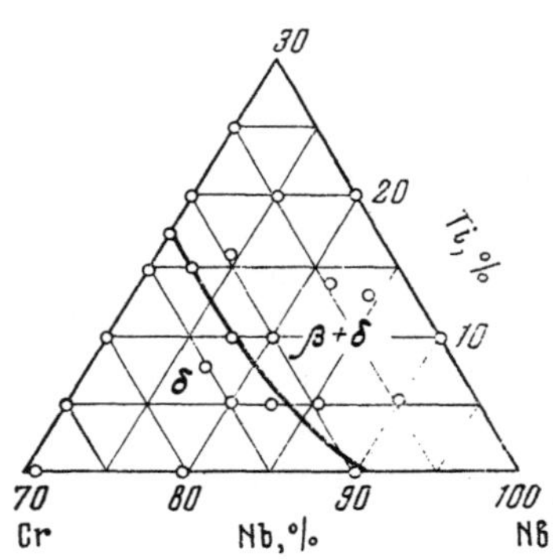


Fig. 65. Isothermal cross section of part of the niobium–titanium–chromium system at 1250°C.

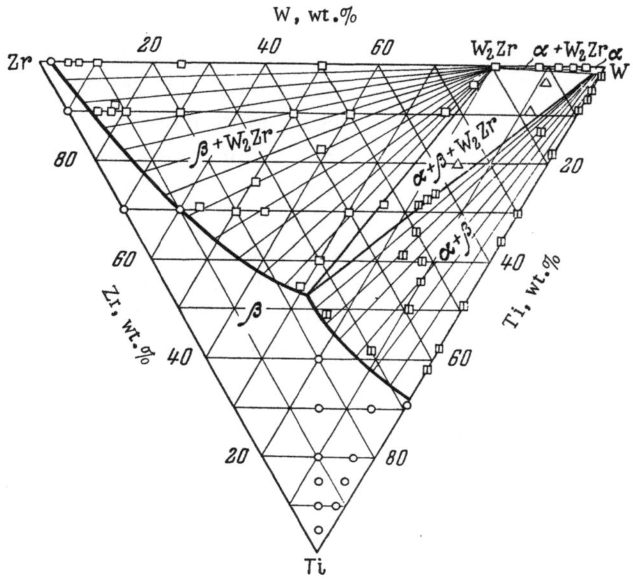


Fig. 66. Isothermal cross section of the tungsten–zirconium–titanium system at 1000°C.

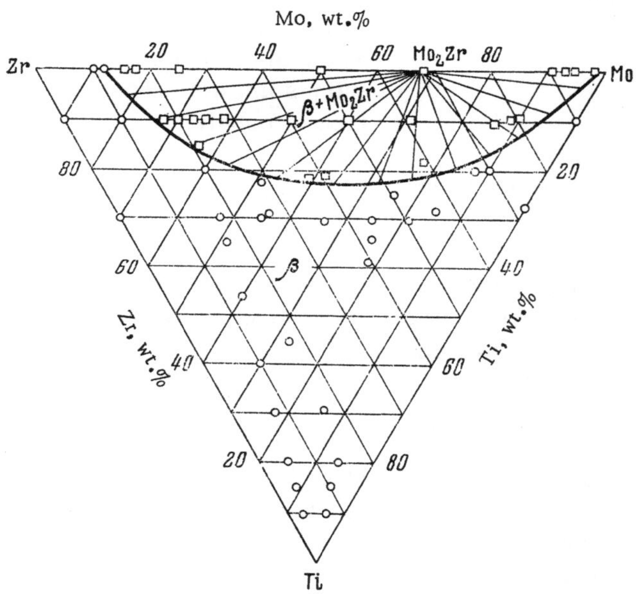


Fig. 67. Isothermal cross section of the molybdenum–zirconium–titanium system at 1500°C.

Figure 64 shows the isothermal cross section of the molybdenum–titanium–chromium system at 1200°C according to the results of [209]. No ternary compounds are formed in the system. At 1200°C there are regions of $\beta$ and ($\beta$ + $TiCr_2$)-phases. At 600°C, in addition to these regions, $TiCr_2$, $\alpha$ + $TiCr_2$, $\alpha + \beta$, and $\alpha$ + $\beta$ + $TiCr_2$ regions occur.

The niobium–titanium–chromium system was studied in [210]. Figure 65 shows the isothermal cross section of part of this system at 1250°C. The tungsten–zirconium–titanium ternary system (Fig. 66) and molybdenum–zirconium–titanium system (Fig. 67) were investigated by E. M. Savitskii and A. M. Zakharov [208, 211].

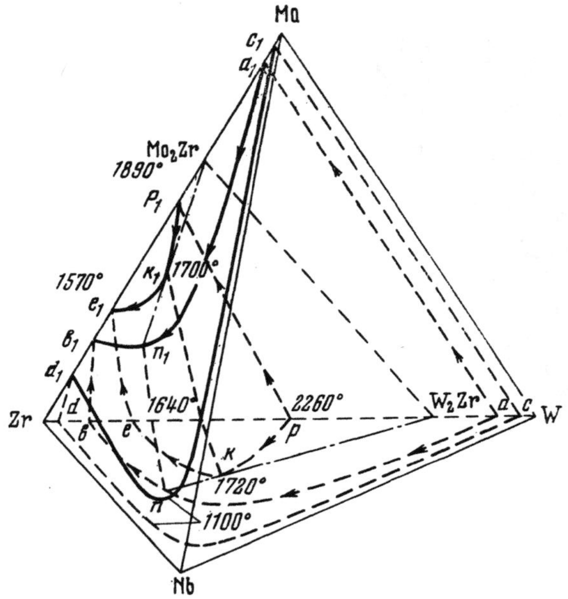


Fig. 68. Polythermal tetrahedron of the quaternary niobium–tungsten–molybdenum–zirconium system.

The niobium–tungsten–molybdenum–zirconium quaternary system was studied in detail by E. M. Savitskii and A. M. Zakharov, using microscopic and x-ray analysis methods [212]. The alloys investigated had a single-phase or two-phase structure. The structure of the single-phase alloys was a ternary solid solution of tungsten, molybdenum, and zirconium in niobium. In the structure of the two-phase alloys an excess phase of $(Mo, W)_2Zr$ was found, this being in equilibrium with the ternary solid solution. The solubility of tungsten, molybdenum, and zirconium in niobium diminished considerably with increase in temperature. The polythermal tetrahedron of the niobium–tungsten–molybdenum–zirconium according to the results of this investigation is shown in Fig. 68.

E. M. Savitskii and collaborators have carried out an investigation of the quaternary system rhenium–ruthenium–osmium–$\alpha$-cobalt. According to provisional results, rhenium, ruthenium, and osmium form continuous quaternary solid solutions with the hexagonal modification of cobalt.

## Interaction of Refractory Metals with Interstitial Impurities

Refractory metals show high chemical activity with respect to gases (nitrogen, oxygen, hydrogen) and carbon. Particularly strong chemical affinity to these elements is found in refractory metals at elevated temperatures.

Table 27 presents the magnitude of the variation in free energy of the interaction of some refractory metals with gases. Titanium has the strongest affinity for oxygen and nitrogen.

Investigations of the oxidation processes of niobium [214-216], tantalum [217, 218], molybdenum [219], and tungsten [220, 221] have shown that at temperatures of up to 500-700°C, oxidation varies with time according to a parabolic function; at higher temperatures, the parabolic function becomes linear.

The intensity of the oxidation of the metals in different temperature ranges is determined primarily by the properties of the oxides formed (Table 28).

Oxides of refractory metals have a relatively high volume compared with the specific volume of the metal base. This circumstance results in cracking or exfoliation of the oxides, the rate of oxidation then sharply increasing. Of the refractory metals, molybdenum and tungsten have a particularly high ratio of the volume of the oxide to the volume of the metal base. The oxides of titanium, zirconium, and hafnium are the most strongly bonded to the metal base. With increase in temperature, the growth of the oxide films is accelerated owing to the weakening of the interatomic bond and the increase in the coefficients of diffusion of the metal and oxygen through the oxide layer.

TABLE 27. Free Energy of the Reaction of Some Refractory Metals with Gases [213]

| Reaction | Maximum work at different temperatures, cal/mole | | | Heat of solution, cal/mole |
|---|---|---|---|---|
| | 20° | 1000° | melting point of the metal | |
| Ti (solid) + $O_2 \rightleftharpoons TiO_2$ (solid) | −208,660 | −175,630 | −147,691 | – |
| Ti (solid) + $0.5O_2 \rightleftharpoons TiO$ (solid) | −135,010 | −112,170 | −97,900 | – |
| 2Ti (solid) + $1.5O_2 \rightleftharpoons Ti_2O_3$ (solid) | −375,304 | −302,640 | −257,450 | – |
| Ti (solid) + $0.5N_2 \rightleftharpoons TiN$ (solid) | −79,843 | −57,450 | −42,200 | – |
| Ti (solid) + $0.5H_2 \rightleftharpoons TiH$ | – | – | – | −20,000 |
| 2Cr (solid) + $1.5O_2 \rightleftharpoons Cr_2O_3$ (solid) | −266,508 | −205,650 | −152,050 | – |
| Cr (solid) + $0.5N_2 \rightleftharpoons CrN$ (solid) | −27,763 | −6,370 | −36,201 | – |
| Cr (solid) + $0.5H_2 \rightleftharpoons [CrH]$ | – | – | – | 37,600 |
| Nb (solid) + $O_2 \rightleftharpoons NbO_2$ (solid) | −192,148 | −150,400 | −64,000 | – |
| Nb (solid) + $0.5H_2 \rightleftharpoons NbH$ | – | – | – | −22,000 |
| Mo (solid) + $O_2 \rightleftharpoons MoO_2$ (solid) | −132,739 | −90,530 | −20,600 | – |
| Mo (solid) + $1.5O_2 \rightleftharpoons MoO_3$ (solid) | −176,301 | −116,210 | −25,910 | – |
| W (solid) + $O_2 \rightleftharpoons WO_2$ (solid) | −130,489 | −93,200 | −1,250 | – |
| W (solid) + $1.5O_2 \rightleftharpoons WO_3$ (solid) | −199,931 | −140,400 | −79,500 | – |

Data on the influence of interstitial impurities on the properties of refractory metals and alloys will be discussed in the next chapter. It should be noted that these impurities are not always harmful. In a number of cases, interstitial impurities may be used in combination with other alloying additions (thermodynamically active elements) in the development of alloys of refractory metals (see Chapter VI).

## 1. Interaction with Oxygen

In the overwhelming majority, refractory metals are quite prone to oxidation when heated in the air. The refractory rare-earth metals (scandium, yttrium), and also titanium, zirconium, and hafnium show the maximum thermodynamic activity toward oxygen. Platinum is the most resistant of the refractory metals to oxidation. Platinum practically does not oxidize when heated in air. The other members of the platinum group oxidize readily. In ruthenium, appreciable oxidation occurs at 100°C, in iridium and rhodium, at 600°C. Osmium oxidizes already at room temperature, and its oxides are very volatile. Ruthenium oxides $RuO_2$ and $RuO_4$ are very volatile.

Of the metals of groups IVa-VIIa, chromium is the only metal which has high oxidation resistance. Most refractory metals oxidize intensely at temperatures above 600°C [222, 230]. Figure 69 shows the variation in loss of weight of some refractory metals with heating temperature [222]. As has been shown [223, 224], the oxidation process consists of two stages: 1) direct reaction of oxygen with the metal, as the result of which a film of oxide is formed; 2) reaction of atoms of oxygen with the oxide film formed, which isolates the metal from direct contact with the oxygen. The first oxidation stage proceeds very rapidly for most metals, even at room temperature, the oxygen compounds formed satisfying the valence rule. Oxidation in the second stage is of substantially different character, since the oxygen atoms then react with the oxide. The kinetics of oxidation in the second stage largely determines the physicochemical properties of the oxide. The phenomenon is complicated by the fact that transition metals form several different oxides with oxygen, as the result of which transition reactions

TABLE 28. Properties of the Oxides of Some Refractory Metals

| Metal | Ratio of vol. of oxide to vol. of metal | Oxide | Melting point, °C | Boiling point, °C | Heat of formation $H_{298}$, kcal/mole | Entropy $S_{298}$, kcal/mole* degree | Oxide structure | Oxide color |
|---|---|---|---|---|---|---|---|---|
| Titanium | — | TiO | 2020 | — | 135.0±10 | 8.3±0.1 | Cubic, NaCl type | Golden |
| | 1.73 | $Ti_2O_3$ | 2130 | — | 375.5±9 | 18.8±0.2 | Rhombohedral $Cr_2O_3$ type | — |
| | — | $Ti_3O_5$ | 1770 * | — | — | 7.2±0.4 | — | — |
| | — | $TiO_2$ | 1850 | 3000 | 219.0±1,5 | 12.0±0.05 | Tetragonal | Yellow |
| Chromium | — | $CrO_3$ | Разл. | — | 137.0±1 | 17.2±2 | Ortho-rhombic | — |
| | 2.07 | $Cr_2O_3$ | 2280 | 3000 | 269.5±2.6 | 19.4±0.2 | Rhombo-hedral | Green |
| Niobium | — | NbO | 1945 | — | 166.0±10 | 12.0±1.5 | Cubic, NaCl type | — |
| | — | $Nb_2O_3$** | — | — | — | — | Cubic (K12) with dis-tortion | — |
| | 2.69 | $NbO_2$ | 1915 | — | 190.9±5 | 13.0±1.5 | Tetragonal, $TiO_2$ type | — |
| | — | $Nb_2O_5$ | 1495 | — | 463.0±8 | 32.5±1.5 | — | — |
| Molyb-denum | — | $MoO_2$ | — | — | 130.0±4 | — | Tetragonal, $TiO_2$ type | Violet |
| | 3.24 | $MoO_3$ | 795 | 1460 | 180.5±2 | 18.7±0.3 | Rhombo-hedral | White |
| Tungsten | — | $WO_2$ | 1270 | — | 134.5±5 | 15±2.5 | Tetragonal, $TiO_2$ type | Brown |
| | 3.35 | $W_2O_5$ | — | — | 335.0±5 | — | — | From blue to violet |
| | — | $WO_3$ | 1473 | — | 200.0±3 | 19.9±0.2 | Distorted $ReO_3$ | Greenish yellow |

*Transformation temperature.

†Stable in neutral atmosphere, but oxidizes in the ordinary atmosphere.

from lower to higher oxides occur in the oxide layer. It is essential for the oxidation process that the atoms of metal and oxygen should be able to diffuse through the previously formed oxide layer. Such penetration of the atoms may occur along pores and cracks in the oxide and also by diffusion through the oxide layer.

According to Kubaschewski and Hopkins [225], in the temperature range of 50-300°C, the effective oxidation mechanism for tungsten, molybdenum, tantalum, and niobium, is that proposed by Cabrera and Mott [226], as refined by Hauffe and Ilschner [227]. These authors consider that the oxidation mechanism of tungsten, molybdenum, niobium, and tantalum is due to the migration of ions through the oxide lattice.

The high rate of oxidation of molybdenum and rhenium is also due to the fusibility and high volatility of their oxides. The melting point of rhenium oxide $Re_2O_7$ is 279°C, that of molybdenum oxide $MoO_3$ is 795°C; the respective boiling points are 363° and 1460°C. The oxidation of rhenium increases sharply at 600°C. If molybdenum and rhenium are kept at temperatures above the melting points of their oxides, total loss of the metal may rapidly follow. Tungsten begins to oxidize intensity at 600°C.

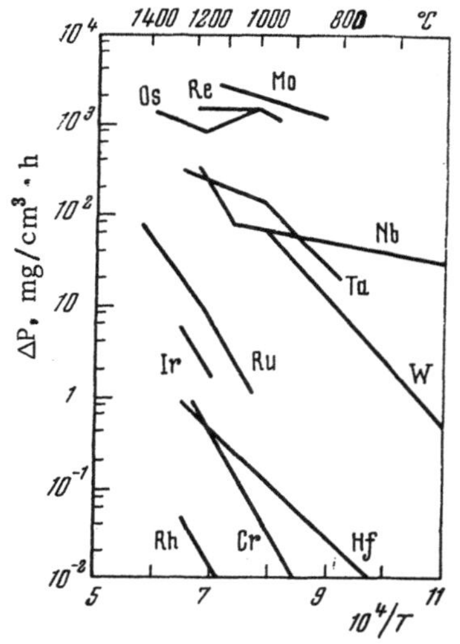


Fig. 69. Oxidizability of refractory metals as a function of the temperature. ΔP) Linear rate of change of weight.

Niobium and tantalum are subject to strong oxidation at temperatures above 400°C, despite the relatively high melting points of their oxides and the firm bonding of the oxides to the metal. Mackintosh [231] considers that this is due to the structural peculiarities of the oxide films. The best resistance to oxidation is preserved by metals having a close-packed cationic or anionic lattice. This in fact explains the high resistance to oxidation of chromium mentioned in [231]; the arrangement of the cations in the chromium lattice is sufficiently close to provide high activation energy for the diffusion of oxygen. Niobium oxides, on the contrary, do not have a close-packed structure; in consequence of the low activation energy of the oxygen ions, active diffusion occurs. According to the results of [232] the activation energy of niobium oxides in pure oxygen is about 10 kcal/mole at 500-600°C.

We shall examine some Me–O phase diagrams. The construction of refractory metal–gas phase diagrams encounters considerable experimental difficulties and requires special apparatus. Such diagrams, constructed for the entire concentration range, are numbered in ones.

During oxidation, oxygen occurs in the metal not only in the form of (lower and higher) oxides, but it also forms solid interstitial solutions, wide ranges of which are found in systems of group IVa metals with oxygen. The solubility of oxygen in $\alpha$-titanium at 600-1700°C is 14.5 at.% [233, 234]. The reaction of oxygen with titanium results in the formation of a number of oxides: $TiO_2$, $Ti_3O_5$, $Ti_2O_3$, and TiO.

The solubility of oxides in group Va metals is much lower. According to A. Seybolt [235], the maximum solubility of oxygen in niobium in the temperature range 775-1100°C varies from 1.4 to 5.5 at.%, and according to [236] the limit saturation of oxygen in solid solution with niobium between 700 and 1550°C varies from 0.7 to 5.5 at.%. According to Gebhardt [237], the temperature dependence of solubility in the range 750-1540°C may be expressed by the equation

$$\ln C_{\max} = k - \Delta H/RT,$$

where $C_{max}$ is the limit solubility of oxygen in solid solution in niobium, at.%; k is 3.829; and ΔH is the heat of solution (7700 cal/g-atom). Above 1540°C and below 750°C, there are deviations from this relationship. The solubility of oxygen in niobium was also studied in [238]; the niobium–oxygen phase diagram according to the results of this work is shown in Fig. 70. It is interesting to note that the rate of oxidation of niobium in air is higher than in oxygen; in [239] this is explained as being due to the catalytic action of nitrogen. In the Nb–O system, the following oxides have been found: $Nb_6O$, $Nb_4O$, NbO, $NbO_2$, and $Nb_2O_5$ [277].

The solubility of oxygen in tantalum at different temperatures and pressures was investigated in [240]. The quantity of oxygen absorbed by tantalum increased continuously with increase in pressure and temperature. The saturation limit at 1500°C was about 3.7 at.%, at 1200°C, about 2.9 at.%, and at 750°C about 1.4 at.% oxygen. In the range 1700-2200°C, a rapid reaction between the pentoxide and metallic tantalum was found, indicating a high rate of diffusion of oxygen in tantalum [241] in this temperature range.

Results obtained by Schoenberg are given in [242]. Six different phases were found in the tantalum–oxygen system: $\alpha$-phase (solid solution based on tantalum with a body-centered cubic lattice), $\beta$-phase ($TaO_4$ having an orthorhombic lattice), $\gamma$-phase (TaO with a structure

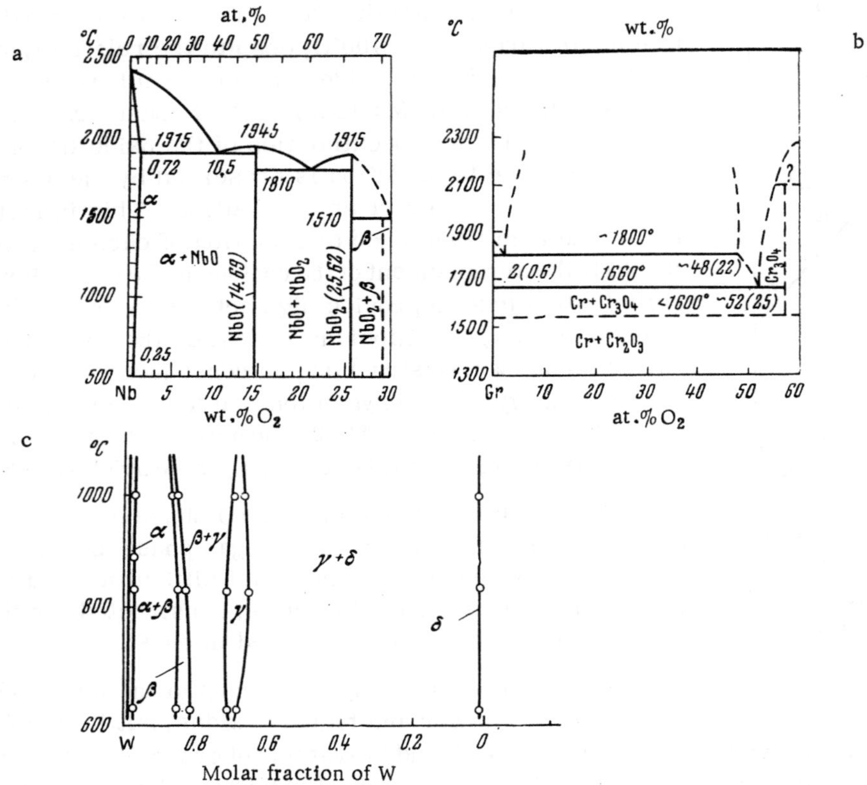


Fig. 70. Phase diagrams of the niobium–oxygen (a), chromium–oxygen (b), and tungsten–oxygen (c) systems.

of NaCl type), δ-phase ($TaO_2$ having a rutile structure), ε-phase (representing an oxide intermediate between $TaO_2$ and $Ta_2O_5$), and ξ-phase (with a structure of $U_3O_8$ type). In [278], the existence of $Ta_2O$, $Ta_6O$, and $Ta_4O$ was established.

Increase in the oxygen content of tantalum in the limits of the solid solution is accompanied by an increase in the modulus of elasticity from 18,000 to 19,840 kg/mm², an increase in strength and at the same time a sharp drop in ductility.

A large number of oxide phases have been found in the vanadium–oxygen system [243]: β (from $VO_{0.18}$ to $VO_{0.33}$), γ ($VO_{0.53}$), δ (from $VO_{0.85}$ to $VO_{1.24}$), ε($VO_{1.27}$), ξ($V_2O_3$), η($V_3O_5$), $\Theta_n(V_2O_{2n-1}$, where $4 \leq n \leq 8$), $\Theta_5(V_5O_9)$, $\Theta(VO_2)$, and $K(V_6O_{13})$.

According to [243], the solubility of oxygen in vanadium at 1400°C is 3.2 at.% (1 wt.%). Rostoker and Jamamoto [244] found that the solubility of oxygen in vanadium varied from 3.2 to 0.8 at.% in the temperature range from 1600 to 900°C.

In the chromium–oxygen system several compounds with different crystal structures are known: $CrO_3$, $Cr_2O_3$, and CrO. At 500°C, $CrO_3$ dissociates into $Cr_2O_3$ (having a rhombohedral structure) and oxygen. The phase diagram of the chromium–oxygen system is shown in Fig. 70b [245].

Tungsten and molybdenum form a large number of oxides. Some of the oxygen compounds of molybdenum and tungsten do not obey the valence rule. Seven oxides of molybdenum are known: $MoO_3$, $Mo_9O_{26}$, $Mo_8O_{23}$, $Mo_4O_{11}$, $MoO_2$, ε-phase, and $Mo_3O$ [246]. $MoO_3$ and $MoO_2$ are the most stable oxides.

In the tungsten–oxygen system, the following oxides are formed: $WO_3$, $W_{20}O_{58}$, $W_4O_{11}$, $WO_2$, and $W_3O$ [247]. The tungsten–oxygen system, according to the results of [248], is shown in Fig. 70c.

For group VIa metals (chromium, molybdenum, tungsten), the insignificant solubility of oxygen in the solid state is characteristic. If the solubility limits are exceeded, the result is the formation of oxides along the grain boundaries. The oxygen dissolved in the solid metal considerably increases the hardness, while the brittle oxides distributed along the grain boundaries often catastrophically reduce the strength and deformability of the metals.

The main features of the interaction of oxygen with refractory metals is also true of alloys based on them. To obtain a complete picture of the interaction of a given alloy with oxygen (or other nonmetal), the corresponding ternary and more complex phase equilibrium diagrams must be known. The construction of such diagrams is quite a complicated matter and is only just beginning, but in its importance it represents a leading problem in the physical chemistry of metals.

## 2. Interaction with Hydrogen

Interaction of group IVa refractory metals with hydrogen has currently been most completely studied. These metals are capable of absorbing large amounts of hydrogen. The quantity of hydrogen retained by group IVa metals depends on the temperature and pressure. The titanium–hydrogen phase diagram was studied by Lenning et al. [249]. A small amount of hydrogen dissolves in hexagonal titanium, as in hexagonal zirconium and hafnium, at temperatures close to room temperature. The solubility increases with increase in temperature and for titanium attains a maximum at 319°C (7.9 at.%). Wide ranges of hydrogen solubility are observed in high-temperature modifications of these metals ($\beta$-phase). In the titanium–hydrogen system, a $\gamma$-phase based on the hydride $TiH_2$ has been found, as well as the hydrides $ZrH_2$, $ZrH$, $Zr_4H$, and $HfH_2$ in the corresponding systems [250, 251, 280, 281].

Group VIa metals also exhibit a considerable sorption capacity for hydrogen, although to a lesser degree than group IVa metals. The process of hydrogen absorption by these metals is activated by preliminary heating [253]. The absorption of hydrogen by vanadium is proportional to the square root of the value of the hydrogen pressure [254]. At lower temperatures, the capacity for hydrogen absorption increases, especially at high pressures. The existence of vanadium hydride with 1.2-2 wt.% hydrogen (38.7-50 at.%) has been established. According to [255], the maximum hydrogen content of vanadium corresponds to the formula $VH_{0.9}$. Vanadium hydride corresponding approximately to this formula occurs in the form of acicular veins along the grain boundaries and is one of the most harmful impurities for group Va metals [251, 252].

The character of the interaction of niobium with hydrogen is approximately similar to that of the vanadium–hydrogen system [279]. Intensive absorption occurs in the range 250-360°C, after which it falls off appreciably (explained in [257] by decomposition of the low-temperature hydride form), and at 560-900°C, niobium again absorbs hydrogen intensely (due to the formation of a more stable hydride). Brauer and Hermann [258] used the x-ray method to determine the range of existence of a solid solution of hydrogen up to 10 at.% in niobium at room temperature. According to these authors, niobium hydride has an orthorhombic lattice with the parameters a = 4.84 Å, b = 4.90 Å, and c = 3.45 Å. Titanium behaves like vanadium and niobium with regard to hydrogen [259]. When vanadium, niobium, and tantalum are heated in hydrogen, the metal becomes rapidly embrittled, owing to the precipitation of brittle hydride along the grain boundaries.

Group VIa metals absorb a very small amount of hydrogen. The hydrides of these metals are unstable. The maximum quantity of hydrogen absorbed by chromium is not more than 0.45 wt.% [260]. The region of existence of an interstitial solid solution does not exceed 0.00001%. The assumed composition of the hydrides is CrH, $CrH_2$, and $CrH_3$.

Molybdenum is inert to hydrogen at room temperature, but absorption of hydrogen by molybdenum is found at elevated temperatures, although in smaller amounts than for chromium [261].

Tungsten absorbs hydrogen in insignificant amounts up to 1500°C. Langmuir [262] found appreciable absorption of hydrogen by heating a tungsten wire in a hydrogen atmosphere at a temperature above 1300°C, which is explained in [253] as being due to the formation of tungsten hydride by an endothermic reaction. V. I. Mikheeva [253] cites results of Dillon [263] on the formation of the compound $WH_2$.

The capacity for absorbing hydrogen is absent in metallic rhenium. V. I. Mikheeva has suggested the possibility of producing rhenium hydrides by successful synthesis of complex rhenium compounds containing hydride hydrogen [253].

The formation of hydrides and very limited solid solutions is characteristic of group VIII metals. Of these metals, the maximum capacity for absorbing hydrogen is found in palladium and its alloys, and is utilized in technology. Among the heavy platinum metals (osmium, iridium, platinum), maximum hydrogen solubility is found in iridium [253]. The palladium–hydrogen system has been studied most completely. The existence of a solid solution based on palladium ($\alpha$-phase with face-centered lattice) and a high-hydrogen $\beta$-phase, also having a face-centered lattice, has been found in this system [264]. It is assumed that the $\beta$-phase is a solid solution based on the compound $Pd_2H$.

## 3. Interaction with Nitrogen

Group IVa metals interact actively with nitrogen. Figure 71a shows the phase diagram of the titanium–nitrogen system [265]. Maximum solubility of nitrogen in $\alpha$-titanium is 7.5%. Two other phases are formed in the system: a $\delta$-phase, homogeneous for nitrogen contents from 11.1 to 22.6 wt.% (it may be regarded as a solution based on the compound $Ti_3N$), and an $\varepsilon$-phase (TiN), formed as the result of the peritectic reaction $\alpha + \delta \rightleftharpoons \varepsilon$ at a temperature of 1050°C. By structure, titanium nitrides are interstitial phases.

The character of the interaction of the other group IVa metals (zirconium and hafnium) with nitrogen is generally similar to the titanium–nitrogen system.

In systems of group Va metals with nitrogen, the regions of existence of interstitial solid solutions are very limited. The limit solubility of nitrogen in niobium is 0.063 wt.% (4.8 at.%). In niobium, nitrogen forms a series of intermediate interstitial phases, the disposition and width of which are shown in the diagram of Fig. 71b [266]: The solid $\alpha$-solution of nitrogen in niobium has a body-centered cubic lattice with a constant differing little from the lattice constant of pure niobium (3.2948 ± 0.0002 Å); the $\beta$-phase has a hexagonal structure and corresponds in composition to the compound $Nb_2N$; the composition of the $\delta$-phase corresponds to NbN (face-centered cubic lattice); the $\gamma$-phase has a composition close to $NbN_{0.75}$ or $Nb_4N$ [242, 267], the compound has a cubic lattice with tetragonal symmetry (c/a = 0.97).

In [267], the coefficient of diffusion of nitrogen in niobium at 500-600°C with the formation of a solid solution is determined as

$$D = 8.05 \cdot 10^{-4} \exp\left(-\frac{3630}{T}\right).$$

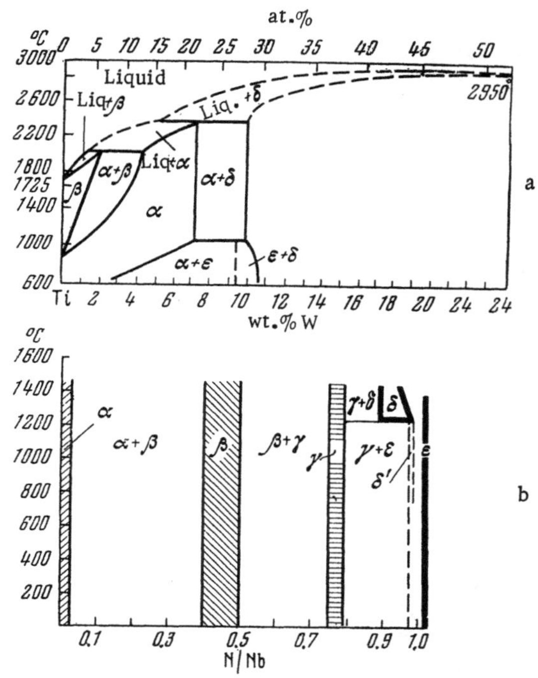


Fig. 71. Phase diagrams of the titanium–nitrogen (a) and niobium–nitrogen (b) systems.

In the same reference, the coefficient of diffusion of nitrogen in niobium with the formation of NbN at 600-900°C is determined by the equation

$$D = 4.5 \cdot 10^{-3} \exp\left(-\frac{5000}{T}\right).$$

The interaction of niobium with nitrogen in the range 400-1600°C obeys a simple parabolic law. The reaction rate depends little on the pressure [242]; at temperatures below 1000°C, a mixture of the nitrides $Nb_3N$ and NbN is formed, and above 1000°C, only NbN. The interaction of niobium with nitrogen was also studied in [282, 283].

In the tantalum–nitrogen system, the existence of four intermediate phases has been found [268]: a $\beta$-phase (approximate composition $TaN_{0.05}$ and arrangement of the tantalum atoms in a body-centered cubic lattice), a $\gamma$-phase (corresponds approximately to $Ta_2N$ and has a close-packed lattice), a $\delta$-phase (with a simple hexagonal lattice, homogeneity range in the limits $TaN_{0.8}$–$TaN_{0.9}$), and an $\varepsilon$-phase (corresponds to the composition TaN). The final product obtained by nitrogenizing was found to be homogeneous with the stoichiometric composition TaN. It may be easily produced at 1000°C provided oxygen is excluded from the reaction. In the presence of oxygen, oxynitrides or a mixture of tantalum nitrides and oxides are formed.

The tantalum–nitrogen system in the concentration range up to 30 at.% nitrogen was also studied by Gebhardt [269]. According to the results of this investigation, at room temperature tantalum dissolves 2 at.% nitrogen, at 2300°C, it dissolves 13 at.%.

The relationship among equilibrium pressure, nitrogen concentration, and temperature in the region of the tantalum solid solution may be expressed by the following equation:

$$C = p^{2.4} \cdot 8 \cdot 10^{4} \cdot e^{-\frac{4.35 \cdot 10^{4}}{T}},$$

where C is in at.% nitrogen, p is in mm Hg, and T is in °K.

Of the group VIa metals, chromium has the highest sorptive capacity for nitrogen. The results of investigations on the interaction of nitrogen with chromium are contradictory, and the phase diagram of the chromium–nitrogen system has not yet been constructed. Nitrogen (above 0.005%) is one of the most harmful impurities for chromium [271]. Evidently, chromium absorbs nitrogen from the air at the ordinary temperature.

The reaction diffusion of nitrogen in chromium commences at 700°C and obeys a parabolic law up to 1200°C. In the interaction of chromium with nitrogen, nitrogen diffuses through the nitride layers to the metal [270]. No diffusion of the metal in the opposite direction is found. At temperatures of from 1600 to 1750°C, the solubility of nitrogen in chromium diminishes from 4.03 to 3.53%. The nitrogen is present in the chromium in solid solution and in the form of CrN and $Cr_2N$ [267]. According to other data [272], an intermediate phase (CrN) of

TABLE 29. Activation Energy of the Diffusion of Nitrogen in Refractory Metals [213]

| Metal | Temperature range, °C | Activation energy, cal/mole |
|---|---|---|
| α-Titanium..... | — | 45,250 ± 2,250 |
| β-Titanium..... | — | 33,800 ± 400 |
| Chromium...... | 500-900 | 10,420 |
| Chromium...... | 900-1200 | 660 |
| Niobium....... | 500-600 | 25,400 |
| Niobium....... | 600-900 | 7,000 |
| Niobium....... | 900-1200 | 10,000 |

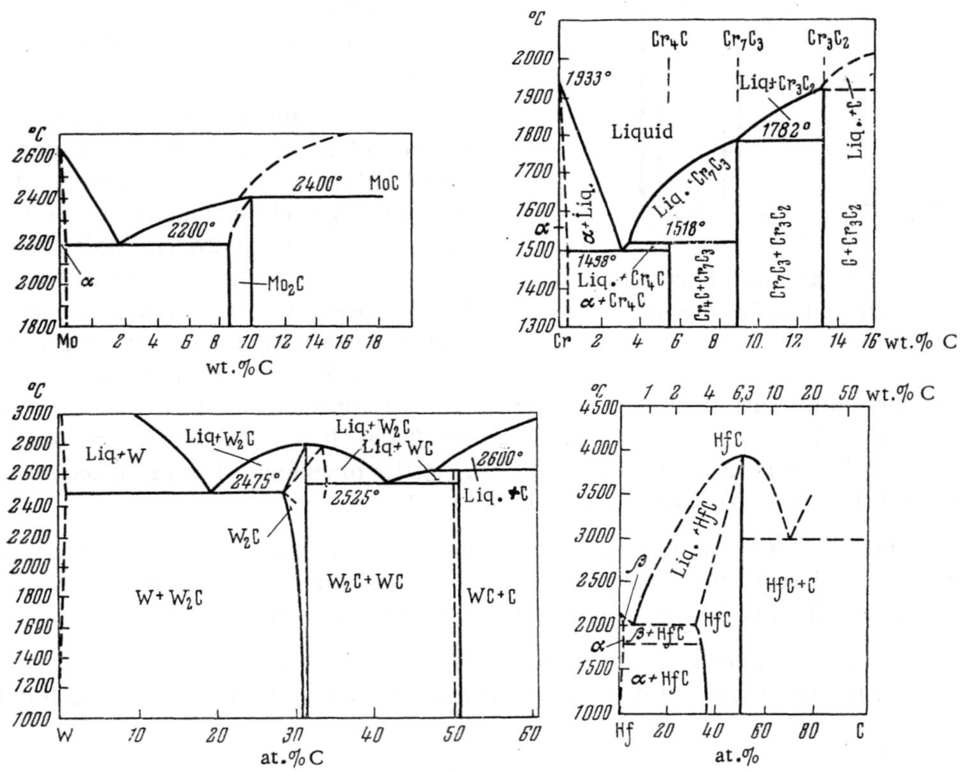


Fig. 72. Phase diagrams of systems of refractory metals with carbon.

variable composition occurs in the chromium–nitrogen system. By their structure, chromium nitrides are interstitial phases. The crystalline structure of $Cr_2N$ is hexagonal (a = 4.806 Å, c = 4.478 Å, c/a = 0.931) [267]. CrN also has a hexagonal crystal structure with the parameters a = 2.751 Å, c = 4.425 Å, c/a = 1.605 [272]. At 1500°C, CrN dissociates.

The phase diagrams of the molybdenum–nitrogen and tungsten–nitrogen systems have not yet been studied adequately. On interaction with nitrogen, molybdenum forms the nitrides $Mo_3N$, $Mo_2N$, and MoN, which are interstitial phases [267]; $Mo_3N$ is stable only up to 600°C, and then dissociates. On interaction of tungsten with nitrogen, the nitrides $W_2N$ (stable up to 400°C) and WN (to 600°C) are formed; $W_2N$ has a face-centered cubic lattice (a = 4.118 Å), WN has a hexagonal structure (a = 2.893 Å, c = 2.826 Å, c/a = 0.977).

Table 29 gives the values of the activation energy of the diffusion of nitrogen in some refractory metals. The diffusion of nitrogen in chromium has the lowest value of activation energy.

## 4. Interaction with Carbon

In addition to gaseous impurities, refractory metals show high chemical activity to carbon, which forms with them interstitial solid solutions and carbides. Carbon has a considerable influence on a number of the physical and chemical properties of refractory metals; the carbon content exercises a decisive influence on the ductile-brittle transition temperature of tungsten and molybdenum; carbides distributed along the grains produce embrittlement of these metals.

In the titanium, zirconium, and hafnium systems with carbon, the compounds TiC, ZrC, and HfC exist with a wide homogeneity range. The solubility of carbon in the $\alpha$- and $\beta$-modifications of these metals increases appreciably with temperature (Fig. 72) [184].

In the interaction of vanadium with carbon, the carbides VC and $V_2C$ are formed; the solubility of carbon in vanadium is very low, being about 0.2 wt.% at 1000°C. In the niobium–carbon system, there are two carbides NbC and $Nb_2C$ with wide homogeneity ranges; the solubility of carbon in niobium does not exceed 0.02 at.%. Carbon is also practically insoluble in tantalum, and forms two carbides with it, $Ta_2C$ and TaC. In the chromium–carbon system, the existence of three carbides has been established: $Cr_3C_2$, $Cr_7C_3$, and $Cr_{23}C_6$. With tungsten and molybdenum, carbon forms the carbides $W_2C$, WC, and $Mo_2C$, MoC, respectively; at 2000°C, the solubility of carbon in tungsten does not exceed 0.05% [287]. As shown by the tungsten–carbon phase diagrams (Fig. 72), $W_2C$ melts at 2750°C and WC decomposes at 2600°C by peritectic reaction, with the formation of liquid phase and graphite [26].

Detailed data on the physical properties of the carbides of refractory metals are given in [184]. Ternary systems of the transition metals with carbon are examined in [284]. The interaction of carbides and nitrides of transition metals is studied in [285].

## LITERATURE CITED


1. N. S. Kurnakov, Selected Works [in Russian], Vols. 1 and 2, Izd. Akad. Nauk USSR (1960, 1961).
2. V. Ya. Anosov and C. A. Pogodin, Basic Principles of Physicochemical Analysis [in Russian], Izd. Akad. Nauk USSR (1947).
3. N. V. Ageev, Nature of the Chemical Bond in Metal Alloys [in Russian], Izd. Akad. Nauk USSR (1947).
4. E. M. Savitskii, Influence of Temperature on the Mechanical Properties of Metals and Alloys [in Russian], Izd. Akad. Nauk SSSR (1957).
5. S. T. Konobeevskii, Nature of the Bonds in Metals. Transactions of a Conference on the Theory of Metal Alloys [in Russian], Izd. MGU (May 1962).
6. Phase Diagrams of Metallic Systems [in Russian], Nos. I–X, VINITI (1959–1966).
7. A. E. Vol, Structure and Properties of Ternary Metal Systems [in Russian], Fizmatgiz (1959)(1) and (1962)(2).
8. M. Hansen and K. Anderko, Structure of Binary Alloys [Russian translation], Parts 1 and 2, Metallurgizdat (1962).
9. V. A. Nemilov, General Metallography [in Russian], Izd. Akad. Nauk SSSR (1947).
10. E. M. Savitskii, V. F. Terekhova, I. V. Burov, O. P. Naumkin, and I. A. Markova, Neorganicheskie Materialy, 1(10):1648 (1965).
11. B. Ya. Pines, Zh. Eksp. Teor. Fiz., 13(11–12):411 (1943).

12. V. I. Danilov and D. S. Kamenetskaya, Zh. Fiz. Khim., 22:69 (1948).
13. D. S. Kamenetskaya, Problems of General Metallography and the Physics of Metals [in Russian], p. 113 (1949).
14. L. S. Darken and R. V. Gurry, Physical Chemistry of Metals [in Russian], Metallurgizdat (1960).
15. O. Kubaschewski and T. Chart, J. Inst. Metals, 93(10):1648 (1965).
16. N. V. Ageev, Zh. Neorg. Khim., 3(3):557 (1958).
17. E. M. Savitskii, Vestnik Akad. Nauk SSSR, No. 2, p. 2 (1964).
18. W. Hume-Rothery, Electrons and Metals [Russian translation], Metallurgizdat (1949).
19. W. Hume-Rothery and G. V. Raynor, Structure of Metals and Alloys [Russian translation], Metallurgizdat (1959).
20. B. W. Mott, Phil. Mag., Ser. 8, 2:259 (1957).
21. K. A. Gschneidner, Rare-Earth Alloys, Van Nostrand, Princeton, New Jersey (1961).
22. E. M. Savitskii, Vestnik Akad. Nauk SSSR, No. 3, p. 1 (1962).
23. A. Nepton, in: Refractory Metals and Alloys [Russian translation], IL, p. 57 (1962).
24. F. Spedding and A. Daan, Production and properties of rare earth elements, in: Rare Earth Metals [Russian translation], IL, p. 298 (1957).
25. W. Gordy and W. J. O. Thomas, J. Chem. Phys., 24:439 (1956).
26. W. P. Sykes, Trans. ASS, 1(18):968 (1930).
27. E. M. Savitskii, V. F. Terekhova, I. V. Burov, I. A. Markova, and O. P. Naumkin, Alloys of Rare Earth Metals [in Russian], Izd. Akad. Nauk SSSR, p. 138 (1964).
28. A. M. Zakharov, Investigation of Phase Diagrams and Properties of Ternary and Quaternary Alloys of Niobium with Tungsten, Molybdenum, and Zirconium (Author's Abstract of Candidate's Dissertation) [in Russian], Moscow (1964).
29. M. A. Tylkina and E. M. Savitskii, in: Rhenium (Transactions of the Second-All-Union Conference on Rhenium Problems, Nov. 19–21 (1962) [in Russian], Izd. Akad. Nauk SSSR, p. 138 (1964).
30. K. A. Gschneidner and J. T. Waber, Paper presented at the Rare Earth Symposium Fall Meeting, Am. Soc. Metals, Chicago (1959).
31. E. Teatum, K. Gschneidner, and J. Waber, Los Alamos Scient. Lab. Rep., p. 2345 (1960).
32. G. Tammann, General Physical Metallurgy [Russian translation], ONTI (1935).
33. A. L. Seifer and Z. M. Ostroukhova, Zh. Neorg. Khim., 7(1):203 (1962).
34. V. M. Vozdvizhenskii, Zh. Fiz. Khim., 37(11):2455 (1963).
35. N. F. Mott and H. Jones, The Theory of the Properties of Metals and Alloys, Dover, New York (1936).
36. S. T. Konobeevskii, Izv. Sektora Fiz.-Tekhn. Analiza, 16(1):19 (1943).
37. E. M. Savitskii and O. Kh. Khamidov, Izv. Akad. Nauk SSSR, Metally, No. 4, p. 116 (1966).
38. E. M. Savitskii and O. Kh. Khamidov, Izv. Akad. Nauk SSSR, Neorg. Materialy, No. 6 (1966).
39. E. S. Makarov, Structure of Solid Phases with a Variable Number of Atoms in the Unit Cell [in Russian], Izd. Akad. Nauk SSSR (1947).
40. G. Shipereit, A. Leserman, and A. Évers, Problemy Sovremennoi Metallurgii, No. 4, p. 99 (1961).
41. E. M. Savitskii, Ch. V. Kopetskii, A. I. Pekarev, and M. I. Novosadov, Zavod. Lab., No. 8, p. 1041 (1961).
42. Am. Metal Market, 71(110):13 (1964).
43. F. Sperner and H. H. Koch, Metall, 18(7):702 (1964).
44. R. Desser and G. Gruber, Problemy Sovremennoi Metallurgii, No. 1, p. 157 (1961).
45. E. M. Savitskii, G. E. Chuprikov, and G. G. Glavin, "Analysis of gases in rhenium," Zavod. Lab., No. 8, p. 957 (1962).
46. Z. M. Turovtseva and L. L. Kunin, Analysis of Gases in Metals [in Russian], Izd. Akad. Nauk SSSR (1959).

47. A. P. Averina, Pribory i Tekhnika Éksperimenta, No. 2, p. 121 (1964).
48. A. P. Averina, Pribory i Tekh. Éksperimenta, No. 5, p. 174 (1965).
49. J. Filibert and K. Crussard, Problemy Sovremennoi Metallurgii, No. 6, p. 95 (1956).
50. I. B. Borovskii and I. P. Il'in, in: Research on Refractory Alloys [in Russian], Izd. Akad. Nauk SSSR, p. 25 (1965).
51. I. B. Borovskii and N. P. Il'in, Dokl. Akad. Nauk SSSR, 106(4):985 (1956).
52. V. G. Kozlova and N. F. Lashko, Zavod. Lab., No. 3, p. 985 (1963).
53. Yu. T. Lukashevich-Duvanova, Slag Inclusions in Iron and Steel [in Russian], Metallurgizdat (1952).
54. K. Agte and I. Vatsek, Tungsten and Molybdenum [in Russian], Izd. "Energiya" (1964).
55. Express Information, Physical Metallurgy and Heat Treatment [in Russian], VINITI, No. 8 (Abstract 26) (1965).
56. T. Berglund, Handbuch d. metallogr. Schlief.-, Polier- und Åtzverfahren, Berlin (1940).
57. V. Éspe, Technology of Electrovacuum Materials [in Russian], Vol. 1, Gosenergoizdat (1962).
58. Refractory Metals and Alloys, 2:17 (April, 1962).
59. A. N. Cottrell, J. Inst. Metals, 90:12 (1962).
60. N. M. Popov, Vestnik AN SSSR, No. 1, p. 39 (1964).
61. N. M. Popov, Nauka i Zhisn', No. 11, p. 54 (1962).
62. N. M. Popov, E. M. Savitskii, and G. L. Tsarev, Dokl. Akad. Nauk SSSR, 162(1):64 (1965).
63. Technique Moderne, 56(3) (1964).
64. E. B. Bas, in: Introduction to the Technology of Electron-Beam Processes, Izd. "Metallurgiya" (1965).
65. W. Hume-Rothery, J. Christian, and V. Pearson, Equilibrium Diagrams of Metallic Systems [Russian translation], Izd. "Metallurgiya" (1965).
66. J. W. Edwards, K. Speiser, and H. L. Johnston, Rev. Scient. Instrum., No. 20, p. 343 (1949).
67. R. G. Ross and W. Hume-Rothery, Less-Common Metals, 5(3):258 (1963).
68. A. F. Tulinov, B. A. Akhmetova, A. A. Puzanov, and A. A. Bednyakov, Zh. Eksp. Teor. Fiz., 2(1):48 (1965).
69. D. Ya. Svet, Temperature Radiation of Metals and Some Substances, Izd. "Metallurgiya" (1955).
70. E. M. Savitskii and M. A. Tylkina, Zh. Neorg. Khim., 3:763 (1961).
71. E. M. Savitskii, Auth. Cert. No. 78001, Byull. Izobretenii, No. 11, p. 54 (1949).
72. E. M. Savitskii, Zavodskaya Laboratoriya, 16(11):1366 (1950).
73. A. M. Borzdyka, Methods of the Hot Mechanical Testing of Metals [in Russian], Metallurgizdat (1955).
74. A. V. Panov, Advanced Scientific, Technical, and Production Experience (Theme 39, No. P-58-93/5) [in Russian], Fil. VINITI (1958).
75. P. G. Strelkov, G. I. Kosourov, and V. N. Samoilov, Izv. Akad. Nauk SSSR, Ser. Fiz., 17(3):383 (1953).
76. B. G. Livshits, Physical Properties of Metals and Alloys [in Russian], Mashgiz (1956).
77. E. M. Savitskii and A. I. Dashkovskii, Izv. Akad. Nauk SSSR, OTN, Metallurgiya i. Toplivo, No. 5, p. 96 (1961).
78. A. I. Dashovskii, A. I. Evstyukhin, E. M. Savitskii, and D. M. Skorov, in: Metallurgy and Physical Metallography of Pure Metals [in Russian], No. 11, Atomizdat, p. 224 (1960).
79. N. D. Kozlova, Yu. V. Emifov, V. V. Baron, and E. M. Savitskii, Transactions of 2nd Conference on the Metallurgy, Physical Chemistry, and Metal Physics of Superconductors [in Russian], Izd. "Nauka" (1966).

80. E. M. Savitskii, I. V. Burov, G. S. Burkhanov, L. M. Litvak, and N. N. Bokareva, Zh. Tekh. Fiz., 7:1310 (1966).
81. A. D. Macvillen and M. K. Macvillen, Titanium [in Russian], Metallurgizdat (1958).
82. E. M. Savitskii, M. A. Tylkina, and A. I. Pekarev, Zh. Neorg. Khim., 4(10):2320 (1959).
83. D. E. Williams and W. H. Rechin, Trans. ASME, 50:1081 (1958).
84. V. N. Eremenko, L. A. Tetr'yachenko, and Z. P. Golubenko, "Dopovidi" Akad. Nauk Ukr.SSR, No. 2, p. 192 (1962).
85. W. Rostoker and A. Jamamoto, Trans. ASME, 46:1136 (1954).
86. O. Carlson, D. Srsh, and A. Eusfice, Reactive Metals, New York, p. 274 (1952).
87. R. P. Elliot, Reactive Metals, New York, p. 293 (1959).
88. H. A. Wilhelm, O. N. Carlson, and J. M. Dickinson, J. Metals, 6:915 (1954).
89. S. Vlodek, Problemy Sovremennoi Metallurgii, No. 2, p. 89 (1961).
90. Metals Handbook, Am. Soc. Metals (1943), p. 1230.
91. C. G. Goetzel, Treatise on Powder Metallurgy, New York, Vols. I and II (1949, 1950).
92. R. Kieffer, K. Seldachek, and G. Braun, Problemy Sovremennoi Metallurgii, No. 4, p. 164 (1959).
93. H. Greenaway, J. Inst. Metals, 80:589 (1951–1952).
94. A. T. Grigor'ev, E. M. Sokolovskaya, A. P. Hefedov, and M. V. Maksilova, Vestnik MGU, Seriya II, Khimiya, No. 16(2), p. 19 (1961).
95. A. T. Grigor'ev and E. M. Sokolovskaya, Vestnik MGU, Seriya II, Khimiya, No. 16(6), p. 3 (1961).
96. W. Trzebiatowski, H. Ploszek, and J. Lubrowski, J. Analyt. Chem., 19:93 (1947).
97. N. V. Grum-Grzimailo and I. A. Popov, Zh. Neorgan. Khim., No. 3, p. 1227 (1958).
98. V. A. Nemilov and N. M. Voronov, Izv. Platinovogo Inst. Akad. Nauk SSSR, No. 12, p. 17 (1935).
99. A. S. Darling, Platinum Metals Rev., 5(2):58 (1961).
100. V. A. Nemilov, Izv. Sektora, Platiny Akad. Nauk SSSR, No. 19, p. 21 (1943).
101. E. Raub and W. Plate, Z. Metallkunde, 47(10):688 (1956).
102. A. A. Rudnitskii, V. A. Nemilov, and R. S. Polyakova, Izv. Sektora Platiny, Akad. Nauk SSSR, No. 25, p. 138 (1950).
103. E. Raub, H. Beeskow, and D. Menzel, Z. Metallkunde, 50(7):428 (1949).
104. M. A. Tylkina, V. P. Polyakova, and E. M. Savitskii, Zh. Neorgan. Khim., 7(6):1471 (1962).
105. A. Makdonado and K. Schubert, Z. Metallkunde, 55(10):619 (1964).
106. M. A. Tylkina, V. P. Polyakova, and E. M. Savitskii, Zh. Neorgan. Khim., 8(3):776 (1963).
107. W. Obrowski and G. Zwingmann, Z. Metallkunde, 53(7):453 (1962).
108. E. M. Savitskii, V. V. Baron, and Yu. V. Efimov, Trudy Inst. Met. im. A. A. Baikova, No. 4, Izd. Akad. Nauk SSSR, p. 230 (1960).
109. E. Pipitz and R. Kieffer, Z. Metallkunde, 46(31):187 (1955).
110. V. V. Baron, Yu. V. Efimov, and E. M. Savitskii, Izv. Akad. Nauk SSSR, OTN, Metallurgiya i Toplivo, No. 4, p. 36 (1958).
111. V. V. Baron, Yu. V. Efimov, and E. M. Savitskii, Izv. Akad. Nauk SSSR, OTN, Metallurgiya i Toplivo, No. 7, p. 70 (1960).
112. R. Kieffer, K. Sedlatschik, and H. Braun, Z. Metallkunde, 50(1):18 (1959).
113. V. N. Eremenko, Ukr. Khim. Zh., 20(3):227 (1954).
114. I. I. Kornilov and R. S. Polyakova, Trudy. Inst. Met. im. A. A. Baikova, No. 11, p. 149 (1957).
115. V. V. Baron, K. N. Ivanova, and E. M. Savitskii, Izv. Akad. Nauk SSSR, OTN, Metallurgiya i Toplivo, No. 4, p. 143 (1960).
116. H. Bückle, Metallforschung, 37:53 (1946).

117. B. I. Krimer and Yu. E. Matveev, Sb. Trudov. Mosk. Inst. Stali, No. 38, Metallurgizdat, p. 420 (1958).
118. V. S. Mikheev and D. M. Pevtsov, Zh. Neorgan. Khim., 3(4):861 (1958).
119. E. M. Savitskii, V. V. Baron, and K. N. Ivanova, Izv. AN SSSR, OTN, Metallurgiya i Toplivo, No. 2, p. 119 (1962).
120. V. N. Svechnikov and V. M. Pan, in: Research on Refractory Alloys [in Russian], Vol. 8, Izd. Akad. Nauk SSSR, p. 47 (1962).
121. A. T. Grigor'ev, V. V. Kuprina, and N. A. Nedumov, Zh. Neorgan. Khim., 4(3):651 (1959).
122. O. H. Schramm, P. Gordon, and A. R. Kaufmann, J. Metals, 2:195 (1950).
123. R. H. Meyers, Metallurgiya, 42:3 (1950).
124. O, Kubaschewski and E. L. Evans, Metallurgical Thermochemistry, Pergamon, New York, 3rd ed. (1958).
125. V. K. Grishin, M. G. Glazunov, A. G. Arakelov, A. V. Vol'dent, and T. S. Makedonskaya, Properties of Lithium [in Russian], Metallurgizdat (1963).
126. R. E. Matt and J. J. Warga, Metalle für die Raumfahrt, Metallwerk Plansee A. G. Reutte/ Tyrol, p. 233 (1965).
127. F. Sauerwald, Z. Anorg. Chem., 258:296 (1949).
128. M. I. Arafonova, V. V. Baron, and E. M. Savitskii, Izv. Akad. Nauk SSSR, OTN, Metallurgiya i Toplivo, No. 5, p. 139 (1959).
129. I. Obinata, I. Takiuchi, and R. Kavanisi, Problemy Sovremennoi Metallurgii, No. 5, p. 116 (1959).
130. F. H. Spedding and A. H. Daane, Metallurg. Rev., 5:19 (1960).
131. E. M. Savitskii and G. S. Burkhanov, in: Titanium and Its Alloys [in Russian], Izd. Akad. Nauk SSSR, pp. 51-60 (1962).
132. C. E. Lundin, Rare-earth metal phase diagrams, Paper presented at the Rare Earth Symposium Fall Meeting, ASM, Chicago (November 1959).
133. E. M. Savitskii and G. S. Burkhanov, Zh. Neorgan. Khim., 7(3):669 (1962).
134. B. Love and J. Wright, Air Development Center Rept. WADC-TR-57-666 (Res. Chem. Inc. Rept.), Part II (March 1959).
135. J. Taylor, Trans. AIME, 209:94 (1957).
136. E. M. Savitskii and G. S. Burkhanov, J. Less-Common Metals, No. 4, p. 301 (1962).
137. M. G. Manrone and J. Z. Briggs, Less-Common Alloys of Molybdenum, Climax Mo. Co. (1962).
138. F. H. Spedding and A. H. Daane, Rare Earths, Wiley, New York (1961).
139. E. M. Savitskii, V. V. Baron, and Yu. V. Efimov, Trudy Inst. Met. im. A. A. Baikova, No. 5, Izd. Akad. Nauk SSSR, p. 166 (1962).
140. E. M. Savitskii, V. V. Baron, and Yu. V. Efimov, Zh. Neorgan. Khim., 7(3):701 (1962).
141. J. Moriarty and N. Baenziger, State Univ., Iowa, Prepublication Data (1959).
142. E. M. Savitskii, V. F. Terekhova, and I. V. Burov, Zh. Neorgan. Khim., 4(6):1462 (1959).
143. E. M. Savitskii, V. F. Terekhova, and I. A. Markova, Zh. Neorgan. Khim., 6(5):1252 (1961).
144. E. M. Savitskii, V. F. Terekhova, and A. V. Kholopov, Zh. Neorgan. Khim., 4(2):435 (1959).
145. A. H. Daane and F. H. Spedding, Iowa State College Rept., p. 530 (1954).
146. W. Obrowski, Naturwissenschaften, 47:201 (1960).
147. E. M. Savitskii, V. V. Baron, and Tao Chu-Chun, Izv. Akad. Nauk SSSR, OTN, Metallurgiya i Toplivo, No. 1, p. 156 (1962).
148. E. M. Savitskii, V. V. Baron, and V. A. Frolov, in: Questions of the Theory and Application of Rare-Earth Metals [in Russian], Izd. "Nauka," p. 138 (1964).
149. A. Hellawell, J. Less-Common Metals, 1:343 (1959).
150. V. Compton and B. Matthias, Acta Crystallogr., 12:651 (1959).

151. V. Compton, Acta Crystallogr., 11:446 (1958).
152. E. M. Savitskii and O. Kh. Khamidov, Izv. Akad. Nauk SSSR, Neorganicheskii Materialy, 1(11):1873 (1965).
153. H. Nowatry, Z. Metallkunde, 452:203 (1954).
154. D. J. Maykuth, H. R. Ogden, and R. I. Jaffee, Trans. AIME, 197:231 (1953).
155. M. Hansen, E. L. Kamen, H. D. Kessler, and D. J. McPherson, Trans. AIME, 191:881 (1951).
156. D. J. Summers-Smith, J. Inst. Metals, 81:73 (1952–53).
157. A. D. McQuillan, J. Inst. Metals, 80:363 (1951–52).
158. V. N. Eremenko, Titanium and Its Alloys [in Russian], Izd. Akad. Nauk Ukr.SSR, Kiev (1955).
159. P. Duwes, J. Inst. Metals, 80:525 (1951–52).
160. V. S. Emel'yanov, Yu. G. Godin, and A. I. Evstyukhin, Atomnaya Energiya, 2(1):42 (1957).
161. D. E. Williams, R. S. Jackson, and W. L. Larsen, Trans. AIME, 224(4):571 (1962).
162. V. N. Svechnikov et al., Questions of the Physics of Metals and Metal Science [in Russian], Izd. "Naukova Dumka," Kiev (1965).
163. M. A. Tylkina, I. A. Tsyganova, and E. M. Savitskii, Zh. Neorgan. Khim., 9(7):1652 (1964).
164. R. F. Domogala, D. J. MacPherson, and M. Hansen, Problemy Sovremennoi Metallurgii, No. 1, p. 83 (1954).
165. G. Braun and E. Rudy, Problemy Sovremennoi Metallurgii, No. 3, p. 164 (1961).
166. B. Giessen, I. Rump, and N. Grant, Trans. AIME, 224(1):60 (1962).
167. A. Taylor, N. J. Doyle, and B. G. Kagle, J. Less-Common Metals, No. 4, p. 265 (1963).
168. W. Köster and W. D. Haehl, Z. Metallkunde, 49:647 (1958).
169. E. M. Savitskii, V. V. Baron, and A. N. Khotinskaya, Zh. Neorgan. Khim., 6(11):2603 (1961).
170. E. M. Savitskii, V. P. Polyakova, M. A. Tylkina, and G. S. Burkhanov, Zh. Neorgan. Khim., 9(7):1645 (1964).
171. L. Menezes, J. K. Poros, and L. A. Read, Sympos. Nature, Occurrence and Effects of Sigma Phase, ASTM (1950).
172. M. V. Nevill and J. W. Downey, J. Metals, 9:1072 (1957).
173. A. E. Dwight and P. A. Beck, Trans. AIME, 215:976 (1959).
174. G. Grube and R. Knabe, Z. Electrochem., 42:793 (1936).
175. E. Raub and N. Mohler, Z. Metallkunde, 45:648 (1954).
176. E. Gebhardt and W. Köster, Z. Metallkunde, 32:262 (1940).
177. E. Friedrich and A. Kussman, Ann. Phys., 6(7):169 (1950).
178. E. Raub and V. Maler, Problemy Sovremennoi Metallurgii, No. 5, p. 114 (1955).
179. M. A. Tylkina, O. Kh. Khamidov, and E. M. Savitskii, Zh. Neorgan. Khim., 9(12):2738 (1954).
180. C. W. Nowarth and W. Hume-Rothery, J. Inst. Metals, 87(8):265 (1959).
181. E. Anderson and W. Hume-Rothery, J. Less-Common Metals, 2(1):19 (1960).
182. E. Anderson and W. Hume-Rothery, J. Less-Common Metals, 2(6):44 (1960).
183. A. Taylor, N. J. Doyle, and B. J. Kagle, J. Less-Common Metals, 4(5):436 (1962).
184. G. V. Samsonov and K. I. Portnoi, Alloys Based on Refractory Compounds [in Russian], Oborongiz (1961).
185. M. A. Tylkina, U. P. Polyakova, and E. M. Savitskii, Zh. Neorgan. Khim., 6:1472 (1961).
186. Feder and Steele, Nature, 205(4971):587 (1965).
187. M. A. Tylkina, U. P. Polyakova, and U. Sh. Shekhtman, Zh. Neorgan. Khim., 8(11):2540 (1963).
188. E. Raub and P. Walter, Festschrift aus Anlass des 100-jahrigen Jubileums der Firma W. C. Heraeus, Hanau, p. 124 (1961).

189. E. M. Savitskii, Izv. Akad. Nauk SSSR, Metally, No. 5, p. 3 (1965).
190. E. J. Rapperport and M. F. Smits, J. Metals, 13(1):86 (1961).
191. A. Taylor, B. Kagle, and N. Y. Dajle, J. Less-Common Metals, 3(4):333 (1961).
192. A. Lina, Problemy Sovremennoi Metallurgii, No. 5, p. 125 (1955).
193. E. M. Savitskii, M. A. Tylkina, and K. B. Povarova, Rhenium Alloys [in Russian], Izd. "Nauka" (1965).
194. Collection: Rhenium, New York (1962).
195. R. I. Jaffee, D. Meikuth, and R. Douglass, Rhenium and Refractory Metals of the Platinum Group [Russian translation], IL (1963).
196. N. V. Grum-Grzhimailo and D. I. Prokof'ev, Zh. Neorgan. Khim., 7(3):596 (1962).
197. P. Chiotti, J. Electrochem. Soc., 101:596 (1954).
198. F. F. Schmidt, H. R. Holden, H. R. Ogden, and R. I. Jaffee, Investigation of the Properties of Tantalum and Its Alloys, WADD Techn. Rept., pp. 61–106 (1961).
199. W. Rostoker, WADC Rept., pp. 59–492 (June 1959).
200. E. M. Savitskii, M. A. Tylkina, and V. P. Polyakova, Zh. Neorgan. Khim., 8(1):146 (1963).
201. M. A. Tylkina, K. B. Povarova, and E. M. Savitskii, Zh. Neorgan. Khim., 5(11):2458 (1960).
202. N. N. Kurnakov and M. Ya. Troneva, Zh. Neorgan. Khim., 5(11):2403 (1960).
203. D. A. Prokoshkin and M. I. Zakharova, in: Researches of Refractory Alloys [in Russian], Vol. 8, Izd. Akad. Nauk SSSR, p. 70 (1962).
204. E. M. Savitskii and A. M. Zakharov, Zh. Neorgan. Khim., 7(11):2575 (1962).
205. E. M. Savitskii and A. M. Zakharov, in: Researches on Nonferrous Metal Alloys [in Russian], Izd. AN SSSR, No. 4, p. 108 (1963).
206. E. M. Savitskii and A. M. Zakharov, Zh. Neorgan. Khim., 9(9):2261 (1964).
207. A. M. Zakharov and E. M. Savitskii, Izv. Akad. Nauk SSSR, Metally, No. 1, p. 151 (1965).
208. A. M. Zakharov and E. M. Savitskii, Izv. Akad. Nauk SSSR, Metally, No. 5, p. 161 (1966).
209. N. V. Grum-Grzhimailo and V. G. Gromova, in: Titanium and Its Alloys [in Russian], Izd. Akad. Nauk SSSR, No. 7, p. 35 (1962).
210. V. N. Svechnikov, Yu. A. Kocherzhinskii, V. I. Latysheva, and V. I. Pan, in: Researches on Refractory Alloys [in Russian], Vol. 8, No. 6, Izd. Akad. Nauk SSSR (1962).
211. A. M. Zakharov and E. M. Savitskii, Izv. Akad. Nauk SSSR, Metally, No. 6 (1966).
212. E. M. Savitskii and A.M. Zakharov, Zh. Neorgan. Khim., 9(10):2424 (1964).
213. B. B. Gulyaev, O. N. Magnitskii, and A. A. Demidov, Refractory Metal Castings [in Russian], Izd. "Maschinostroenie" (1964).
214. E. A. Gulbransen and K. F. Andrew, Trans. AIME, 105:4 (1958).
215. W. D. Klopp, C. T. Sims, and R. I. Jaffee, Trans. ASM, 51:282 (1959).
216. J. V. Cathcart, J. J. Campbell, and G. R. Smith, J. Electrochem. Soc., 105:442 (1958).
217. E. Gebhardt and H. D. Seghezzi, Z. Metallkunde, 50:245 (1959).
218. K. Hauffe, Z. Elektrochem., 63:819 (1959).
219. E. S. Jones, J. E. Mosher, R. Speiser, and J. W. Spretnak, Corrosion, 14:27 (1958).
220. R. Kieffer and F. Kölbi, Z. Anorg. und Allgem. Chem., 262:229 (1950).
221. W. W. Webb, J. T. Norton, and G. Wagner, J. Electrochem. Soc., 103:107 (1956).
222. R. I. Jaffee, Problemy Sovremennoi Metallurgii, No. 6, p. 115 (1960).
223. V. I. Akharov, Oxidation of Metals at High Temperatures [in Russian], Metallurgizdat (1945); Trudy Instituta Fizika Metallov, No. 20, p. 229, Izd. Akad. Nauk SSSR (1958).
224. N. A. Shishakov, Electron-Diffraction Examination of Oxide Films [in Russian], Izd. Akad. Nauk SSSR (1953).
225. O. Kubaschewski and B. Hopkins, in: Protection of Refractory Metals from Oxidation [Russian translation], IL, p. 11 (1962).
226. N. Cabrera and N. F. Mott, Repts. Progr. Phys., 12:163 (1948–49).
227. K. Hauffe and B. Ilschner, Z. Electrochem., 53:382 (1954).
228. A. Jenkins, J. Inst. Metals, 82:213 (1954).

229. W. Kunn and W. Knorr, Techn. Mitt. Krupp, 14(4) (1956).
230. G. Pew, Problemy Sovremennoi Metallurgii, No. 5, p. 90 (1958).
231. A. Mackintosh, Problemy Sovremennoi Metallurgii, No. 4, p. 78 (1957).
232. D. W. Bridges and W. M. Fassel, J. Electrochem. Soc., 103:326 (1956).
233. E. S. Bumps, H. D. Kessler, and M. Hansen, Trans. ASME, 45:1008 (1953).
234. A. E. Jenkins and H. W. Worner, J. Inst. Metals, 80:157 (1951–52).
235. A. U. Seybolt, Trans. AIME, 200:770 (1954).
236. R. Bryant, J. Less-Common Metals, 4:62 (1962).
237. E. Gebhardt and R. Rothenbacher, Z. Metallkunde, 54(12):689 (1963).
238. R. Elliott, Problemy Sovremennoi Metallurgii, No. 5, p. 128 (1960).
239. G. Mayo, V. Shepherd, and A. Thomas, in: Protection of Refractory Metals from Oxidation [Russian translation], IL, p. 115 (1962).
240. E. Gebhardt and G. Preisedanz, in: Niobium and Tantalum, No. 2 [Russian translation], IL, p. 51 (1956).
241. R. Vasilevskii, in: Niobium and Tantalum, No. 2 [Russian translation], IL, p. 47 (1956).
242. F. E. Bacon and P. M. Moanfeldt, Columbium and Tantalum, Wiley, New York, p. 347 (1963).
243. J. Stringer, J. Less-Common Metals, 8:1 (1965).
244. W. Rostoker and A. S. Yamamoto, Trans. ASM, 47:1022 (1955).
245. V. S. Mozgovoi and A. M. Samarin, Izv. Akad. Nauk SSSR, OTN, Metallurgiya i Toplivo, No. 10, p. 1529 (1950).
246. A. U. Seybolt and H. T. Sumsion, Trans. AIME, 197:292 (1953).
247. V. I. Arkarov and Yu. D. Kozmanov, Fizika Metallov i Metallovedenie, 2(3):566 (1956).
248. F. Marion and C. Choain-Maurin, Chimie et Industrie, No. 5, p. 88 (1962).
249. G. A. Lenning, C. M. Graighead, and K. F. Jaffee, J. Metals, No. 3, p. 367 (1954).
250. W. Trzebiatowski and B. Stalinski, Roczn. Chem., 30:691 (1956).
251. S. S. Sidhu and J. C. McGuire, J. Appl. Phys., 23:1257 (1952).
252. G. T. Hahn, A. Gilbert, and R. I. Jaffee, Proc. Refractory Metals Symposium, Chicago, p. 23 (1962).
253. V. I. Mikheeva, Hydrides of the Transition Metals [in Russian], Izd. Akad. Nauk SSSR (1960).
254. L. Kirschfeld and A. Sieverts, Z. Electrochem., 36:123 (1930).
255. M. J. Trzeriak, D. F. Dilthly, and M. W. Mollet, U. S. Atomic Energy Commiss., BMI, 1112, 32 (1956).
256. V. Roberts and G. Rodgers, Problemy Sovremennoi Metallurgii, No. 3, p. 110 (1957).
257. E. A. Gulbrandsen and K. F. Andrew, J. Electrochem. Soc., 101:560 (1954).
258. G. Brauer and R. Hermann, Z. Anorgan. und Allgem. Chem., 274:12 (1953).
259. E. A. Gulbrandsen and K. F. Andrew, Rev. Metallurgie, 51:101 (1954).
260. G. T. Huttig and F. Broadkorb, Z. Anorgan. und Allgem. Chem., 144:341 (1925).
261. A. Sieverts and K. Brüming, Arch. Eisenhüttenwes., 7:611 (1934).
262. I. Langmuir, J. Amer. Chem. Soc., 37:451 (1915).
263. T. J. Dillon, Proc. Phys. Soc. A, 41:546 (1929).
264. G. Wilkinson and J. M. Briminham, J. Amer. Chem. Soc., 77:3421 (1955).
265. A. E. Palty, H. Margolin, and T. P. Nielsen, Trans. ASM, 46:312 (1954).
266. G. V. Samsonov and V. I. Konstantinov, Tantalum and Niobium [in Russian], Metallurgizdat, p. 264 (1959).
267. G. V. Samsonov, Refractory Compounds. Reference Book on Properties and Application [in Russian], Metallurgizdat, p. 397 (1963).
268. A. Nönberg, in: Niobium and Tantalum [Russian translation], IL, p. 72 (1956).
269. E. Gebhardt, H. D. Seghezzi, and E. Fromm, Z. Metallkunde, 52(7):464 (1961).
270. R. Blix, J. Phys. Chem., 3:229 (1929).

271. E. M. Savitskii, Izv. Akad. Nauk SSSR, OTN, Metallurgiya i Toplivo, No. 5, p. 52 (1960).
272. J. Electrochem. Soc., 103(6):347 (1956).
273. E. M. Savitskii, V. F. Terekhova, and N. A. Birun, Zh. Neorgan. Khim., 6(8) (1961).
274. E. M. Savitskii, V. P. Polyakova, and M. A. Tylkina, Palladium Alloys [in Russian], Izd. "Nauka" (in press).
275. I. I. Kornilov and R. S. Polyakova, Izv. Akad. Nauk SSSR, OTN, Metallurgiya i Toplivo, No. 1, p. 85 (1960).
276. M. O. Teslyuk, Discussion on papers presented at the Conference "Nature of Metallic Phases and the Character of the Chemical Bond in Them" [in Russian], Izd. "Nauka," p. 17 (1965).
277. S. Steeb and J. Renner, Z. Metallkunde, 56(8):531 (1965).
278. S. Steeb and J. Renner, J. Less-Common Metals, 9(3):181 (1965).
279. C. Wainwright, A. J. Cook, and B. E. Hopkins, J. Less-Common Metals, 6(5) (1964).
280. P. Cotterill, Hydrogen Embrittleness of Metal, Pergamon, New York (1961).
281. C. M. Schwartz and M. W. Mallet, Trans. ASM, No. 46, p. 640 (1954).
282. R. A. Pasternak and B. Evans, Trans. Met. Soc. AIME, No. 233, p. 1194 (1965).
283. R. A. Pasternak, B. Evans, and B. Bergsnov-Hansen, J. Electrochem. Soc., 113(7):731 (1966).
284. E. I. Gladyshevskii, N. M. Popova, and T. F. Fedorov, Zh. Neorgan. Khim., 8(8):1997 (1963).
285. H. Bittner, H. Goretzki, F. Benesovsky, and N. Nowotny, Monatshefte für Chemie, 94(3):518 (1963).
286. E. Goudremont, Special Steels [Russian translation], Metallurgizdat, Vols. I and II (1959, 1960).
287. H. J. Goldschmidt and J. A. Brand, J. Less-Common Metals, 5:181 (1963).

CHAPTER IV

# THE PROBLEM OF THE COLD BRITTLENESS OF REFRACTORY METALS AND ALLOYS

In addition to the oxidizability and poor weldability of refractory metals, their cold brittleness forms the principal obstacle to their utilization in the new technology. Transition from the ductile to the brittle state occurs in refractory metals in a certain temperature range. Group Va metals and particularly group VIa metals having a bcc lattice show a tendency to lose ductility on cooling (Fig. 73). Group VIa metals and primarily tungsten have the highest ductile–brittle transition temperature. Iron is the classic example of a cold-brittle metal.

Some hexagonal metals such as zinc, cadmium, magnesium, ruthenium, and osmium (the last two are refractory) are also found to be cold brittle. Cold brittleness is found in metallic compounds and in some semiconductor elements (germanium, silicon).

Thus, the cold brittleness problem goes beyond the scope of refractory metals and becomes one of the difficult problems of the physical chemistry, physical metallurgy, metal physics, and the mechanics of materials. It is of very great importance for the national economy.

This chapter examines the principal current theoretical concepts of the cold brittleness problem.

## Cold Brittleness of Metals and the Interatomic Bond

Currently, it is possible to distinguish four approaches to the study of the nature of cold brittleness; they may be grouped roughly in the following categories: mechanical, energy, metal physics, and physicochemical approaches.

The mechanical or phenomenological approach starts out from the relationship between torsional strength and tensile strength which varies with temperature (with reduction in temperature, the torsional strength increases much more rapidly than the tensile strength). This approach is being followed up by the N. N. Davidenkov school [1, 2].

The energy approach is connected with the calculation of the energy of fracture. In the USSR, it is being followed up by V. S. Ivanov and collaborators [3].

The metal physics approach starts from the decisive part played by crystal defects [4-7]. It is being followed up in non-Russian and Russian metal physics institutes, in particular by V. I. Trefilov and collaborators. From the standpoint of modern metal physics concepts the formation of cracks which result in brittle fracture is caused by the resistance to motion of dislocations [4]. A very important development in this line of approach is the progress from qualitative descriptions, which predominate in the literature on this question, toward the establishment of quantitative relationship in clarifying the question as to the kind of dislocation structure that must be present (form and number of dislocations and other defects, and their arrangement in the metal or alloy) to ensure the necessary minimum ductile properties at ordinary and low temperatures.

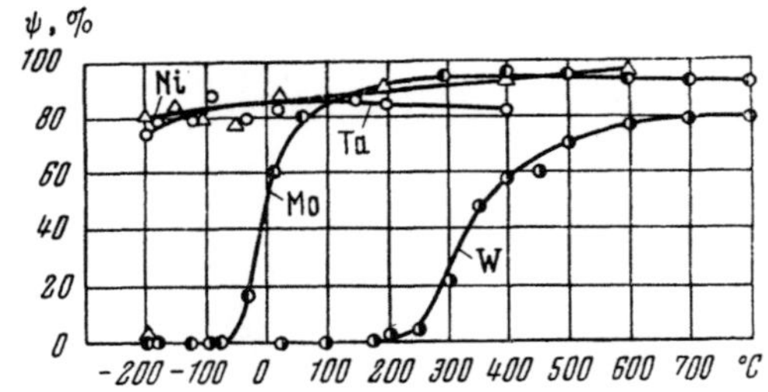


Fig. 73. Dependence of ductility (relative reduction in area at fracture ψ) of refractory bcc metals and nickel on test temperature [18].

The physicochemical theory of ductility (plasticity) starts from the decisive influence exerted by the peculiarities of the metallic bond and purity of metals and alloys (interstitial impurities) on their plasticity. It is being developed in the Laboratory of Alloys of Refractory and Rare Metals of the A. A. Baikov Institute of Metals, Academy of Sciences of the USSR, and in a number of other Russian and non-Russian scientific research institutions [6-10].

It would not be correct to compare with each other all the above-mentioned approaches to the study of the nature of cold brittleness. On the contrary, they must be complementary to each other and underline the importance of the many-sided research on this extremely complex phenomenon. Ductility, like the strength, of crystalline solids, including of course metals and alloys, is a function of a large number of factors, the most important of which are the nature of the interatomic bond, macro- and microstructure, different kinds of defects, rate of strain, presence or absence of notches, state of the surface scale factor, etc. Ductility is thus not merely a property but also a state of a material.

We shall examine in more detail the most important hypotheses of the physicochemical theory of ductility. The nature of the interatomic bond is one of the most important factors which determine the behavior of a material under the action of external forces. The concept of high ductility follows from the very definition of a metal. The metallic state, the metallic character of the interatomic bond favor most the ductile state of matter, since only the metallic bond, i.e., the presence of free electrons, permits movement of the atoms relative to each other without disturbing the continuity of the body. All the age-old experience of man confirms this. It is here not out of place to recall that precisely the forgeability of metals, coupled with their hardness and strength, was the most important condition for the emergence of the bronze and iron ages and of modern technical progress. In his time, Lomonosov wrote: "Metals may be defined as lustrous solids which may be forged" [11]. All the subsequent developments of science and technology have confirmed and are daily confirming Lomonosov's brilliant generalization.

The brittlest materials are those having a covalent bond, in which the valence electrons belong to a given atom and its closest neighbors. Such a type of bond is found in diamond, silicon, germanium, and semiconductor compounds at ordinary and low temperatures. The covalent bond, once ruptured, is difficult to restore. Crystals having an ionic bond, for example, metal oxides and natural salts, show a similar behavior. However, when they are heated to relatively high temperatures (close to the melting point), with the application of stresses in accordance with a system of hydrostatic nonuniform compression, covalent and ionic crystals assume the property of plastic deformation [6].

Temperature is one of the most important external thermodynamic factors determining the physicochemical state of a substance and consequently its properties. The results of systematic research into the influence of temperature on the mechanical properties of various metallic substances (monomorphic and polymorphic metals, metallic compounds, and multiphase alloys) are given in [6]. Increasing the temperature usually results in weakening the effect of interatomic and intercrystalline bonding forces, and consequently in increasing the ductility. It is well known that in pure metals having a fcc structure, brittle fracture is not observed even when they are cooled to temperatures close to absolute zero. The same applies to alkali metals having a bcc structure [1, 6]. In group Va and VIa monomorphic refractory metals having a bcc lattice, the type of crystal lattice does not alter and the metallic character of the bond persists when the temperature is reduced to close to absolute zero. Consequently, from the physical standpoint, there is no reason for the existence of temperature brittleness zones in the case of such metals. Brittleness is a property of metals of commercial purity and is due to the difficulties of producing them in a pure form and preserving that purity in their subsequent treatment. This hypothesis was expressed ten years ago [6]. Subsequent development of the physical metallurgy and physical chemistry of very pure metals has fully confirmed the dependence of their ductility on the attainable degree of purity [7, 9, 10, 14]. The supposition of some writers that cold brittleness of metals having a bcc lattice is their natural characteristic is not sufficiently founded and is contradictory to the nature of the metallic state of a substance. Obviously, cold brittleness is inherent in bcc metals only if impurities forming interstitial solid solutions are present in them, and are arranged in the form of nonmetallic inclusions along the grain and block boundaries. Experimental data (see Chapter III) confirm that metals such as tantalum, niobium, and molybdenum in a state of high purity, and also potassium and sodium, do not show any signs of cold brittleness down to helium temperatures [6-10]. Tungsten having a low carbon content shows high ductility in liquid nitrogen [9], while tungsten perfectly free from carbon and other interstitial impurities would indisputably be deformed even in liquid helium. We also do not share the view that refractory metals ought always to be cold brittle because of the presence in them of some proportion of covalent bond. There is nothing new in this assertion since it is well known that real substances are most often characterized by a mixed interatomic bond and the presence of a covalent component follows from the very definition of transition metals. What is important is that the metallic bond always predominates in metals, in fact it determines their mechanical and electrical properties. The covalent component evidently determines their particular sensitivity to impurities. One of the principal problems of the physical chemistry of metals at the present time is to progress from qualitative conceptions to the quantitative determination of electron distribution in metals and alloys. As already stated in Chapter II, for this purpose, it is necessary to resort to modern physical methods, such as investigation of the distribution of electron density by means of Fourier series [12], study of x-ray emission spectra [13], neutron diffraction determination of spin superstructure, resonance methods, experimental determination of the Fermi surface in metals [14-16], measurement of various physical properties, etc. For us, physical metallurgists, there ought to be more contacts on these questions with the physical institutes.

## Methods of Determining the Ductile – Brittle Transition Temperature

In metals, the transition from the ductile to the brittle state, characterized by the critical brittleness temperature, is connected with a change in the mechanism of fracture. Above the critical temperature, fracture occurs after considerable plastic deformation produced by shear. Below the transition temperature, fracture proceeds by the propagation of a brittle crack along cleavage planes or along the grain boundaries without previous deformation or after slight plastic deformation [17].

We shall consider briefly the method of determining the ductile–brittle transition temperature (or more exactly, temperature range).

Curves characteristic of the transition of metallic materials from the ductile to the brittle state are shown in Fig. 74. The ductility parameter is plotted along the axis of ordinates and the temperature along the axis of abscissas. Curve A is characteristic of the ideal transition to the brittle state, curve B of the gradual transition frequently observed experimentally. In the latter case, in the determination of the transition temperature, various writers

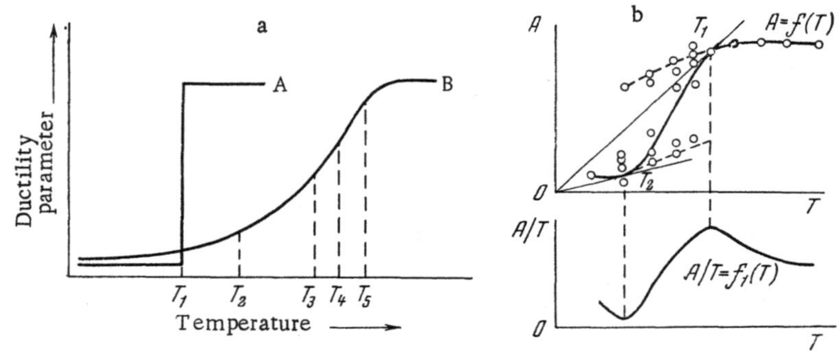


Fig. 74. Characteristic curves of the ductile–brittle transition temperature (a) [19] and graphical determination of the ductile–brittle transition temperature zone (b) [21].

permit a certain arbitrariness. As cold brittleness threshold, it is possible to assume the temperature at which the ductility parameter attains an arbitrarily selected low value or at which the ductility parameter is half-way between the upper and lower levels, etc. It would obviously be more correct to indicate the limit temperatures confining the argon (zone) of transition from the ductile to the brittle state. When tested in this temperature zone, specimens of bcc metals and alloys undergo a certain plastic deformation and a definite degree of work hardening. An exact determination of the temperature range of the transition zone, based on the methods of mathematical statistics, was proposed in [20]. The limit temperatures of the transition zone may be readily determined by the graphical method [21]. Tangents are drawn from the origin of the coordinates to the smooth ductility curve; the points of contact $T_1$ and $T_2$, corresponding to the maximum and minimum of the transformed curve, may be assumed to be the upper and lower limits of the transition range (Fig. 74b). By graphical differentiation of the transformed curve, it is possible to construct a curve having one maximum, corresponding to the most rapid change in ductility [21].

Brittleness zones are revealed mainly by the ductility values obtained in mechanical tests involving tensile stresses: elongation or reduction in area in tensile testing, impact toughness, deflection in bend tests, etc. [6].

The severest conditions are created in testing notched specimens for impact toughness. The highest ductile–brittle transition temperature is found in this method of testing. Steel is regarded as brittle if its impact toughness is less than 2 kgm/cm$^2$. The dependence of the critical brittleness temperature of recrystallized molybdenum on the method of testing is examined in [22]. According to results of Semchyshen and Barr, for recrystallized molybdenum rods, the transition temperature in impact testing is approximately 350-450°C higher than in tensile testing [23]. According to E. M. Savitskii and V. A. Trapeznikov [24], the ductile–brittle transition temperature of chromium may be determined from temperature curves of hardness and reduction in length of specimens subjected to uniaxial compression. However, like other characteristics of the resistance of meterials to plastic deformation (yield point, ultimate strength, fracture), hardness by itself cannot be used as a measure of ductility [6]. Wood and Westbrook [19] developed a special apparatus for determining the transition temperature of thick wire heated in oil or other medium while being wound on a steel drum.

The value of the fracture energy of a specimen tested for impact strength and notch sensitivity are perhaps the most objective and severe criteria of the cold brittleness of materials. An interesting method of determining the cold brittleness threshold is by constructing the curve of temperature dependence of the percentage of tough zone in the fracture of a specimen after toughness impact tests at different temperatures [25].

The influence of the nature of the stress on the transition temperature is in agreement with the variation in the ratio of the tangential stresses to normal stresses: in tension it is 1:1, in torsion 1:2, and in compression 1:4. According to [23], the transition temperature for commercial chromium in tension, torsion, and compression tests is 355°, 250°, and 0°C. As is well known, in bending, the maximum stress acts on the outer surface of the specimen, and therefore to avoid random results due to contamination, the surface must first be polished.

The basic parameters of the ductile–brittle transition of materials are brought out distinctly by subjecting notched specimens to tension. The introduction of the notch in tension gives rise to two effects. The rate of strain is increased in a localized portion of the specimen and a triaxially stressed state is produced, leading to an increase in the ratio of normal stress to tangential stress. Notches, cracks, and other discontinuities in the metal result in stress concentration. The notch sensitivity of the material increases, if their deformability decreases, since plastic deformation results in stress relief. The notch sensitivity of a material is defined as the ratio of the tensile strength of notched specimens to the tensile strength of smooth specimens. Another criterion of the notch sensitivity of materials may be the resistance to crack propagation $G_c$. This parameter is defined as the energy necessary for the propagation of a brittle crack when the crack becomes unstable, and it may be determined from the approximate equation $G_c = 5\gamma$ [26], where $\gamma$ is the specific surface energy.

Thus, the ductile–brittle transition temperature is not a constant physical characteristic of a metal, but depends on a number of factors: purity of the metal, rate of strain and system of deformation, the surrounding gaseous medium, etc.

## Factors Affecting the Ductile – Brittle Transition Temperature

### 1. Interstitial Impurities and Structure

The study of the influences of impurities on the ductile state of a material is one of the most important problems of the physicochemical theory of plasticity.

In examining the influence of the degree of purity on the temperature of the transition to the brittle state, it is important to know not only the total content and chemical composition of the impurities, but also the character of their interaction with the metal base and their structural arrangement. Questions of the influence of chemical composition and structure must therefore be examined in their interrelationship. According to modern views, a real metal or alloy is a physicochemical system of crystals of the base metal, having impurities, alloying elements and also crystal defects distributed in their interior and along their boundaries. The mechanical and other physical properties of such a system, in addition to the nature of the metal base, are primarily determined by the properties and relative arrangement of the structural components, metalloid and metallic impurities, and crystallographic defects. This arrangement may be favorable or unfavorable for the plastic deformation of the crystals of a metal or alloy [27]. There is now no doubt that the degree of purity of a metal is one of the principal factors affecting the ductile–brittle transition temperature, interstitial impurities exercising the maximum influence.

The mechanism of the embrittlement of refractory bcc metals by interstitial impurities is associated with the distinctive features of the interaction of refractory metals with nonmetals. As already pointed out in Chapter III, the interstitial impurities (nitrogen, carbon, hydrogen, oxygen) form with refractory metals very limited ranges of solid solutions, the solubility diminishing from group IVa to group VIa. The solubility of the interstitial impurities

TABLE 30. Influence of Interstitial Impurities on the Ductility–Brittle Transition Temperature of Group Va Metals [10, 28, 30–32]

| Metal | Content of interstitial impurities, wt.% | | | | Ductile–brittle transition temperature, °C |
|---|---|---|---|---|---|
| | hydrogen | nitrogen | oxygen | carbon | |
| Niobium, arc melted, recryst. at 1200°C | 0.0001 | 0.065 | 0.0317 | 0.009 | −185 |
| The same, electron-beam melted (recryst. at 1200°C) | 0.0001 | 0.011 | 0.029 | 0.016 | −200 |
| The same, wire (annealed in a vacuum of $5 \cdot 10^{-6}$–$1 \cdot 10^{-7}$ mm Hg) | 0.0001 | 0.002 | 0.0008 | 0.006 | 1.4°K |
| Vanadium, iodide | 0.005 | 0.0005 | 0.015 | 0.015 | −40 |
| The same, electron-beam melted | 0.0001 | 0.005 | 0.01 | 0.02 | −196 |
| Tantalum, cermet | — | 0.006 | 0.008 | 0.0014 | −200 |
| The same, electron-beam melted | — | 0.0008 | 0.003 | 0.003 | −250 |

TABLE 31. Influence of Oxygen on the Ductile–Brittle Transition Temperature of Molybdenum [27]

| Quality of molybdenum | Oxygen content, % | Ductile–brittle transition temperature, °C |
|---|---|---|
| Commercial | 0.02 | 300 |
| Purified | 0.002 | 25 |
| Specially pure | 0.0001 | Below −196 |

found in group VIa is exceptionally low. If the solubility limits are exceeded, the result is the formation of compounds, precipitated on the grain boundaries and severely embrittling the metal. If the content of interstitial impurities does not exceed the solubility limit, metals of groups Va and VIa are ductile at low temperatures.

Thorough purification of refractory metals from interstitial impurities lowers the ductile–brittle transition temperature considerably.

An investigation of the temperature dependence of the strength and ductility of niobium in the range 1.4-300°K showed that niobium of high purity (impurity content: oxygen < 0.0008%, nitrogen 0.002%, hydrogen 0.0001%, carbon 0.006%, iron 0.025%, titanium 0.015%), retained a relatively high degree of ductility (elongation 8-10%, reduction in area 50-60%), even near the absolute zero, at 1.4°K [28]. Interstitial impurities may be arranged in the following order for group Va metals in descending order of their influence on the production of brittleness: hydrogen, nitrogen, oxygen, carbon. Table 30 shows the influence of interstitial impurities on the ductile–brittle transition temperature of group Va metals. For rhenium, the most harmful impurity is oxygen, which affects its mechanical properties by a factor of approximately 100 more strongly than substitutional elements [29].

The amount of interstitial elements which can remain in solid solutions based on group VIa metals is not more than 0.0001% [10]. In view of their low solubility in group VIa metals, oxygen, nitrogen, carbon, and hydrogen, even in very small quantities, are present in the form

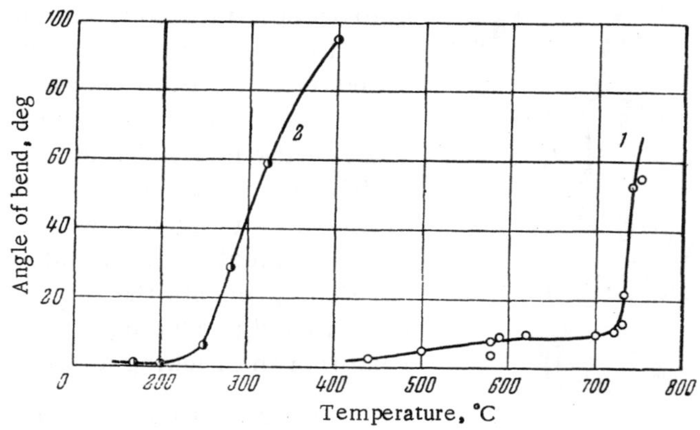


Fig. 75. Ductile–brittle transition temperature of tungsten versus interstitial impurity content. Impurity content, %: 1) 0.029 C, 0.085 $O_2$, 0.014 $N_2$, 0.0004 H; 2) 0.011 C, 0.02 $O_2$, 0.012 $N_2$, 0.0001 H.

of oxides, nitrides, carbides, and hydrides, which also produce cold brittleness in these metals. The influence exerted by oxygen is illustrated for molybdenum by the data of Table 31.

A reduction by two orders in the oxygen content of molybdenum results in a reduction in the brittleness threshold from 300°C to a temperature below −196°C. Nitrogen has approximately the same influence as oxygen on the transition temperature of molybdenum (reduces the brittle strength and shifts the ductile–brittle transition zone to a region of higher temperatures), but in a lesser degree [33]. Carbon in a content of 0.001-0.01% has the strongest effect on the solid-brittleness threshold of molybdenum compared with nitrogen and oxygen. Nitrogen evidently has the strongest influence on the temperature threshold of cold brittleness in chromium [34]. For a given content of interstitial impurities the transition temperature for group VIa metals in the recrystallized state is approximately 300°C higher than in the deformed state. The cause of this is the precipitation, along the grain boundaries of the refractory metal, of comparatively coarse accumulations of nonmetallic compounds occurring during the cooling of the specimen or article after recrystallization annealing, or not working by pressure.

In collaboration with A. M. Zakharov, we have studied the effect of the total content of interstitial impurities on the ductile–brittle transition temperature range of tungsten (Fig. 75). For tungsten of commercial purity, the ductile–brittle transition temperature lies in the range 700-500°C. A reduction in impurity content by approximately half considerably reduces the brittle temperature threshold of tungsten. For very pure single-crystal tungsten produced by zone refining using an electron beam with special purification by removal of carbon, the transition temperature is below −196°C [35, 36].

The distribution of nonmetallic compounds (characteristic of high hardness and brittleness) in the form of film or shells, enveloping the grain boundaries of the metal is particularly detrimental. In this case, the bond between the crystallites is greatly weakened, resulting in brittle fracture of the material along the grain boundaries. For example, in molybdenum a content of 0.006% oxygen is sufficient to surround all the crystal boundaries with molybdenum oxides; for tungsten, this content is still lower. Figure 76 shows the microstructure of cast tungsten annealed in an insufficient vacuum ($10^{-3}$ mm Hg) for two hours at a temperature of 1000°C. The layer of nonmetallic inclusions bordering the grains is distinctly visible. With a low impurity content (of the order of 0.01-0.001%) and a finely granular structure, the film of nonmetallic inclusions may be very thin and may not be revealed by ordinary

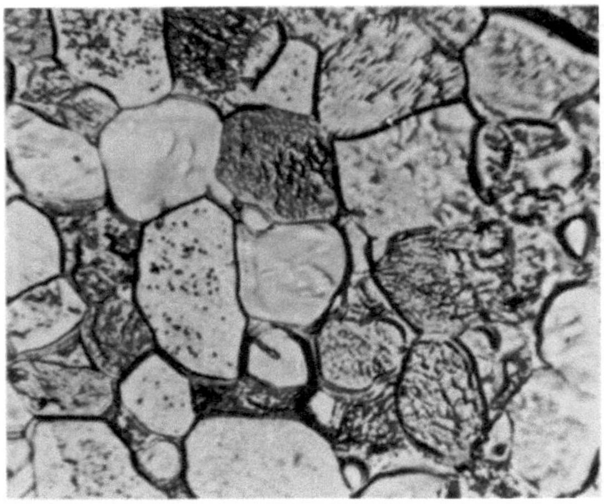

Fig. 76. Microstructure of cast-annealed tungsten having a high interstitial impurity content (×500).

microscopic methods. In molybdenum, the distribution of nonmetallic impurities along the grain boundaries has been shown by fractographic examination of the cast metal [37], and also by a study of the internal friction of polycrystalline molybdenum. In this case, the bond between the crystallites is greatly weakened, resulting in embrittlement of the metal. The brittle, hard envelope prevents the flow of the metal, obstructing slip and producing stress concentration, resulting in cracking and fracture of the metal.

A study of the individual effect of interstitial impurities on the structure and properties of these metals is important in elucidating the nature of cold brittleness of group VIa metals.

Work conducted in the Refractory and Rare Metal Alloys Laboratory of the A. A. Baikov Institute of Metallurgy confirms that carbon is the most harmful impurity in molybdenum and especially in tungsten. Even in highly pure tungsten, obtained by electron-beam melting in a vacuum produced by oil-vapor pumps, an excess phase, identified as $W_2C$ may be detected [38]. Saturation of tungsten by carbon results in an abrupt increase in the amount of dispersed carbides, and at the same time in an increase in the dislocation density of approximately one order (from $1.2 \cdot 10^6$ to $1 \cdot 10^7$ cm$^{-2}$). This is associated with the relief of internal stresses produced by the introduction of hexagonal carbide $W_2C$ into the tungsten bcc lattice. The presence of a large number of dislocation sources and obstacles to the movement of dislocations (carbides) creates favorable conditions for the occurrence of microcracks, and the inception of brittle fracture. According to [35], interstitial impurities may be arranged in the following descending order of the degree of their effect on the ductile–brittle transition temperature of tungsten: carbon, oxygen, nitrogen, and hydrogen. It is possible to arrange the interstitial impurities in an "embrittlement series" also for molybdenum. The oxides and nitrides are characterized by lower thermodynamic stability and higher vapor tension than the carbides of molybdenum and tungsten [39, 40]. Oxygen and nitrogen, therefore, can be removed comparatively easily in the vacuum melting of tungsten and molybdenum, owing to the decomposition or evaporation of the corresponding compounds. Carbon, however, may be removed in melting only by the reaction of the tungsten or molybdenum carbides with oxygen or hydrogen, an excess amount of carbon remaining in the metal in the form of carbides, as is convincingly shown by microscopic and electron-microscopic examinations [36, 38].

Figure 77 shows the microstructure of molybdenum as revealed by the emission microscope [41]. The excess phase (molybdenum carbide) is a stress raiser and is consequently the source of microcracks. The solubility of carbon in solid molybdenum, however, is evidently still higher than in solid tungsten. To produce ductile tungsten, therefore, requires a more thorough purification from carbon than in the case of molybdenum.

The influence of carbon on the ductile–brittle transition temperature of molybdenum is shown in Fig. 78. A decrease in the carbon content of molybdenum from 0.017 to 0.004% reduces the transition temperature from 120° to −50°C [42].

Table 32 shows the effect of interstitial impurities on the ductile–brittle temperature of tungsten and molybdenum.

TABLE 32. Influence of Interstitial Impurities on the Ductile–Brittle Transition Temperature of Group VIa Metals [10, 35, 42, 45]

| Metal | Impurity content, wt.% | | | | Dilute–brittle transition temperature, °C |
|---|---|---|---|---|---|
| | carbon | oxygen | nitrogen | hydrogen | |
| Molybdenum, cermet, vacuum sintered | 0.047 | 0.02-0.01 | 0.02 | 0.0006 | +200 |
| The same, vaccum arc-melted (deformed) and recrystallized) | 0.014 | 0.005-0.003 | 0.0006 | 0.001-0.0005 | 0 |
| The same, electron-beam zone vacuum melted | 0.012 | 0.001 | 0.0001 | 0.0001 | −196 |
| Tungsten, cermet, vacuum welded | 0.040 | 0.023 | 0.002 | 0.0003 | +500 |
| The same, vaccum arc-melted | 0.030 | 0.004 | 0.001 | 0.0002 | +200 |
| The same, electron-beam zone melted after | | | | | |
| one pass | 0.024 | 0.001 | 0.001 | 0.0001 | +100 |
| two passes | 0.020 | 0.001 | 0.001 | 0.0001 | +20 |
| The same, electron-beam zone melted (initial blank subjected to preliminary decarburization) | 0.001 | 0.001 | 0.001 | 0.0001 | −196 |

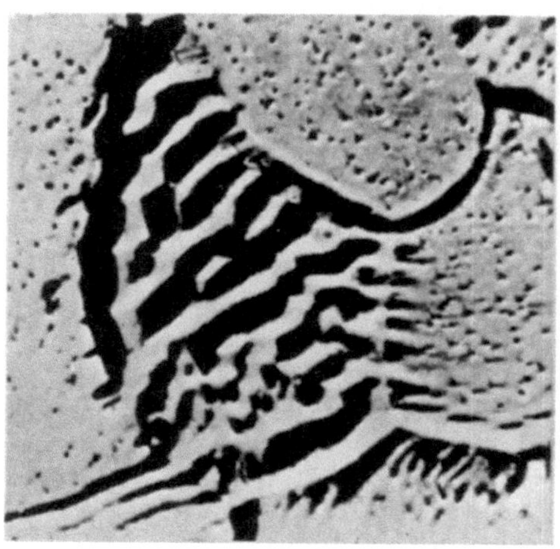

Fig. 77. Microstructure of molybdenum containing carbon.

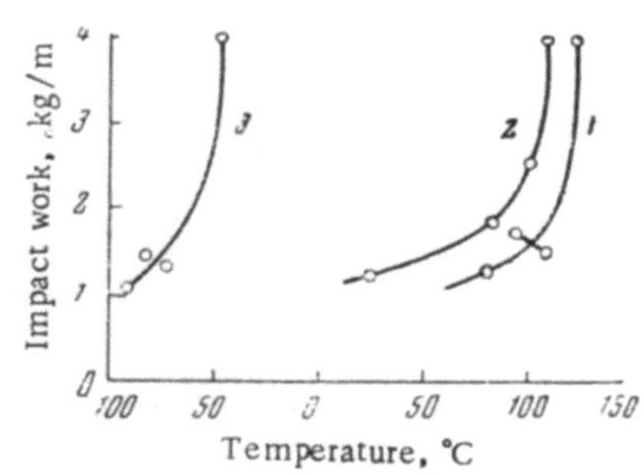


Fig. 78. Ductile–brittle transition temperature of molybdenum versus carbon content (%). 1) 0.017; 2) 0.008; 3) 0.004.

According to [43], the residual carbon content in deoxidized molybdenum should not exceed 0.005% to ensure a cold-brittleness temperature not above room temperature.

The examination of the fracture of various metals shows that fracture may occur along the grain boundaries and through the grain, but as a rule crack inception and propagation proceed along the grain boundaries. Inception of fracture along the grain boundaries in the case of chromium, molybdenum, and tungsten was demonstrated in [37]. The surface of the fracture of arc-melted molybdenum specimens was studied in the optical and electron microscope in [44]. The fracture of fine-grained molybdenum occurred by simultaneous cleavage and crack propagation along the grain boundaries. Coarsely crystalline molybdenum fractured

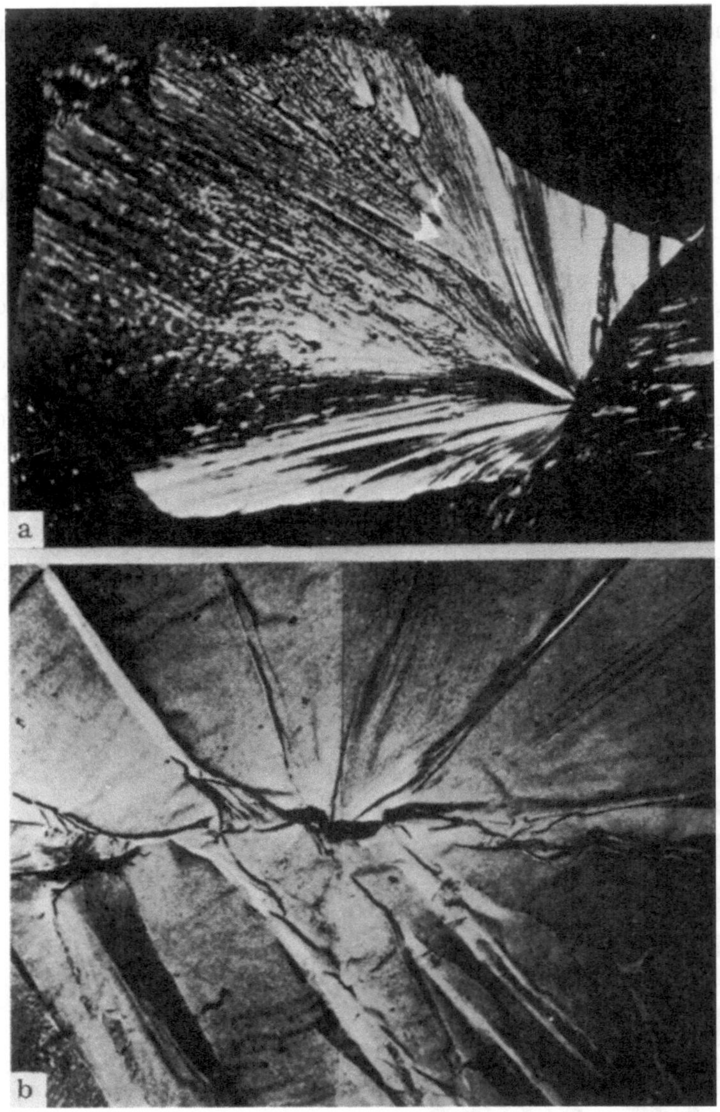

Fig. 79. Fracture inception in coarsely crystalline molybdenum: a) optical fractograph (×150); b) electron-microscope fractograph (×1300).

by cleavage through the grain, but crack inception was observed on the grain boundary. As shown in Fig. 79, the lines of fracture start from one point; a network of the carbide $Mo_2C$ on the grain boundaries of molybdenum has been found by electron-microscope and x-ray methods. Available data show the existence of complex nonmetallic inclusions (oxycarbides, oxynitrides, etc.) on the grain boundaries of refractory metals [46].

Very interesting are results on the effect of purity on the ductility of $\alpha$-iron having a bcc lattice, isomorphic with group Va and VIa metals. The temperature dependence of the ductility of $\alpha$-iron of commercial purity is similar to the temperature dependence of the ductility of group Va and VIa metals. It has been found that the impurity content level and grain size considerably affect the ductile–brittle transition temperature of $\alpha$-iron [47, 48]. The

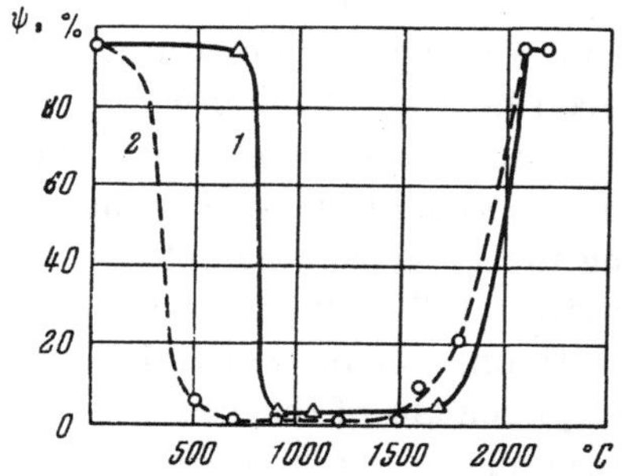


Fig. 80. Temperature range of carbide (1) and oxide (2) brittleness in molybdenum.

results of these investigations show that by electron-beam zone purification in a vacuum, and also by grain refinement, the transition temperature of α-iron may be reduced to temperatures close to absolute zero.

Speisel and Wulf [49] cite interesting results obtained by Beisch on the effect of nonmetallic boundary inclusions (oxides and carbides) on the brittleness of molybdenum wire, from which oxygen and nitrogen and partly carbon (in the form of CO) had been removed by vacuum annealing at 2100°C. Heating the molybdenum to a temperature of 900-1800°C followed by slow cooling produced brittleness, owing to the precipitation of carbides on the grain boundaries. A second heating reconverted the carbides into a solid solution, and after quenching the molybdenum became ductile again. Oxide precipitation on the grain boundaries, as the result of heating the molybdenum wire in moist hydrogen, resulted in abrupt embrittlement. Conversion of the oxides into solid solution followed by quenching again imparted high ductility to the molybdenum. The results of [49] on the oxide and carbide brittleness of molybdenum are shown in Fig. 80.

As is well known from the practice of the production of cermet tungsten, an increase in the annealing temperature increases the ductile–brittle transition temperature. In a recent investigation [50], it was shown that such a relationship is associated with the redistribution of impurities on annealing with increase in temperature, the impurities dissolve, and on subsequent cooling segregate along the grain boundaries, producing embrittlement of the metal. The much higher ductile–brittle transition temperature of group VIa metals than that of group Va metals is evidently due to the higher solubility of interstitial impurities in group Va metals. Values are given in [10] for the maximum solubility of interstitial impurities in group Va and VIa metals (Fig. 81). The maximum solubility of interstitial impurities in group Va metals is 3-4 orders higher than their solubility in group VIa metals. The insignificant solubility of interstitial impurities in group VIa metals is due to the characteristics of their electronic structure. According to Robins [52], minimum solubility occurs when the electron concentration of the bond is half the value of the coordination number. This condition is found in group VIa metals which have six valence electrons although, as we know, the coordination number in the bcc lattice is eight. For bcc metals, Robins assumes the value of the effective coordination number to be twelve (first- and second-order closest neighbors are considered). In an electron concentration of six, the resonance electronic bond becomes very stable. The interstitial impurities concentrate around defects characteristic of a lower coordination number compared with the ideal lattice. These concepts are sketchy, and not generally acceptable. If the interstitial impurity content does not exceed the solubility limit, group Va and VIa metals will be ductile at low temperatures. In this case, the temperature threshold of brittleness is shifted abruptly to the low-temperature region. In a condition of highest purity, group Va metals are unsaturated solid solutions and are ductile at low temperatures (below −196°C), while group VIa metals having the same level of impurity content may be supersaturated solid solutions and may contain inclusions of a second phase in the form of oxides, nitrides, and carbides. As already pointed out, far-reaching purification of group VIa metals (molybdenum, tungsten) abruptly lowers the ductile–brittle transition temperature (for data on chromium, see [51]).

Since the abrupt embrittlement of metals is caused primarily by interstitial impurities dispersed along the grain boundaries, the length of the grain boundaries is of considerable importance. The high sensitivity of refractory metals to increase in grain size was noted in

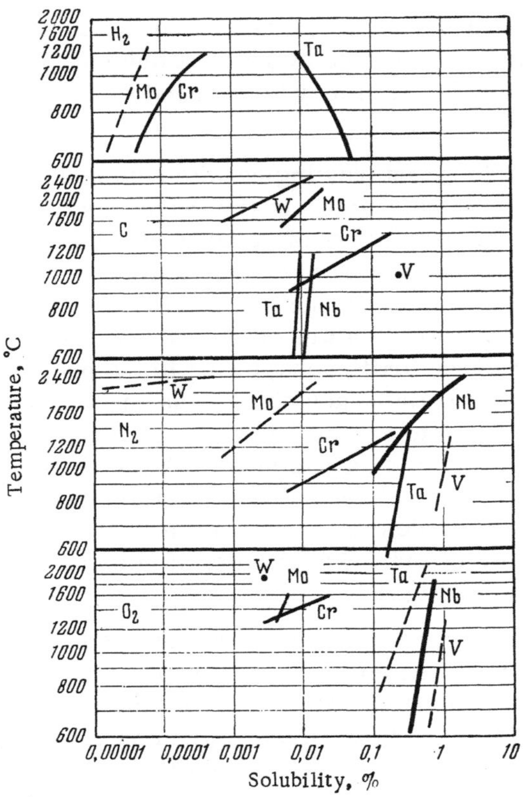


Fig. 81. Values of maximum solubility of interstitial impurities in refractory metals.

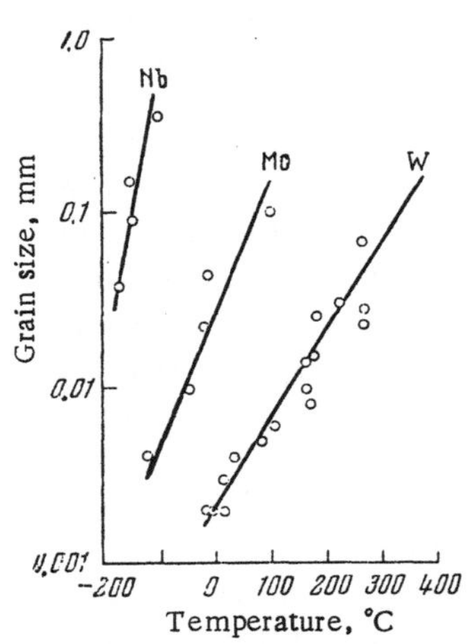


Fig. 82. Ductile–brittle transition temperature of some refractory metals as a function of average grain size.

[53]. The area of intergranular surface per unit volume is an important factor in refractory metals having a body-centered lattice. Therefore, for any given interstitial impurity content, the area of the intergranular surface should preferably be a maximum for reducing the relative concentration of interstitial impurities [27]. In a fine-grain structure, the harmful influence of interstitial impurities is diminished, because of the reduction in their concentration on the grain boundaries. Figure 82 shows the influence of grain size on the transition temperature of some group Va and VIa refractory metals [54]. The transition temperature increases in proportion to the logarithm of the grain diameter. Such a linear relationship is evidently due to the dependence of the yield point on grain size, which for bcc metals may be expressed by means of Petch's formula [55]

$$\sigma_T = \sigma_0 + kd^{-\frac{1}{2}},$$

where $\sigma_T$ is the lower yield point, d is the grain diameter, $\sigma_0$ is the frictional stress in the lattice, and k is a constant.

Evidently, the dimensions of the subgrains and of the regions of coherent scattering show, although to a lesser degree than the grain size, a definite influence on the ductility of metals [56, 57]. It is necessary to take into account also the influence of angular disorientation of the subgrains; the greater the angular disorientation of the blocks, the higher is the yield point of the material.

Experience of cermet production shows that the harmful influence of interstitial impurities is diminished as the result of plastic deformation of refractory metals. The reduction in the transition temperature by work hardening of molybdenum was examined in [18]. Work hardening increases the stress at which fracture of molybdenum occurs. The presence of a deformation structure with elongated grains is especially important for group VIa metals, since in these metals, the interstitial impurity content almost always exceeds the limit of their solubility. The favorable effect of work hardening is more noticeable in coarse-grain material, in which the ductile–brittle transition temperature is higher than in the same material, but with fine grain. Evidently, the reduction in the ductile–brittle transition temperature as the result of work hardening is due to the rupture of the coarse films along the grain boundaries. In group Va and VIa metals,

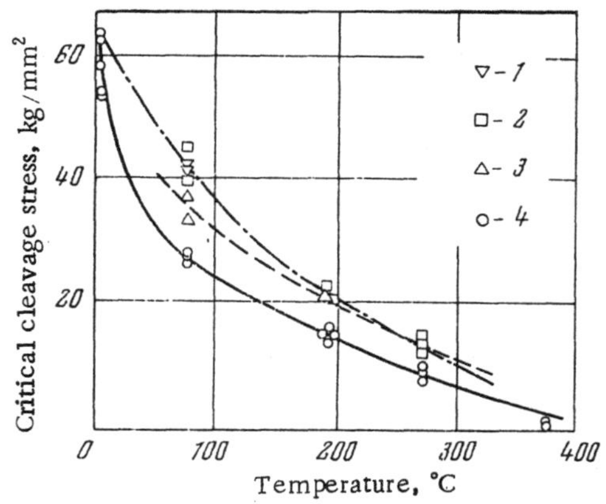


Fig. 83. Dependence of the critical cleavage stress of a zone-refined molybdenum single crystal on temperature [76]. 1) Starting material; 2) after one zone melting pass; 3) after three passes; 4) after six passes.

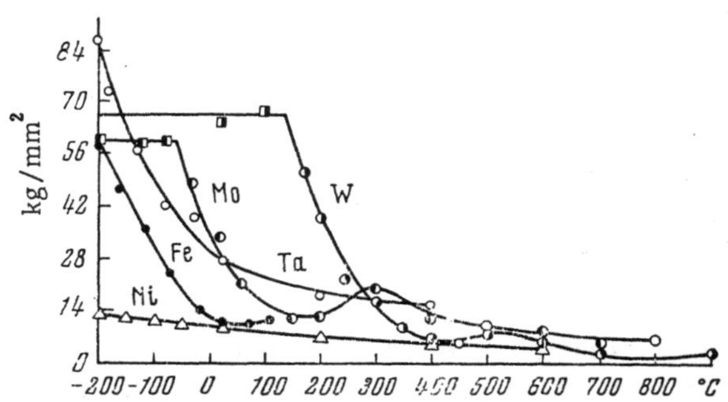


Fig. 84. Dependence of yield point of refractory metals on test temperature.

the presence of a deformation structure permits an increase in the amount of interstitial impurities.

So far, we have examined the influence of interstitial impurities on the brittleness of refractory metals for impurity contents exceeding the solubility limits in the metal base, when the impurities occur in the form of nonmetallic inclusions along the grain boundaries. Interstitial impurities in solid-solution form also considerably impair the ductile properties of refractory metals, mainly as the result of the retardation of dislocation movement, and the reduction in the resistance to crack propagation along effective atomic planes. According to existing concepts, plastic deformation occurs by multiplication and movement of dislocations. The interaction of dislocations with interstitial impurities was studied by Cottrell [58, 59]. A reduction in temperature creates unfavorable conditions for the detachment of dislocations from impurity "atmospheres," which results in an increase in the yield point. The stresses necessary for the detachment of dislocations are high and lead to cracking. The activation energy of the flow process is 0.5-1.5 eV [18].

The presence of nonmetallic impurities along the grain boundaries, as well as the accumulation of impurities along the boundaries of the mosaic structure blocks and around dislocations (Cottrell's atmospheres) render plastic flow very difficult and greatly increase the critical cleavage stress. This is shown in Fig. 83 for molybdenum of different degrees of impurity. The increase in yield point of bcc metals in the region of the ductile–brittle transition temperature is shown in Fig. 84. The existence of a narrow ductile–brittle transition range for refractory bcc metals and the strong temperature dependence of the yield point in the region of this transition are without doubt interrelated phenomena. According to the dislocation theory, the existence of a strong temperature dependence of the yield point for bcc metals, as well as the presence of a lower and upper yield point on the stress–strain curve, are the result of the interaction of dislocations with impurities precipitated on the boundaries and dissolved in the crystals of the metal base. As the result of the slight plastic deformation preceding the upper yield point, grouping of dislocations occurs, resulting in their accumulation on the grain boundaries, producing the stress concentration necessary for the inception of slip in neighboring grains [18, 60, 61]. The stresses increase with decrease in temperature, and the formation of microcracks becomes possible. If a crack attains the initial dimensions, satisfying the Griffiths condition, brittle fracture occurs. The nonmetallic impurities on the grain boundaries act like a notch; they increase the stress concentration and lower the resistance of the metal to brittle fracture. Decrease in the distance between the barriers where the dislocations are accumulating, due to grain refinement, diminish the yield-point inflection and increase the resistance of the metal

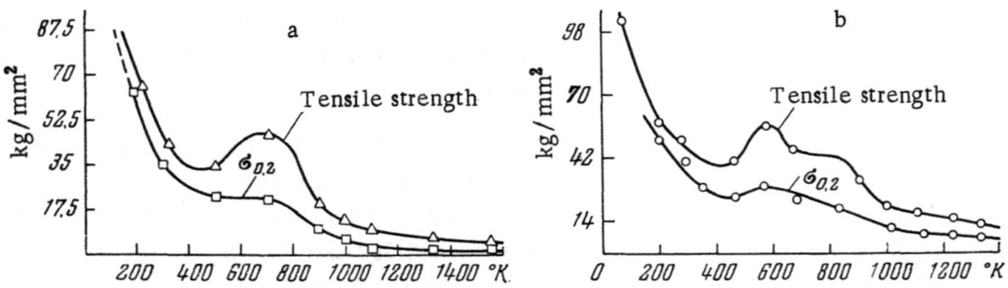


Fig. 85. Temperature dependence of tensile strength and yield point for vanadium (a) and tantalum (b).

to fracture. This effect occurs on decrease in the impurity content. In a metal of high purity, the yield point inflection may not be observed at all or is very slight at low temperatures. According to the Cottrell theory [58, 59] and experimental data obtained by various authors, the temperature dependence of the yield point and the ductile–brittle transition in metals having body-centered cubic and hexagonal lattices is more sharply pronounced than in metals having face-centered cubic lattices. According to Cottrell, this difference is explained by the fact that in bcc metals, the interstitial impurities produce tetragonality of the lattice and consequently interact with linear and screw dislocations. In metals having a fcc lattice, the impurities produce symmetrical distortion of the lattice and react weakly with dislocations (reaction occurs only with linear dislocations). According to [60], the difference in the temperature dependence of the yield points of bcc and fcc metals is also due to the greater activation energy of the diffusion processes of impurity atoms in fcc lattices compared with bcc lattices. It should be noted that in some cases, a strong dependence of yield point on temperature and a sharply defined ductile–brittle transition temperature are also found in metals having a face-centered cubic lattice, for example in iridium.

According to [76], the temperature dependence of the critical cleavage stress of molybdenum purified six times (having the minimum impurity content) increases smoothly down to 77°K, then increases more rapidly, attaining at 4.2°K values obtained for less pure molybdenum (see Fig. 83, one pass). According to Lowly et al. [76], the sharp drop in yield point with temperature, found for the purest material, may be due to the combined action of Peierls-Nabarro mechanisms [63] and Cottrell systems.

As already pointed out, the existence of upper and lower yield points is characteristic of refractory metals. The yield point "tooth" is more distinctly pronounced in metals of commercial purity at low temperatures. With increase in temperature, it becomes less marked and disappears altogether at a certain temperature. According to Cottrell, the upper yield point corresponds to the increased stress necessary for separating the dislocations from impurity atmospheres. The dislocations liberated from the impurities move more freely, which in stress–strain diagrams appears as a stress reduction after the yield point "tooth."

Interaction of impurities with dislocations depends on the rate of movement of the dislocations and the diffusion of the impurity atoms. At certain temperatures, the rate of diffusion of impurity atoms may be so high that, in plastic flow, they may succeed in following the dislocations, i.e., they may occupy positions of minimum energy. Such migration of the atoms increases the stress necessary for the movement dislocations [64]. The result is strain aging. Figure 85 shows the temperature dependence of tensile strength and yield point for vanadium and tantalum [65, 66].

It is pointed out in [70] that dislocations are capable of moving along with the impurity atmospheres forming round them. In the high-temperature range, the rate of diffusion of interstitial atoms increases to an extent such that interaction with dislocations ceases. Johnston and Gilman [68] consider that the yield point is not determined by the stresses necessary for separating the dislocations from a Cottrell atmosphere; in their view, the yield point is determined by the stress necessary for the movement of newly formed dislocations. According to Johnston and Gilman, the occurrence of a flow area when body-centered cubic metals are subjected to tension is due to the increase in the number of dislocations with increase in the degree of deformation. Cottrell atmospheres play a part in the initial stage of deformation up to the upper yield point.

The view is also expressed in [69] that the yield point phenomenon in bcc metals is associated with the rate of movement of dislocations. Assuming that the movement of dislocations and the propagation of elastic waves are controlled by the same factors, the authors of [69] connect the lower yield point with the relation $(E/p)^{1/2}$, representing the longitudinal velocity of the wave (E is the elastic modulus, p is the density of the metal):

$$\sigma_y = \text{const}\left(\frac{E}{p}\right)^{1/2}\frac{T_{mp}}{T}.$$

Thus, from the standpoint of the dislocation theory, the temperature dependence of the yield point and the related transition of metals to the brittle state are determined by the following factors:

a) the barrier effect dependent on the extent of the grain boundaries and impurity concentration, and also on the extent and angular disorientation of the subgrain boundaries [58];

b) the resistance of the crystal lattice to the movement of dislocations;

c) Cottrell atmospheres retarding the movement of dislocations.

The assumption has also been made that dislocation "forests" also serve as barriers to the passage of dislocations [70].

To make an overall assessment of contribution of dislocation concepts to the solution of the problem of cold brittleness, the progress achieved must be regarded as quite modest, particularly with regard to the indication of practical methods of combating cold brittleness.

## 2. Alloying Metals

As a rule, the alloying of refractory metals with substitutional elements leads to an increase in the ductile–brittle transition temperature of these metals, which is evidently due to the additional distortions produced in the crystal lattice by alloying. For example, when commercial tantalum is alloyed with niobium, the transition temperature is increased from 300 to 800°C. The substantial reduction in the transition temperature observed when refractory metals are alloyed with small additions of thermodynamically active elements (zirconium, titanium, lanthanum, cerium, yttrium) is due to their refining action, and not to their alloying action [8, 71, 72].

The influence of substitutional elements on the ductile–brittle transition temperature of vanadium and niobium is shown in Fig. 86 [10]. A considerable reduction in the temperature of the commencement of brittleness is found in molybdenum when alloyed with small amounts of zirconium, titanium, lanthanum, and cerium [10, 71]; with increase in the deoxidizer content as solid solution in the molybdenum (as also in other refractory metals), the critical cold-brittleness temperature increases.

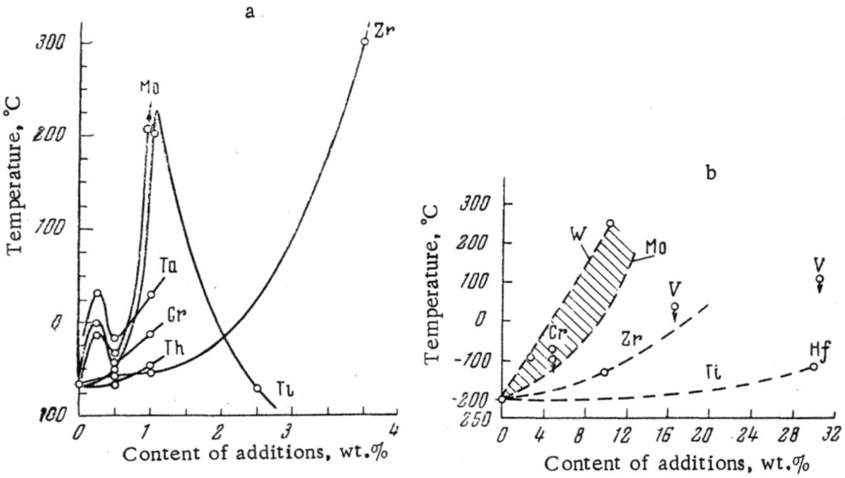


Fig. 86. Brittleness threshold (temperature at which the reduction in area is less than 10%) of vanadium (a) and niobium (b) as a function of various additions.

The ductile–brittle transition temperature for alloys of the tantalum–tungsten system was studied in [74].

The effect of a reduction in the cold-brittleness temperature during deoxidation by thermodynamically active additions is also observed in tungsten, but here the data are rather more scanty. The question of the deoxidation of chromium by cerium, yttrium, and other rare-earth metals is examined in [53, 75, 103].

The influence of substitutional alloying elements on the ductile–brittle transition of chromium in bending tests was investigated by the authors of [77].

A review of investigations dealing with the interaction of alloying elements with the base metal is given in [75, 78]. The considerable increase in the ductile–brittle transition temperature on alloying refractory metals of commercial purity is a serious obstacle to the production of new alloys based on refractory metals. Such a situation will remain until the refractory metals, representing the base of the alloys (chromium, molybdenum, tungsten), are obtained in the state of the necessary freedom from interstitial impurities for utilizing the advantages of strengthening by alloying. Such a procedure was adopted for titanium alloys; at first titanium was brittle on account of interstitial impurities and became still more brittle when alloyed; subsequently, after development of the method of purification from oxygen and especially from hydrogen, titanium acquired the properties of high ductility inherent in all pure metals, and at the present time, constructional and heat-resistant alloys of titanium are highly alloyed compositions. Niobium and its alloys are currently in a similar situation. Niobium which has been purified by electron-beam vacuum remelting permits of a high degree of alloying while retaining the necessary minimum commercial ductility. By alloying, the tensile strength of niobium may be increased by 500% [27]. With regard to the alloying of group VIa metals, rhenium is at present an exception (evidently also ruthenium and osmium), since it reduces the transition temperature of tungsten, molybdenum, and chromium, and considerably increases their deformability [53, 75]. The influence of rhenium on the transition temperature of tungsten and molybdenum is shown in Fig. 87 [78]. With increase in the rhenium contents, the critical ductile–brittle transition temperature of molybdenum is reduced (from room temperature) to below the temperature of liquid nitrogen. Similar results were obtained in [79] and in the work of the Refractory and Rare Metal Alloys Laboratory of the A. A. Baikov

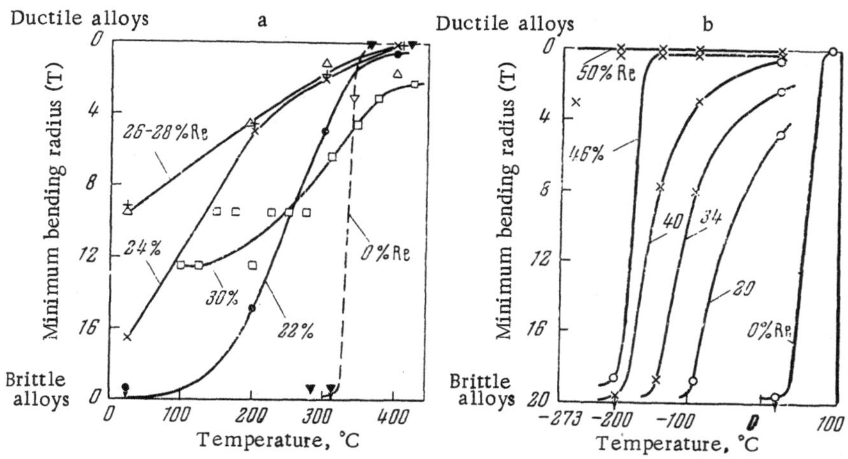


Fig. 87. Ductile–brittle transition temperature of tungsten (a) and molybdenum (b) as a function of the rhenium content.

Institute of Metallurgy [80]. In the case of tungsten, the ductile–brittle transition temperature is reduced to room temperature. The ductile–brittle transition of molybdenum alloys occurs in a narrow temperature range; in tungsten–rhenium alloys, the transition zone occupies a wide temperature range.

The reduction in the ductile–brittle transition temperature and the increase in ductility on alloying tungsten and molybdenum with rhenium is due firstly to the redistribution of oxygen and the formation of complex oxides which, owing to high surface tension, are arranged in the form of globules along the boundaries of the grains and in their interior [78], and secondly to the occurrence of an additional deformation mechanism, i.e., twinning [53, 72-84]. The authors of [85] attribute the influence of twinning on ductility to a decrease in stress concentrations on the grain boundaries. At the same time, a number of investigators have pointed out that twinning may be the source of crack inception, especially in iron and steel [1]. It is quite important that the function of twinning in the cold brittleness of bcc metals should be clarified.

Alloys of group VIa metals with rhenium are prone to strain aging [86], usually the result of interaction between moving dislocations and impurities forming interstitial solid solutions. The increase in the strain-aging effect on alloying group VIa metals with rhenium led the authors of [85] to assume the possibility of increasing the solubility of interstitial impurities in the range 400-700°C. This decreases the concentration of impurities on the grain boundaries and is evidently one of the causes of the increase in ductility of alloys of chromium, molybdenum, and tungsten with rhenium.

The mechanism of the influence of rhenium on the ductility of group VIa metals, however, has not yet been finally clarified. According to data obtained in our laboratory, it has been established that, in general, the hexagonal group VII and VIII metals (rhenium, ruthenium, osmium) are capable of increasing the ductility of group VIa metals [8]. This is also confirmed by work of other investigators, in particular Jaffee et al. [85].

It is pointed out in [87] that the addition of osmium increases the ductility of tungsten and molybdenum, although to a lesser extent than rhenium. Tungsten–osmium alloys become ductile only in the heated condition. The mechanism of the influence of osmium is not yet clear. Bearing in mind that osmium, like rhenium, is a metal of hexagonal structure, it may be assumed that in the present case, twinning is one of the causes of the increase in ductility. According to Jaffee [78] osmium is the only metal of the platinum group which has an influence

TABLE 33. Composition Ranges in Chromium, Molybdenum, and Tungsten Alloys with Metals of the Platinum Group in Which Twinning Was Observed at Room Temperature [78]

| Alloying element | Limits of content of alloying elements in the metals, wt.% | | | | | |
|---|---|---|---|---|---|---|
| | chromium | | molybdenum | | tungsten | |
| | as cast | homogenized | as cast | homogenized | as cast | homogenized |
| Ruthenium | — | 1—5 * | 5—12.5 | 7.5—12.5 | 5—10 | 5—33 * |
| Rhodium | — | 1—5 * | 10 | 0.1—33 * | 5—33 * | 5—33 * |
| Palladium | — | 1—5 * | 0.1—33 * | 0.1—33 * | — | — |
| Osmium | — | 1—5 * | 7.5—12.5 | 7.5—12.5 * | 5—10 | 5—33 * |
| Iridium | 1—5 * | — | 5—33 | 0.1—33 * | 5—10 | 5—33 * |
| Platinum | 5 | — | 5—10 | 5—10 | 5 | 5—33 * |

* Not twinning.

(very small) on the redistribution of oxides in molybdenum. It is noted that ruthenium (also a metal with a hexagonal structure) lowers the ductile–brittle transition temperature of chromium, especially at a content of 20% [91]. In [78] data of Klopp, Holden, and Jaffee are cited, showing that the alloying of chromium, molybdenum, and tungsten by metals of the platinum group in certain concentrations promotes twinning at room temperature (Table 33).

It is pointed out in [78], that, as a rule, metals of the platinum group do not reduce the transition temperature of group VIa metals, despite some improvement in commercial ductility at elevated temperatures.

## 3. Rate of Strain

The rate of strain has a substantial effect on the ductility characteristics and resistance to deformation of refractory metals [6]. An increase in the rate of strain of refractory metals is accompanied by a shift in the brittleness threshold to higher temperatures and by some widening of the temperature range of the transition zone. The dependence of yield point and plastic flow of molybdenum on the rate of strain was studied in [89]. An increase in the rate of strain increases the upper and lower yield points and also the stress level of plastic flow up to the commencement of necking during elongation. At a definite rate, the resistance to deformation may attain values corresponding to the resistance to fracture. The value of the yield point increases correspondingly in the transition region. The influence of the rate of strain on the resistance to deformation of tungsten is shown in [62].

The influence of the rate of strain on the transition temperature may be expressed by the influence of the rate of strain on the yield point under tension [90]:

$$\sigma_T = \left[\varepsilon e^{\frac{Q}{RT}}\right]^n,$$

where $\sigma_T$ is the yield point, $\varepsilon$ is the rate of strain, n is the exponent of the rate of strain determined at constant temperature, and Q is the activation energy.

According to the generally accepted mechanical model, brittle fracture is caused by an increase in the yield stress to the level of fracture stress. As follows from the above-mentioned equation, an increase in the rate of strain increases the yield point and consequently the transition temperature. Figure 88 shows the experimentally determined dependence of the ductile–brittle transition temperature of chromium, molybdenum, and tungsten on the rate of strain.

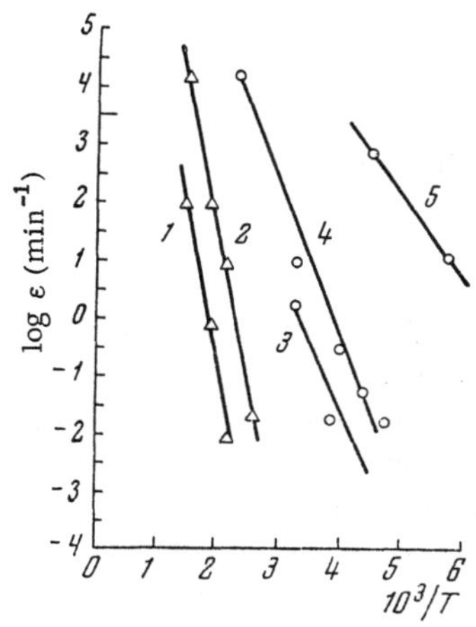


Fig. 88. Experimental dependence of the transition temperature of chromium, molybdenum, and tungsten on rate of strain. 1) Tungsten (recrystallized); 2) tungsten (rolled); 3) chromium (bent); 4) molybdenum (tension); 5) molybdenum (twisted).

An increase in the rate of strain leads to irregularity in the stressed state, an increase in the resistance to deformation, and a drop in the ductility of metals.

L. D. Sokolov [92], drawing inferences from data on the rate function of the resistance to deformation of different metals, remarks that this dependence appears to be more pronounced in metals having a hexagonal or tetragonal crystal structure, and to a lesser degree in metals having fcc or bcc lattices.

There is considerable interest in investigating the behavior of a material under strain, the rate of which is commensurate with the speed of sound. It has been found that in deformation by explosion, additional structural changes occur in the metal, compared with those occurring at the ordinary rates of strain [62]. These additional changes are mechanical twinning, anomalous slip processes, phase conversions, increased formation of vacancies, and dislocated atoms. The effect of the shock wave on the ductile–brittle transition temperature of refractory metals has not been studied in detail. Experiments made in this direction on steel showed a sharp increase in yield point and a decrease in elongation [93]. According to data of [62], an increase in the ductile–brittle transition temperature of a metal is determined by the intensity of the shock waves. The action of the shock substantially increases the hardness of metals; the action of a shock wave (300 kbar) increases the hardness of iron by a factor of more than four, and that of niobium by a factor of two [93]. In the action of shock waves of low intensity, the hardening effect is approximately the same as in strain at ordinary rates.

The inference that the ductile–brittle transition temperature is directly proportional to the rate of strain of the material (in the region of ultrahigh rates) is in our view incorrect. High rate of strain of a metal or alloy is accompanied by a high heating effect, and consequently also by softening processes. The need to examine the temperature-rate factor in the deformation of metals and alloys is emphasized in [94, 95]. It has been pointed out that in explosive deformation, as the power of the charge increases, at some moment a softening effect appears, due to the liberation of heat [96]. The heating in explosive deformation is not only due to dynamic ductile deformation, but to an instantaneous ductile deformation by hydrostatic compression [96]. Evidently, in the region of superhigh rates of strain, there is some critical value of the rate for each metal or alloy, which corresponds to the maximum transition temperature, and above this temperature, the transition begins to fall, owing to the heat effect of deformation and evidently to the special features of the stressed state and the nature of this process.

## 4. Surrounding Medium

The surrounding medium usually exercises a very strong chemical effect on the behavior of metals and alloys during their plastic deformation. This applies particularly to refractory metals which are characterized by high chemical activity. The surrounding medium may cause dissolution of the surface layer and boundary layers of the deformed body, corrosion of these layers, a change in the chemical composition of the substance and, due to adsorption, may facilitate or, on the contrary, retard the deformation process. The process of plastic deformation considerably increases the effect of the surrounding medium on the metal. Corrosion is

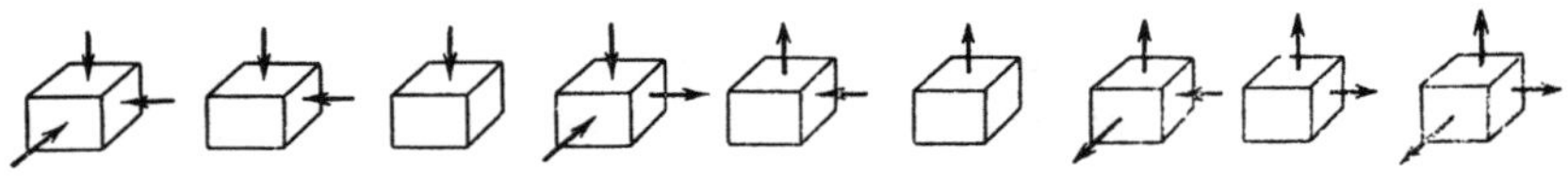

Fig. 89. System of principal stresses in order of their influence in decreasing the ductility of deformed materials.

usually accelerated under the influence of the stressed state. "Loosening" of the metal in the deformation process accelerates the penetration of gases. Corrosion by damaging the grain boundaries, the change in chemical composition under the influence of a corrosive surrounding medium result in such profound structural changes that plastic deformation becomes impossible owing to the intense embrittlement of the material. So-called brittleness or hot-shortness zones are formed [6]. These circumstances render it necessary to reduce the deformation temperature as much as possible and to carry out deformation in an inert medium or in protective envelopes. It must be emphasized that on open storage in the air, even at the ordinary temperature, there is selective absorption of gases from the surrounding atmosphere by refractory metals, for example, hydrogen by titanium, nitrogen by chromium, etc. These processes, however, have not yet been sufficiently examined.

## 5. Stressed State System

The system of the stressed state is also one of the most important factors in determining the behavior of a metal during deformation [2]. The systems of stressed state are shown in Fig. 89 [94]. The ductility of a metal increases with increase in the part played by compressive stresses in the system of principal stresses. The most unfavorable system for plastic deformation is a system of triaxial tension, and the most favorable is a system of nonuniform hydrostatic compression. In deformation according to a system of triaxial tension, even highly ductile materials may experience brittle fracture. On the contrary, transition to the triaxial system of nonuniform hydrostatic compression greatly facilitates the deformation conditions and renders possible the deformation of refractory metals of commercial purity and metallic compounds (pressing through a die), which undergo brittle fracture on deformation by other stressed state systems [6]. The influence of the stressed state system on the plasticity of substances was convincingly shown by N. S. Kurnakov and collaborators in their well-known experiments with marble, red sand, and natural salts [101]. These brittle substances are plastically deformed by the addition of lateral pressure, the deformability increasing with increase in the lateral pressure [102].

The favorable influence of the system of hydrostatic compression on the ductility of brittle materials, including refractory metals, is accounted for by the following physicochemical causes:

a) hydrostatic compression prevents intercrystalline deformation which rapidly results in fracture of polycrystalline material and increases intracrystalline shear;

b) hydrostatic compression compacts the metal and restores the intracrystalline and intercrystalline bonds, ruptured in the process of plastic deformation;

c) the harmful influence of nonmetallic inclusions and liquid phases are rendered wholly or partly harmless (nonmetallic inclusions are dangerous in the case of tensile stresses; in the action of compressive stresses, the nonmetallic inclusions are crushed and exert no influence on plasticity);

d) hydrostatic compression reduces the embrittling effect of tensile stresses produced as the result of nonuniform plastic deformation.

As previously pointed out, the question of the notch sensitivity of refractory metals is closely bound up with the influence of the stressed state system on the plasticity of the material. All-round notching of a specimen results in a system of triaxial tensile stress in the zone of the notch. A system of triaxial tensile stresses in the specimen considerably increases the ductile–brittle transition temperature of molybdenum. In [98] it is shown that the strength of notched specimens of molybdenum rapidly falls at a temperature of about 50°C, while the strength of smooth specimens continues to increase on further reduction in temperature. The premature embrittlement of notched specimens of molybdenum may be attributed to the concentration of tensile stresses in the region of the notch. The elimination of microcracks, scratches, and other possible defects acting as stress raisers from the surface of specimens (by mechanical polishing followed by electropolishing) considerably increases the technological ductility of refractory metals. According to [99], tungsten rods subjected to careful electropolishing may be bent through 180° at room temperature. According to [98], some refractory metals may be arranged in the following descending order of notch sensitivity: tungsten, molybdenum, niobium, tantalum. The criterion for the assessment of the notch sensitivity of these metals was the temperature at which the ratio of the tensile strengths of notched and smooth specimens rapidly decreased. Chromium exhibits exceptionally high sensitivity to surface condition. Thus, filling the surface of an electropolished chromium specimen raised its transition temperature from −60 to +120°C [34].

### 6. Scale Factor

The dimensions of a deformed specimen also exercise a substantial influence on ductility, strength, and nature of the fracture of a material. The smaller is the volume of the specimen, the greater is the amount of plastic deformation before fracture.

It has been established experimentally that bodies of small dimensions have a higher strength than bodies of large size of the same materials. In some cases, the experimental values of the strength of bodies of small dimensions attain and even surpass the theoretical strength. Usually, an appreciable increase in tensile and compressive strength begins to appear for specimens having a diameter of less than 1 mm. The principal causes of this are the increase in the function of surface energy with decrease in a diameter of the specimen, change in the character of stress distribution, and also the increase in the probability of dangerous defects and inclusions, with which the fracture process commences in specimens of large dimensions. The dependence of strength on the linear dimensions of tested bodies is fairly well described by the following equation [6]:

$$P = \frac{\alpha}{l} + P_0$$

where P is the strength and $l$ the linear dimensions of the specimen, $P_0$ is the strength of specimens of large cross-sectional dimensions, and $\alpha$ is a proportionality factor.

When examining the influence of the scale factor on ductility, it is also necessary to take into account the stressed state system, the treatment temperature, surface condition of the specimen, the type of fracture (tough or brittle), etc. The question of the influence of the scale factor calls for further investigation.

## Methods of Combating Cold Brittleness in Refractory Metals

Existing methods of combating cold brittleness in refractory metals may be divided into chemical and physical methods.

The chemical methods include purification from interstitial impurities at various stages in the technological production of the pure metals, deoxidation and modification of the metals in vacuum melting, and creation of the optimum structure by alloying with substitutional elements. A few examples will be given of the considerable increase in ductility of refractory metals as the result of thorough purification from interstitial impurities or neutralization of their harmful effect by interaction with alloying additions, thermodynamically more active than the base metal.

In [8], the possibility was pointed out of considerably increasing the ductility of chromium by refining with yttrium or other rare-earth metals, which combine with the nitrogen in chromium to form nitrides. The point is that the heat of formation of yttrium nitride (72 kcal/mole) is almost three times that of the formation of chromium nitride (25 kcal/mole). Experimental work carried out in the Refractory and Rare Earth Metal Alloys Laboratory of the A. A. Baikov Institute of Metallurgy [76] and in other scientific research institutions have confirmed the efficiency of this method. Recently, in the Institute of the Physics of Metals of the Academy of Sciences of the Ukrainian SSR, V. I. Trefilov et al. produced chromium which after treatment with rare-earth metals showed high ductile properties [104]. Specimens in the form of plates withstood bending through 180°C at room temperature.

Rare-earth metals have also been found to have a beneficial influence on the ductility of molybdenum. Deoxidation of molybdenum by rare-earth metals lowers the brittle–ductile transition temperature by approximately 300°C [105]. Molybdenum, deoxidized by lanthanum boride, after previous forging and rolling at 700°C, could be rolled at room temperature into a strip 0.15 mm thick (total degree of deformation 98%) [71]. Molybdenum single crystals, purified from interstitial impurities by electron-beam zone refining in a vacuum, exhibit phenomenal ductility [7].

Group IVa and Va metals (titanium, zirconium, hafnium, vanadium, niobium, and tantalum) become fairly ductile at low temperatures as the result of electron-beam vacuum remelting, since they dissolve much smaller amounts of interstitial impurities than do group VIa metals. As was found recently, the bulk of the oxygen is removed in the form of lower oxides. For group VIa metals, a promising method is that known as the duplex process: electron-beam melting with an active deoxidizer for purification from interstitial impurities followed by arc melting with modifiers for grain refinement. With the object of combating the contamination of the metal by carbon during melting, a promising method is that of producing the vacuum not by means of oil-vapor pumps but by absorption of the gases in a closed space by means of chemically active getter substances, such as zirconium, boron, various ion exchange resins, etc.

Perhaps the most interesting example of the efficiency of chemical methods of combating cold brittleness in refractory metals is the considerable increase in ductility of tungsten by its high purification from carbon (to 0.001%), as was recently carried out in our laboratory by a thermocyclic treatment in a medium containing hundredths of one percent of oxygen [36]. A tungsten single crystal, produced by electron-beam zone refining in a vacuum from ingots decarburized by the above-mentioned methods, was bent through 180°C in liquid nitrogen. Carbon, which forms refractory carbides with tungsten, is evidently the most harmful impurity and one difficult to remove, whereas oxygen, nitrogen, and hydrogen are almost completely removed by vaporization on melting or by heating the tungsten in a vacuum. The numerous carbide inclusions act as obstacles for the movement of dislocations, and also as nuclei for the inception of microcracks; such inclusions raise the ductile–brittle transition temperature of tungsten by hundreds of degrees [38]. To produce tungsten free from carbon and to preserve this purity on further treatment of the semifinished products and during the service life of finished articles is a combined problem for chemists, metallurgists, technologists, and users.

It will be clear from this example that the fight for the ductility of refractory metals must be conducted from the first stages of conversion of the crude ore and continued throughout the entire chemical, metallurgical, and technological cycle.

Reference has already been made in the foregoing to the considerable increase in ductility of tungsten and molybdenum by alloying with rhenium. To this may be added the increase in ductility of chromium by alloying with hexagonal ruthenium.

The rule of the ductility of high-temperature modifications of polymorphic metals points to a way of producing ductile alloys based on polymorphic metals which are brittle under ordinary conditions [6, 48]. For this purpose, it is necessary to add to the polymorphic metals, such metals as stabilize the ductile high-temperature modification at the ordinary temperature. A well-known example of combating cold brittleness by the chemical method which is of enormous industrial value is the stabilization of the high-temperature $\gamma$-phase of iron by nickel, manganese, chromium, and other additions. Stabilization of the austenite fcc structure must be regarded as one of the cardinal methods of combating the cold brittleness of $\alpha$-iron which, as we know, has a bcc structure, typical of a cold-brittle metal.

The *physical* methods of combating the cold brittleness of refractory metals include deformation by a system of nonuniform hydrostatic compression, including back pressure (extrusion); production of the optimum deformation texture, optimum dislocation structure, and subgrain structure by combined heat and mechanical treatment and work hardening; application of vibration, ultrasonics, purification from dissolved gases, plastic deformation in a vacuum, etc.

Of all these physical methods of combating cold brittleness, the one most widely used until recently was work hardening of metals produced by deformation below their recrystallization temperature. Currently, the treatment of refractory metals by a system of nonuniform hydrostatic pressure, including the use of back pressure (extrusion), is becoming more widely used. Methods of increasing the ductility of refractory metals by combined heat and mechanical treatment are as yet only in the initial stages of development. It is currently already possible to make recommendations regarding the purity of the metals, grain size, optimum form of intergranular boundaries, and necessary deformation structure, but it is not yet possible to make recommendations regarding the optimum fine structure and the methods of achieving it.

The examples of increasing the ductility of metals which have been considered are evidence of the fruitfulness of the physicochemical approach to the solution of the cold-brittleness problem.

Some success is being achieved in solving the problem of cold brittleness of refractory metals, but before the final solution of this problem is reached much remains to be done not only on the scientific plane but also on the practical plane. The high sensitivity of refractory metals and alloys to interstitial impurities in the production and treatment of refractory metals and alloys calls for exclusive attention to the cultivation and technical level of production. It is necessary to emphasize the need for preventing cold brittleness in refractory metals throughout the entire production cycle, and the importance of utilizing the means of the chemical technology of pure substances, vacuum metallurgy, and heat treatment for producing and preserving a high degree of purity of the metal. It is important not only to produce the pure refractory metal having a low ductile—brittle transition temperature, but to preserve the purity attained also in the subsequent treatment of the semifinished products and in the service life of the finished articles.

## LITERATURE CITED

1. N. N. Davidenkov and G. N. Chuchman, in: Research on Refractory Metals [in Russian], Vol. 2, Izd. Akad. Nauk SSSR, p. 9 (1957).
2. N. N. Davidenkov, in: Research on Refractory Metals [in Russian], Vol. 4, Izd. Akad. Nauk SSSR, p. 13 (1959).
3. V. S. Ivanova and Yu. I. Ragozin, Izv. Akad. Nauk SSSR, Neorganicheskie Materialy, 1(10):1684 (1965).
4. I. A. Oding, Theory of Dislocations in Metals and Its Application [in Russian], Izd. Akad. Nauk SSSR (1959).
5. V. I. Trefilov, in: Physical Nature of the Brittle Fracture of Metals [in Russian], Kiev, Izd. "Naukova Dumka," p. 22 (1965).
6. E. M. Savitskii, Influence of Temperature on the Mechanical Properties of Metals and Alloys [in Russian], Izd. Akad. Nauk SSSR (1957).
7. E. M. Savitskii, Izv. Akad. Nauk SSSR, Metally, No. 5, p. 3 (1965).
8. E. M. Savitskii, Izv. Akad. Nauk SSSR, OTN, Metallurgiya i Toplivo, No. 5, p. 52 (1960).
9. E. M. Savitskii and G. P. Tsarev, Izv. Akad. Nauk SSSR, Metally, No. 2, p. 84 (1965).
10. G. T. Hahn, A. Gilbert, and R. I. Jaffee, Proc. Refractory Metals Sympos., Chicago, p. 23 (April 1962).
11. M. V. Lomonosov, Complete Works [in Russian], Vol. 1, Izd. Akad. Nauk SSSR (1950).
12. N. V. Ageev, Nature of the Chemical Bond in Alloys [in Russian], Izd. Akad. Nauk SSSR (1947).
13. M. I. Korsunskii and Ya. E. Genkin, Izv. Akad. Nauk SSSR, Seriya Fizicheskaya, 28(5):832 (1964).
14. Collection: Ultrasonic Metals [in Russian], Izd. "Metallurgiya" (1966).
15. J. M. Ziman, Theory of Solids, Cambridge University Press, New York (1965).
16. Collection: Electronic Structure of Transition Metals and the Chemistry of Their Alloys [in Russian], Izd. "Metallurgiya" (1966).
17. Express Information, Metallovedenie i Termoobrabotka [in Russian], VINITI, No. 4, p. 1 (Abstr. 12) (1965).
18. J. G. Bechtold and E. T. Bessel, in: Molybdenum [Russian translation], IL, p. 159 (1962).
19. D. L. Wood and J. H. Westbrook, Reprint from Proceedings ASTM, 63:759 (1963).
20. E. M. Shvetsova, Zavod. Lab., 4(6):694 (1935).
21. G. I. Pogodin-Alekseev, Properties of Metals under Impact Stress [in Russian], Metallurgizdat (1953).
22. R. I. Jaffee, Problemy Sovremennoi Metallurgii, No. 6, p. 115 (1960).
23. M. Semchyshen and R. Q. Barr, in: Refractory Metals and Alloys [in Russian], Izd. "Metallurgiya," p. 149 (1966).
24. E. M. Savitskii and V. A. Trapeznikov, in: Research on Heat-Resistant Alloys, Vol. II, Izd. Akad. Nauk SSSR, p. 141 (1957).
25. A. P. Gulyaev and V. N. Nikitin, Zavod. Lab., 30(7):885 (1964).
26. Express Information, Metallovedenie i Termoobrabotka, VINITI, No. 4 (Abstr. 13) (1965).
27. E. M. Savitskii, G. S. Burkhanov, and Ch. V. Kopetskii, Izv. Akad. Nauk SSSR, Metallurgiya i Gornoe Delo, No. 6, p. 12 (1963).
28. I. A. Gindin and Ya. D. Starodybov, Fiz. Metallov i Metallovedenie, 15(5):736 (1963).
29. E. M. Savitskii and G. E. Chuprikov, Izv. Akad. Nauk SSSR, OTN, Metallurgiya i Toplivo, No. 4, p. 137 (1962).
30. A. L. Eustice and O. N. Carlson, Trans. AIME, 221:238 (1961).
31. J. H. Bechtold, Acta Metallurgica, 3:249 (1955).
32. A. G. Imgram, F. C. Holden, H. R. Ogden, and R. I. Jaffee, Trans. AIME, 221:517 (1961).

33. V. N. Platt, in: Molybdenum [Russian translation], IL, p. 246 (1949).
34. H. L. Vane, F. Henderson, S. T. Johnstown, and N. J. Luat, Problemy Sovremennoi Metallurgii, No. 6, p. 115 (1958).
35. E. M. Savitskii and G. L. Tsarev, Izv. Akad. Nauk Belorussian SSR, Seriya Fiz.-Tekhn. Nauk, No. 2, p. 84 (1965).
36. E. M. Savitskii and G. L. Tsarev, in: Growth and Imperfections of Metal Crystals [in Russian], Izd. "Naukova Dumka," p. 290 (1966).
37. G. T. Hahn, A. Gilbert, and R. I. Jaffee, Refractory Metals and Alloys, II (Metallurgical Society Conferences, Vol. 17), New York, pp. 23–63 (1963).
38. N. K. Popov, E. M. Savitskii, and G. L. Tsarev, Dokl. Akad. Nauk SSSR, 162(1):64 (1965).
39. G. V. Samsonov and K. I. Portnoi, Alloys Based on Refractory Compounds [in Russian], Oborongiz (1961).
40. O. Kubaschewski and E. L. Evans, Metallurgical Thermochemistry, Pergamon, New York, 3rd ed. (1958).
41. E. B. Bas, L. Preuss, and W. Epprecht, Z. Metallkunde, 48(12):516 (1957).
42. R. I. Jaffee, J. Metals, 410 (May 1960).
43. H. Braun, M. Samchyshen, and R. Q. Barr, Metallwerk Plansee A. G. Reutte/Tyrol, 35 (1965).
44. A. Gilbert, W. R. Warke, and B. A. Wilcox, J. Less-Common Metals, No. 7, p. 222 (1964).
45. E. M. Savitskii, G. S. Burkhanov, Ch. V. Kopetskii, and G. E. Chuprikov, in: "Growth of Crystals" [in Russian], Izd. "Nauka," p. 308 (1965).
46. E. Rudy, F. Benesovsky, and L. Toth, Z. Metallkunde, 54(6):345 (1963).
47. G. S. Bazinskii and É. Slisvik, Problemy Sovremennoi Metallurgiya, No. 4, p. 109 (1957).
48. E. M. Savitskii, Metall. i. Term. Obrabotka, No. 7, p. 3 (1965).
49. G. S. Speisel and J. Wulf, in: Molybdenum [Russian translation], IL (1962).
50. R. C. Koo, Trans. Met. Soc. AIME, 227(1):280 (1963).
51. D. J. Maykuth and R. I. Jaffee, Problemy Sovremennoi Metallurgii, No. 1, p. 83 (1958).
52. D. A. Robins, J. Less-Common Metals, 1:396 (1959).
53. E. M. Savitskii, Rare Metals and Alloys [in Russian], Moscow, Izd. Doma Tekhniki (1959).
54. L. L. Seigle and M. D. Dickinson, Proc. Refractory Metals Sympos., Chicago (April 1962).
55. N. J. Petch, Atomic Mechanism of Fracture [Russian translation], Metallurgizdat (1965).
56. R. C. Koo, J. Less-Common Metals, 3(5) (1961).
57. A. J. Opinsky et al., Proc. Refractory Metals Sympos., Chicago (April 1962).
58. A. H. Cottrell, Dislocations and Plastic Flow in Crystals, Oxford University Press, New Jersey (1953).
59. A. H. Cottrell and B. A. Bilby, Proc. Roy. Soc. A, 62:49 (1949).
60. E. T. Wessel, J. Metals, 9:930 (1957).
61. N. Mott, The Rutherford Memorial Lecture, Atomic Physics and the Strength of Metals (1962).
62. G. A. Verbraak, Z. Metallkunde, 55(12):723 (1964).
63. G. Schoeck, Acta Metallurgica, 9:382 (1961).
64. A. H. Cottrell, Trans. ASME, 46A:13 (1964).
65. J. W. Pugh, J. Metals, 10(5):335 (1958).
66. J. W. Pugh, J. Metals, 9(10):1243 (1957).
67. U. R. Hibbard, in: Mechanism of the Strengthening of Solids [in Russian], Izd. Metallurgiya, p. 9 (1965).
68. W. G. Johnston and P. S. Gilman, J. Appl. Phys., 30:129 (1959).
69. L. L. Seigle and Dickinson, Refractory Metals and Alloys, Vol. 17, p. 65, Chicago (April 1962).

70. A. Seeger, Phil. Mag., No. 46, p. 1194 (1955).
71. E. M. Savitskii, V. V. Baron, Yu. V. Efimov, and V. A. Frolov, in: Questions of the Theory and Application of Rare-Earth Metals [in Russian], Izd. "Nauka" (1964).
72. E. M. Savitskii, Metallovedenie i Termoobrabotka Metallov, No. 9, p. 19 (1961).
73. R. T. Begley and J. H. Bechtold, J. Less-Common Metals, 3(1):1 (1961).
74. F. Schmidt, W. Klopp, D. Maykuth, H. Ogden, and R. I. Jaffee, WADD TR 61-106 (1961).
75. E. M. Savitskii, Izv. Akad. Nauk SSSR, OTN, Metallurgiya i Toplivo, No. 5, p. 52 (1960).
76. R. Smith and A. Hendrickson, in: Superpure Metals [Russian translation], Izd. "Metallurgiya," p. 78 (1966).
77. E. P. Abrahamson and N. J. Grant, Trans. ASME, 50:705 (1958).
78. R. I. Jaffee, D. J. Maykuth, and R. U. Douglass, Rhenium and Refractory Metals of the Platinum Group [Russian translation], IL (1963).
79. E. M. Savitskii and Tao Chu-Chun, Fizika Metallov i Metallovedenie, 14(1):148 (1962).
80. E. M. Savitskii, M. A. Tylkina, and K. B. Povarova, Rhenium Alloys [in Russian], Izd. Akad. Nauk SSSR (1965).
81. P. A. Flinn, in: Mechanism of the Strengthening of Solids [in Russian], Izd. "Metallurgiya," p. 25 (1965).
82. E. M. Savitskii, M. A. Tylkina, and K. B. Povarova, Dokl. Akad. Nauk SSSR, 119(2):274 (1958).
83. M. A. Tylkina, K. B. Povarova, and E. M. Savitskii, Zh. Neorgan. Khimii, 5(11):2458 (1960).
84. D. J. Maykuth, F. C. Holden, and K. J. Jaffee, Rhenium, Amsterdam, p. 114 (1962).
85. T. G. Booth, R. I. Jaffee, and E. L. Salkovitz, Metallwerk Plansee, A. G. Reutte/Tyrol (1965).
86. A. Gilbert, C. N. Reid, and G. T. Gahn, AIME Meeting, New York (February 1964).
87. R. Kieffer, K. Seldachek, and H. Braun, Problemy Sovremennoi Metallurgii, No. 4, p. 164 (1959).
88. C. S. Wukusick, Evaluation of Chromium–Ruthenium Alloys, General Electric (June 25, 1965).
89. R. Orava, Trans. Met. Soc. AIME, 230(7):1614 (1964).
90. J. H. Bechtold, J. Metals, 5:1969 (1953).
91. J. H. Bechtold, E. T. Wessel, and L. L. France, Refractory Metals and Alloys, Interscience, New York, p. 25 (1961).
92. L. D. Sokolov, Resistance of Metals to Plastic Deformation [in Russian], Metallurgizdat (1963).
93. N. N. Davidenkov, The Impact Problem in Physical Metallurgy [in Russian], Izd. Akad. Nauk SSSR (1938).
94. S. I. Gubkin, Theory of the Working of Metals by Pressure [in Russian], Metallurgizdat (1947).
95. S. I. Gubkin and E. M. Savitskii, Izv. Sektora Fiz.-Khim. Analiza, 14:235 (1941).
96. E. S. Atroshchenko, P. O. Pashkov, and I. M. Rodinskaya, Fizika Metall. i Metalloved., 19(4):619 (1965).
97. P. W. Bridgman, Studies in Large Plastic Flow and Fracture, McGraw-Hill, New York (1952).
98. E. J. Imgram, F. S. Holden, H. R. Ogden, and R. I. Jaffee, Problemy Sovremennoi Metallurgii, No. 6, p. 80 (1961).
98a. E. M. Savitskii, Dokl. Akad. Nauk SSSR, 89(1):85 (1963).
99. J. Stephens, Some Problems of Refractory Metals and Alloys [Russian translation], IL, p. 89 (1963).
100. G. S. Zhdanov, Solid State Physics [in Russian], Izd. MGU (1962).
101. N. S. Kurnakov, Collected Works [in Russian], Vols. I and II, Izd. Akad. Nauk SSSR (1960–61).

102. S. I. Gubkin, Plastic Deformation [in Russian], Vol. 2, Metallurgizdat (1960).
103. E. M. Savitskii, V. F. Terekhova, I. V. Burov, I. A. Markova, and O. P. Naumkin, Alloys of the Rare-Earth Metals [in Russian], Izd. Akad. Nauk SSSR (1962).
104. Yu. V. Mil'man, A. P. Rachek, and V. I. Trefilov, in: Scientific Transactions of the Metal Institute of the Academy of Sciences, Ukrainian SSR [in Russian], Kiev, No. 20, p. 3 (1960).
105. E. M. Savitskii, V. V. Baron, and Tao Chu-Chun, Izv. Akad. Nauk SSSR, OTN, Metallurgiya i Toplivo, No. 1, p. 151 (1962).

CHAPTER V

# SINGLE CRYSTALS OF REFRACTORY METALS AND ALLOYS

The significance of single crystals is growing in science and the new technology [1], which make ever increasingly severe demands on the purity and structural perfection of crystals, since many of the physical and chemical properties of solid substances are determined to a considerable extent by these factors.

The outstanding scientific achievements in solid state physics, progress in radioelectronics and other fields of the new technology are largely due to development work on the growing of single crystals and the study of their structure and properties. This work has been done mainly on single crystals of semiconductor substances and also on metals of low and moderate refractoriness. Only in recent years has any great attention been paid to single crystals of refractory metals.

As is well known, millionths or even milliardths of some impurities may exert a substantial influence on the electrical and other properties of semiconductors and metals. It has been found that single crystals of a high degree of purity possess a number of unique physical properties which are absent in polycrystalline metals of commercial purity. For example, filamentary single crystals of iron ("whiskers"), having a perfect crystal structure, have yield points and tensile strength which are approximately 2-3 orders higher than those of polycrystalline iron of commercial purity. Single crystals of molybdenum, tungsten, iron, and other cold-short metals, in a state of high purity, lose the brittleness usually inherent in the same metals of commercial purity and assume phenomenal ductility down to temperatures of liquid helium. Even some time ago, it was found that molybdenum and tungsten in a state of high purity became superconductors and had a critical temperature of transition to the superconducting state of 0.9 and 0.05°K, respectively. Single crystals of refractory metals have been found to be stable when used in the plasma of cerium, potassium, and other alkali metals. Considerable anisotropy has been found in the thermionic emission of different faces of tungsten and molybdenum single crystals. The discovery of special chemical and physical properties of metal single crystals continues.

The study of single crystals of refractory metals of a high degree of purity is of considerable interest, since it offers the possibility of assessing more correctly the various physicomechanical properties bound up with the nature of a metal when the effect of impurities and grain boundaries is excluded.

The high technological ductility of single crystals of refractory metals makes it possible to obtain from them various semifinished products which have found extensive application in electronic engineering instruments and devices, for which it is imperative to use metals of high degree of purity capable of giving service at high temperatures while being ductile at low temperatures and resistant to thermal fluctuations and vibrations. The high thermionic emission and high stability in cesium plasma and other specific media render single crystals of refractory metals and alloys extremely promising materials for cathodes of thermionic emission converters.

This chapter examines the production methods, structural properties, and applications of single crystals of refractory metals and alloys. As will be shown in the following, the purification and growth of single crystals of refractory metals and alloys by the method of vacuum zone refining, using an electron beam, is one of the effective means of increasing their ductility.

## Methods of Growing Metal Single Crystals

We shall review briefly the existing methods of growing metal single crystals and of estimating their purity. The commonest methods of growing metal single crystals are pulling from the melt [2], recrystallization [3, 4], and zone refining [5-8]. Methods are being developed for growing single crystals by the crystallization of fine powders, crystallization in a melting vessel, from the gaseous phase by the method of thermal dissociation of halogen compounds [9, 10], by electrolysis, distillation, the plasma method [11], and other ways. Reviews of methods of producing single crystals of metals and other substances having low and moderate melting points are given in [3, 4, 12].

The method of pulling single crystals with a seed from the melt (Czochralski method) has been extensively used in recent years for growing large single crystals of germanium, silicon (of a diameter of up to 80 mm), and semiconductor compounds of various compositions for semiconductor triodes, transistors, and capacitors [13].

The essence of the method is to immerse a single-crystal seed to a shallow depth in the melt and then slowly pull it out. The temperature of the seed is somewhat lower than the temperature of the metal and during the pulling operation, the melt gradually crystallizes on the plane surface of the seed. The crystallographic orientation of the crystallizing melt corresponds to the crystallographic orientation of the seed. In the absence of a single-crystal seed, a polycrystalline seed is used, which is immersed in the melt and pulled, the temperature of the melt being raised until a neck is formed. The temperature of the melt is then lowered again to the previous value. Such a method ensures the production of a single crystal. For the production of single-crystal seeds, Czochralski proposed a method of drawing the melt into a capillary in which the melt crystallized in single-crystal form [2].

Crystal pulling from the melt is usually carried out in a vacuum or in an inert atmosphere.

The method proposed by A. V. Stepanov for producing sheets, rods, tubes, and other products directly from the melt [14-16] is of practical interest. A former, in the shape of a horizontal plate with a slot, the periphery of which corresponds to the cross section of the article to be made, is placed on the surface of the melt. Using a suitable seed, a part of definite cross section is pulled directly from the melt through the slot [17].

Using this method, ribbons and tubes of aluminum alloys and inorganic salts, as well as germanium, have been produced [18].

The application of the Czochralski method to refractory metals is as yet only in its initial stages of development [19]. By using the method of pulling from the melt, theoretically a single crystal of any diameter may be produced. The chief difficulty is the lack of a chemically stable material for making crucibles in which chemically active refractory metals may be kept for some time without contamination. Melting is carried out in the metal itself. The use as source of a heat of a directional electron beam makes it possible to melt only a definite portion of the metal, so that the melt does not come into contact with the material of the crucible. The directional electron beam is formed by means of an electromagnetic system, permitting the production of a tubular convergent electron beam, forming an annular heating zone. Annular

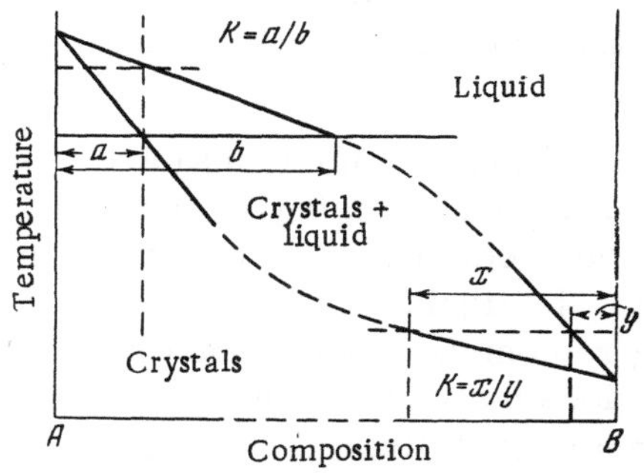


Fig. 90. Portions of the phase diagram of a binary system (schematic).

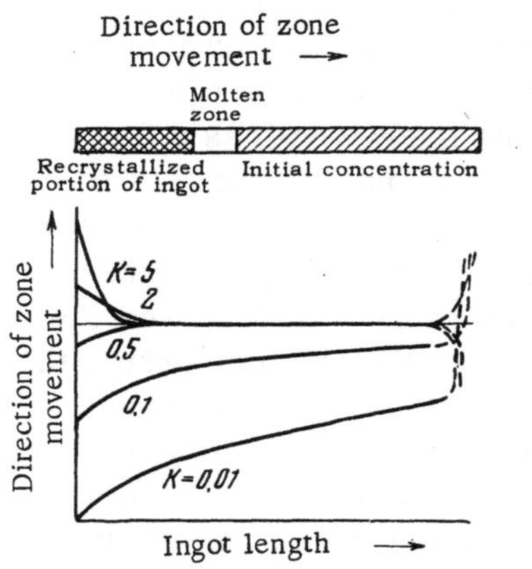


Fig. 91. Distribution of impurities with different values of the distribution coefficient K along an ingot undergoing zone refining.

heating of the molten metal enables crystals to be pulled by means of a seed from the center of the liquid bath [19]. Further development of this method must also be directed to the production of single-crystal tubes, plates, and other sections from refractory metals and alloys, a development which would be of undoubted interest for the new technology.

The method of recrystallization anneal after a slight ("critical") degree of plastic deformation has been found very suitable for growing single crystals of low and moderate refractoriness, and especially for metals having a polymorphic transformation in the solid state, for example iron [20], uranium [21], and the rare-earth metals [22]. Single crystals of the refractory rare-earth metals, scandium with a melting point of 1530°C and yttrium with a melting point of 1500°C, were first produced by this method [22]. In this method, use is made of the extremely rapid growth of the metal grains in the "critical" deformation zone by the mechanism of collective recrystallization [22] and polygonization stage only.

The method of zone melting or zone recrystallization is one of the most effective and extensively used methods for the separation, extreme purification, and the growing of single crystals of various materials. This method is often called zone refining. It consists in melting a zone (short length) of a solid ingot and the slow movement of the zone along the ingot in one direction. Not merely one but several zones may be advanced at the same time. Purification occurs as the result of the different solubilities of the impurities in the liquid phase (higher solubility) and solid phase (lower solubility); in addition, in zone refining in a vacuum, the low-melting impurities and those having a high vapor tension evaporate.

An important characteristic of the process of separating substances by crystallization is the distribution coefficient K, representing the ratio of the impurity concentrations in the solid and liquid phases: $K = C_{sol}/C_{liq}$.

The distribution coefficient may be determined from the phase diagram. Since the refining is applied to almost pure metals, it is of interest to consider the limit portions of the phase diagram in which the liquidus and solidus curves may be represented by straight lines. Figure 90 [24] shows a typical portion of a phase diagram, in which the liquidus and solidus curves may be represented by straight lines. If the impurities increase the melting point, $K > 1$, if they reduce it, $K < 1$. The values of K usually vary in the limits from 0.001 to 10.

Unfortunately, few phase diagrams have been published in which the liquidus and solidus lines are determined with sufficient accuracy in the regions of very low concentrations. For a definite value of K in the region of low concentrations, Pfann proposes to extrapolate the curve of K versus impurity concentration to zero concentration; an experimental method of

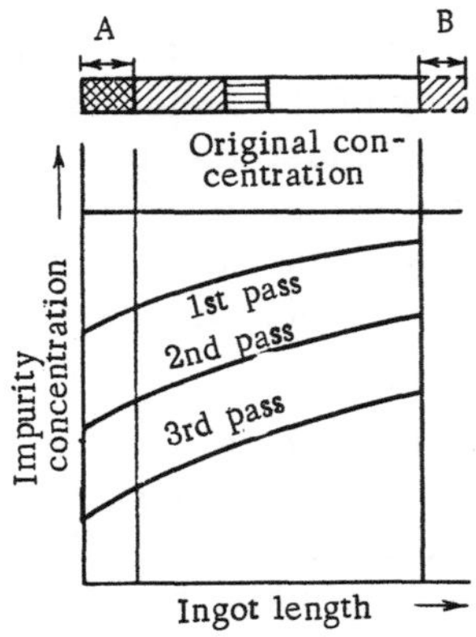


Fig. 92. Impurity distribution along an ingot (schematic) as a function of the number of passes in zone refining. A) Initial portion, equal to the zone length; B) final portion to be removed.

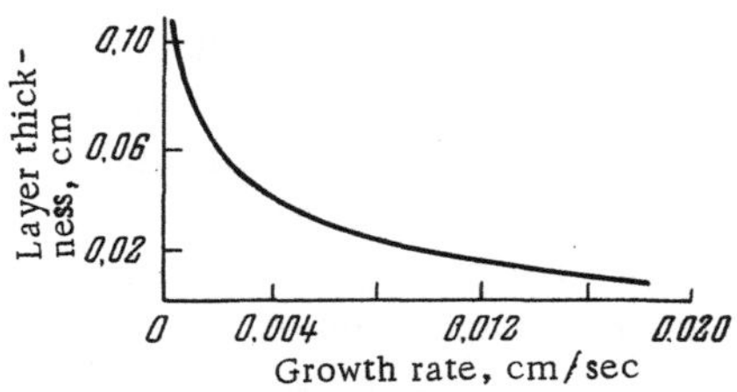


Fig. 93. Thickness of enriched layer as a function of growth rate.

determining K has also been proposed. Thermodynamic calculations are also used for making a rough estimate of K [15].

Impurities with K < 1 ("formed" impurities) are concentrated in the melt and in zone refining are carried by the zone to that part of the specimen which crystallizes last. During the movement, the liquid zone on the fusion front is continuously being replenished by the starting material. Its composition does not vary so sharply, therefore, as it would do in normal crystallization. When the zone reaches the end of the ingot, it crystallizes according to the law of normal crystallization, resulting in an abrupt increase in the impurity content of the ingot end. Impurities with K > 1 ("reverse" impurities) are moved in a direction opposite to the movement of the zone, increasing the content of the solid phase. Thus, the final portion of the ingot is purified. Purification from reverse impurities proceeds with less effect. The impurity distribution along the ingot length as a function of the distribution coefficient may be expressed graphically as shown in Fig. 91, on the assumption that all other factors influencing the effective purification have optimum values [5]. Additional passes lower the initial portion of the curve, raise the end and reduce the length of the horizontal portion. Figure 92 shows schematically the influence of the number of passes on impurity concentration. Distribution curves may be determined experimentally or calculated theoretically by the method proposed by Pfann [5]. With increase in the number of passes, the efficiency of zone refining is reduced.

## Specific Features of the Zone-Refining Process

The efficiency of zone refining also depends on the width and the rate of movement of the zone. At a very low rate of movement of the zone, equilibrium may be attained in the ideal case, owing to diffusion processes in the solid–liquid state. For purification to occur, the crystallization rate should be much higher than the diffusion rate in the solid state; at the same time, the rate of movement of the molten zone should be low enough for the impurity atoms separating on the crystallization front to be carried off by diffusion, convection, and mechanical mixing. The dependence of the thickness of the impurity-enriched layer on the crystallization rate is shown in Fig. 93 [25, 26].

The increased impurity concentration near the crystallization front results in concentration supercooling (which is the cause of the high growth rate of the crystals) and the trapping by the crystal of liquid having an increased impurity content. The result is the formation of what are called impurity entrapment bands [27]. Zone refining may be intensified by forced mixing of the melt in the zone. Good results are given by electromagnetic mixing. In vertical zone refining of refractory metals, mixing is effected by rotation of the ingot. The zone mixing speeds generally used are 0.1-10 mm/min.

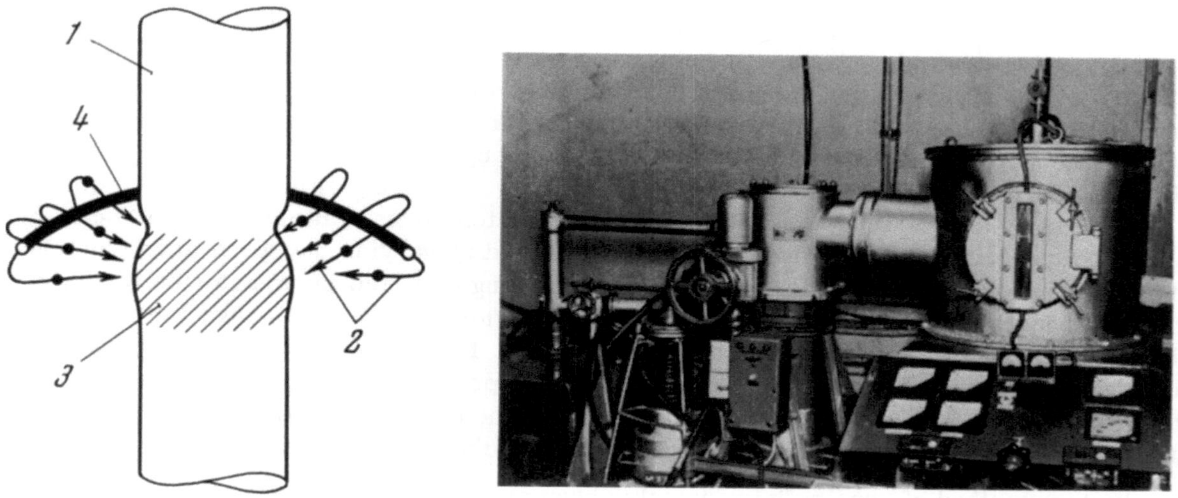


Fig. 94. Diagram of electron-beam zone refining and general view of apparatus for growing single crystals of refractory metals. 1) Cylindrical specimen; 2) electrons, the movement of which is accelerated by an electrostatic field between cathode and specimen; 3) molten zone; 4) cathode.

The zone width also has a considerable influence on the efficiency of purification. A narrow confined zone with sharp boundaries between the liquid and solid phases ensures the best refining. The zone width is determined by focusing the heat energy and by the physical properties of the purified material on which the concentration of heat energy depends (melting point, specific heat, thermal conductivity, etc.). A narrow molten zone can be produced more readily in a metal having a high melting point and low thermal conductivity. In this respect, the method of zone refining is convenient for the purification and single-crystal production of refractory metals.

To avoid contamination, the material to be purified is isolated from the atmosphere and is placed in a quartz tube or steel sheath. The process is usually carried out in a vacuum or protective atmosphere. In horizontal zone refining, the boats are made of a material which does not react with the molten zone of material, for example, graphite for germanium and silicon. Depending on the physical and chemical properties of the material to be purified, plastics, metals, in particular cooled copper tubes, quartz, alundum, etc., are used for the production of the boats.

Highly reactive substances (including refractory metals) are subjected to noncrucible melting with vertical arrangement of the ingot; the molten zone is supported by surface tension forces and may move both upward and downward along the ingot.

Sometimes, what is known as magnetic suspension of the zone is used. Direct current is passed along the horizontally fixed specimen. At the position of the molten zone a magnetic field of such strength is applied perpendicularly to the specimen that the lifting force produced exactly balances the force of gravity in the molten zone [28].

Heating for the production of molten zones may be carried out by various methods – electrical resistance heaters, high-frequency currents, liquid or gaseous fuel flames, electric discharge (arc, bombardment by electrons), and radiation (focusing the radiation of an incandescent lamp or solar energy). Figure 94 shows diagrammatically zone heating by electron bombardment, and a general view of an apparatus for the vacuum electron-beam zone refining of refractory metals developed at the Refractory and Rare Metal Alloys Laboratory of the A. A. Baikov Institute of Metallurgy [7, 77].

The fusion of a narrow zone of metal is carried out by electron bombardment of the specimen, acting as anode. A stand is set up on a water-cooled plate for fixing the specimen, which is secured by molybdenum clamps with tantalum springs, allowing the specimen to elongate freely during heating. An annular cathode of tungsten wire is fixed in steel holders, to which the current for heating the cathode is supplied by means of flexible bars. The bracket with the holders is moved vertically along a lead screw. Focusing of the electrons emitted by the cathode is by means of two parallel molybdenum plates situated at a distance of 4-5 mm apart. The rate of movement of the cathode is regulated by varying the speed of an electric motor. The anode circuit is fed by a rectifier, consisting of a step-up transformer and eight parallel-connected-type KRM-110 kenatrons. The rectifier provides half-wave rectification with a voltage of 8 kV and a maximum current of 0.8 A. Regulation of the working current and voltage is by smooth variation of the voltage in the primary winding of the step-up and heating transformers from 0 to 220 V. The circuit of the electric supply apparatus ensures a smooth regulation of the melting temperature and precludes unexpected current overloads.

Another method of zone refining has also been proposed [29] in which three or more guns, arranged symmetrically in a horizontal plane around the rod to be refined, produce a narrow fused zone. In this method, the guns are stationary while the rod makes a gradual vertical movement and rotates on its axis. Such a system ensures the protection of the purified metal from contamination by the cathode material.

Recently, a start has been made in the case of automatic control of zone-refining conditions by control of the thermionic emission current using a dc regulator [30]. The dc regulator protects the electric supply unit of the apparatus in the event of the formation of a glow discharge during intense liberation of gas from the specimen.

The vacuum system of the apparatus comprises a type VN-2 backing pump, a type N5 oil-vapor diffusion pump, liquid nitrogen trap, and shield. A vacuum of down to $8 \cdot 10^{-6}$ mm may be produced in the system. Of considerable significance is the controlled heating of the walls of the working chamber for their degasification.

A substantial drawback in the use of oil-vapor pumps is the contamination of the refractory metal due to oil vapor and cracking products finding their way into the working chamber, thus considerably reducing the purifying effect. The authors' experience has shown that use of a nitrogen trap provides considerable protection for the purified metal from carburization. For a more radical solution of the problem, however, oil-less pumping systems must be used (sorption, electric discharge, getter-ion pumps, etc.). There is a published account of the development of an electron-beam zone-refining apparatus having a pumping system, comprising a zeolite sorption pump with liquid nitrogen cooling and a getter-ion pump [31]. Pre-pumping of the working chamber is by backing pump and zeolite pump. Figure 95 shows the results of a mass-spectrometer analysis of the composition of the residual gases in vacuum systems using different pumping methods [31]. When oil-vapor pumps are used, a large quantity of heavy masses [54-200] is found, differing approximately by the atomic weight of carbon, and indicating the presence of a large quantity of carbon-containing compounds in the gaseous medium, and oil-cracking products on the heated metal surface. The presence of heavy masses is rapidly reduced on pumping with a getter-ion or absorption pump.

In the Refractory and Rare Metal Alloys Laboratory of the A. A. Baikov Institute of Metallurgy, single crystals of almost all refractory metals have been produced: vanadium, niobium, tantalum, molybdenum, tungsten, titanium, zirconium, ruthenium, and rhenium. As shown by the authors' researches, the effect of purification and the degree of perfection of the single crystals produced primarily depend on the purity of the initial ingot. For producing single crystals of refractory metals, therefore, the metal used is first subjected to vacuum remelting.

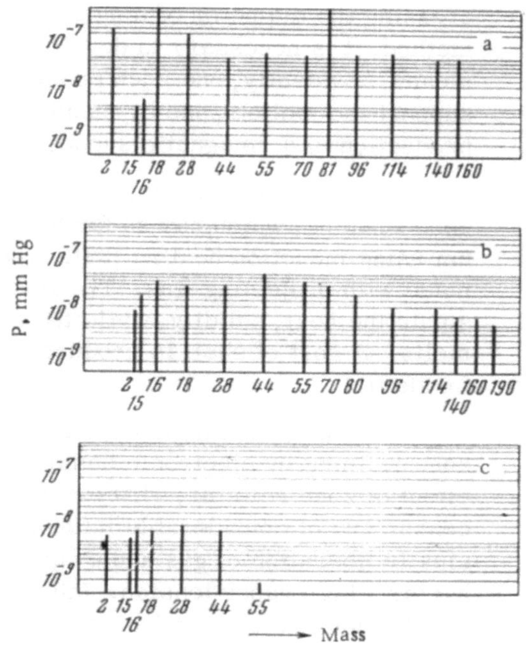


Fig. 95. Composition of gases in the vacuum system a) using an oil-vapor pump without nitrogen trap ($P = 10^{-6}$ mm Hg); b) the same with nitrogen trap ($P = 4 \cdot 10^{-7}$ mm Hg); c) using a getter-ion pump ($P = 3.3 \cdot 10^{-7}$ mm Hg).

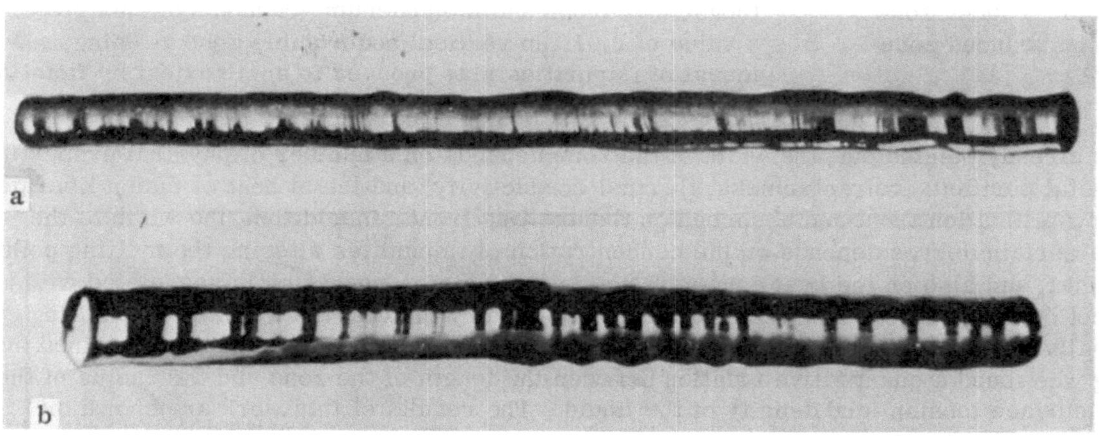


Fig. 96. Single crystals of tungsten (a) and molybdenum (b).

In vertical noncrucible zone refining, the ratio of the surface tension to density of the melt should not be less than 100:1 [32], since the "melting" zone is prevented from spreading out by surface tension forces. The maximum diameter of single-crystal rods produced by this method is, approximately, for molybdenum 18-20 mm, tungsten 10-12 mm, with a length of the order of 200-250 mm. Figure 96 shows single crystals of tungsten (diameter 10 mm) and molybdenum (16 mm) produced by electron-beam zone refining. The stable fusion zone is characterized by vertical lines at the crystallization front (Fig. 97b) [24]. For further in-

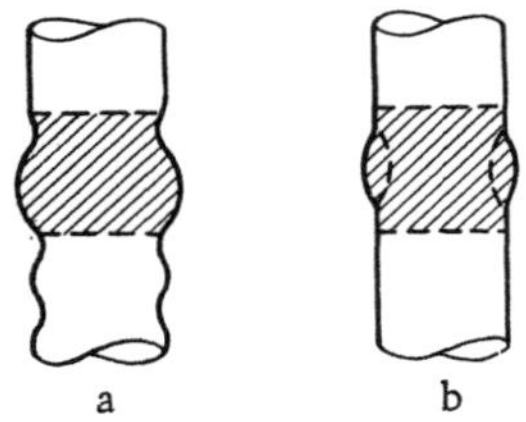


Fig. 97. Vertical fusion zone (upward direction of movement of zone). a) Unstable zone (there is a risk of rupture close to the crystallization front); b) stable zone.

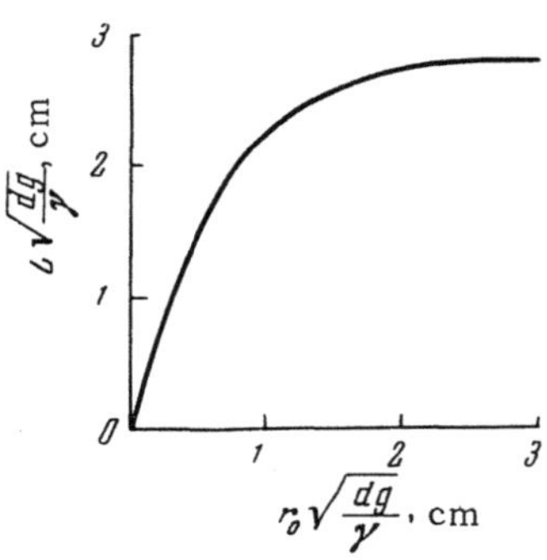


Fig. 98. Graph showing dependence of reduced zone length L on reduced zone radius $r_0$.

creasing the diameter of single crystals produced by this method, it is necessary to develop a number of additional means. Evidently, one such promising means may be to support the molten zone by means of a magnetic field (the production of a "magnetic support").

A noteworthy development is the "cage" method of zone refining, which has so far been applied only to titanium using induction heating [33]. The refined specimen is made with projecting fins. The fins, which strongly dissipate the heat, solidify, while the core of the specimen melts. A sort of shell is produced, with the fused zone inside it.

In producing single crystals of refractory metals by zone refining by electronic bombardment, in addition to the effect of zone refining, what are known as secondary processes are of considerable influence. These include principally liberation of gas and evaporation of impurities having a high vapor tension. Interstitial impurities, which exert a considerable influence on the properties of refractory metals, are removed by degassing. Many metallic impurities accompanying tungsten, molybdenum, and other refractory metals have a high vapor pressure and purification from them occurs largely by evaporation. Purification from carbon occurs not only by degassing (liberation of CO, $CO_2$, and $CH_4$) but also by the zone-refining effect. An approximate estimate of the coefficient of distribution in tungsten–carbon and molybdenum–carbon systems gives a value of 0.01. In vertical noncrucible zone refining, removal of impurities may proceed to some extent by flotation or gravitational effects.

As already pointed out, the width of the zone depends on a number of physical properties of the metal purified: refractoriness, thermal conductivity, and latent heat of fusion liberated on the crystallization front and absorbed on the melting front. In addition, the width of the zone to a certain degree depends on the concentration of impurities altering the melting point of the metal, and also on the heat concentration in the fusion zone, which is associated with the cross section of the ingot. In the refinement of refractory metals, the width of the melting zone usually does not exceed the diameter of the rod. Schadler [34] cites results obtained by Heywang who found a quantitative relation between the length of the zone and the radius of the rod, and surface tension and density of the liquid. The results of this work are shown in graphical form in Fig. 98.

Work is currently being carried out on the production of single crystals of refractory metals in tubular form. Evidently, the most promising method of producing single crystals in the form of tubes, sheets, and other sections, and also for producing large single crystals 20-100 mm in diameter, will be the Czochralski method using electron-beam heating, and perhaps the Bernouilli method (fusion of powders on a seed).

## Production of Molybdenum and Tungsten Single Crystals

As already pointed out, the quality of the single crystals produced from refractory metals (degree of purity, perfection of structure) is largely determined by the purity of the starting

TABLE 34. Electrical Conditions for Zone Refining of Tungsten and Molybdenum

| Material | Diameter, mm | | No. of passes | Voltage, V | | Current, A | |
|---|---|---|---|---|---|---|---|
| | ingot | cathode | | cathode heating | accelerating | cathode heating | on the cathode |
| Molybdenum | 12–15 | 24–27 | 2–3 | 4.5–8.0 | 4500–8000 | 50–80 | 0.7–1.0 |
| Tungsten | 8–10 | 20–22 | 2–3 | 4.5–8.0 | 7000–10000 | 60–90 | 0.8–1.5 |

ingots, especially with regard to interstitial impurities. Therefore, the starting ingots used for the production of tungsten single crystals are vacuum-melted cermet rods, or rods produced from vacuum-melted tungsten, and for molybdenum single crystals, vacuum arc-melted molybdenum rods [37]. The initial vacuum is not more than $5 \cdot 10^{-5}$-$6 \cdot 10^{-6}$ mm Hg. The total number of fusion zone passes should not be less than two. Table 34 gives the electrical working conditions for zone refining of tungsten and molybdenum.

After termination of the growing of the tungsten or molybdenum single crystals, it is cooled in a vacuum. During cooling, internal stresses are set up in the single crystals.

For reducing the internal stresses, the top end of the rod may be "cut off" by melting it across on the last pass, or the rod may be heated by an electric current for reducing the rate of cooling.

It has been possible in zone refining to grow single crystals of large diameter by increasing the power of the feed current and by using additional heating of the refined specimen. Furthermore, the application of a constant electric field promotes better mixing of the liquid zone, whereby the zone-refining effect is increased owing to the distribution coefficient approaching the equilibrium value more closely. The rate of travel of the molten zone is 0.5-3.0 mm/min.

When electron-beam zone-refining apparatus is used, precautions should be taken to protect the personnel from x rays and luminous radiation.

The tungsten and molybdenum rods produced after zone refining are tested for single-crystal quality by the method of electrolytic surface etching. The electrolyte used is a 5 to 10% solution of NaOH in water. The rod under test is the anode; the annular cathode of nickel strip 20 mm wide is suspended horizontally in the bath. The current density is not less than 1 A/cm$^2$. The tested rod, fixed by one end, is lowered into the cathode ring and is held for several seconds in the electrolytic bath. The rod is then fixed by the other end and is lowered into the cathode ring and again held for several seconds in the electrolytic bath. The alkali is washed from the rod by holding the latter in cold running water. The single-crystal rod etched in this way ought to have on its surface alternating dull and bright longitudinal bands. Testing of single-crystal quality is also done by taking Laue photographs; the orientation of the crystallographic planes of the single crystal is determined by x-ray analysis.

## Estimation of Purity and Degree of Perfection of Metal Single Crystals

The estimation of the purity and degree of perfection of the single crystals obtained is a complicated problem. The answer to this problem involves an entire combination of methods developed by modern physics and chemistry. In particular, it is necessary to emphasize the importance of physical methods of analysis, since the classical analytical ("wet") methods of

TABLE 35. Presumed Content of Interstitial Impurities in Solid Solutions Based on Metals of Groups Va and VIa after Cooling at a Moderate Rate [35]*

| Metal | Impurity content, % | | | |
|---|---|---|---|---|
| | hydrogen | carbon | nitrogen | oxygen |
| Vanadium | 1 | 0.1 | 0.5 | 0.3 |
| Niobium | 0.9 | 0.01 | 0.03 | 0.1 |
| Tantalum | 0.4 | 0.07 | 0.1 | 0.02 |
| Chromium | 0.00001–0.0001 | 0.00001–0.0001 | 0.00001 | 0.00001 |
| Molybdenum | 0.00001 | 0.00001–0.0001 | 0.0001 | 0.0001 |
| Tungsten | Not found | 0.00001 | 0.0001 | 0.0001 |

*The data in the table are based on an estimate of the solubility equilibrium of the interstitial impurities at a temperature at which the diffusion coefficient $D = 10^{-11}$ $cm^2/sec$.

chemistry are now often found to be insufficiently accurate for determining the impurities in materials of semiconductor purity, i.e., containing impurities of the order of one millionth of one percent. Matters are further complicated by the fact that it is not the number of nines after the decimal point which is important, but the determination of the residual harmful impurities (boron in semiconductor silicon, gadolinium and europium in uranium, gaseous impurities in metals).

The solubility of interstitial impurities in group VIa metals does not exceed 0.0001% (Table 35). In large quantities, oxygen, nitrogen, and carbon form in metals of this group second phases (oxides, nitrides, carbides), which are distributed over the crystal faces and produce cold brittleness. The accuracy ($10^{-2}$–$10^{-3}$%) of the principal method used for determining the gas content of metals (vacuum melting) no longer satisfies modern requirements of science and technology.

The problem of increasing the accuracy of analysis by a further one or two orders is an urgent one. Therefore, considerable attention is now being paid to other physical methods of analysis, radioactivation, mass-spectrometric, and spectral-isotope methods.

E. M. Savitskii and G. E. Chuprikov [36] have shown that the influence of the oxygen content on the hardness and electrical resistance of rhenium is approximately 100 times greater than that of the content of metallic impurities in solid solution (tungsten, molybdenum, etc.). For estimating the purity of single crystals, the method of residual electrical resistance is used, i.e., the ratio of the specific electrical resistances of single crystals determined at 300°K (room temperature) and 4.2°K (liquid helium) is determined. The method is based on the Mathiessen rule, according to which the electrical resistance of metals consists of two components, a component dependant on the scattering of electrons by the thermal vibration of the atoms, and a component determined by the scattering of electrons at impurities and crystal defects. Since, at the temperature of liquid helium, thermal scattering of electrons is small, the value of the resistance of a single crystal is mainly determined by the impurities. The above-mentioned ratio has the value of several tens for metals of commercial purity, several thousands for pure metals, and several tens of thousands for specially pure metals [37]. The drawbacks of the method include high sensitivity to work hardening and other defects of the specimens and to the experimental conditions (soldering of contacts, etc.). In order to exclude the influence of contacts, a start has recently been made in the use of the noncontact method of measuring residual resistance.

The residual resistance method is quite an approximate method of estimating the purity of a material. Near the temperature of liquid helium, the electrical resistance depends to a considerable extent on the diameter of the specimen. In this range, for the purest material, the free path of an electron is longer than the diameter of the specimen [38]. In measurements of specific resistance of single-crystal tungsten in low temperature ranges, it has been shown that the magnetic field, produced by the current of the specimen, has an influence on the resistance of tungsten at low temperatures [39]. In addition, the value of the resistance in liquid helium may be abruptly reduced by the presence of superconducting impurities (tungsten carbide, rhenium nitride, etc.) in the investigated metal. Evidently, the enumerated errors also explain basically the divergence in the values of K ($\rho_{300°K} / \rho_{4.2°K}$), obtained by different workers, and also the poor reproducibility of the experiments. In addition to the indirect method of estimating the purity of single crystals, the determination of the oxygen, hydrogen, and nitrogen content of single crystals and initial specimens is therefore determined by the vacuum-melting method. It has been found that in zone refining, the content of gaseous impurities is reduced from hundredths of one percent in the initial specimens to thousandths and even tens of thousandths of one percent in single crystals.

The commercial bend test or bending through 90 or 180°C at room temperature is quite effective in estimating the purity of single crystals of cold-brittle metals for interstitial impurity content. The angle of bend and the number of bends which a single crystal withstands without the appearance of cracks is in good relationship to the purity. Another purity characteristic is the reduction in cross-sectional area on elongation [40] and particularly the ductile–brittle transition temperature for cold-brittle metals; the lower the interstitial impurity content of the single crystal, the lower is the transition temperature. Of course, all the metallographic methods of checking the purity of metals using optical and electron microscopy, as well as x-ray analysis methods, are fully applicable to the inspection of single crystals. A recent achievement which must be mentioned is the examination by transmission in the electron microscope of thin metal films, instead of replicas [41, 52], for ascertaining the nature and distribution of nonmetallic inclusions in single crystals.

By using vacuum-melted molybdenum and a liquid nitrogen trap in zone refining, the carbon and oxygen content of single crystals may be reduced to 0.012 and less than 0.001 wt.%, compared with 0.014 and 0.015 wt.% of the initial material, respectively. A considerable reduction in the nitrogen and hydrogen content is also found. Zone refining without a liquid nitrogen trap results in a certain increase in the carbon content of molybdenum single crystals. As already pointed out, this phenomenon is due to the entry of oil and its cracking products into the working vacuum chamber of the apparatus. Spectral analysis of molybdenum and tungsten single crystals showed that the content of other impurities (iron, copper, silicon, etc.) was less than 0.01 wt.%.

To elucidate the effect of zone refining, the degree of purity for different sections along the length of a single crystal was determined by the residual resistance method and microhardness measurements. The measurements showed that the purity of the middle portions of the specimens was somewhat higher than the purity of the end portions. It was found that in the production of single crystals of refractory metals by zone refining, purification occurs to a greater extent by degasification and evaporation of the impurities in the melting of the zone of metal in a high vacuum, and to a lesser extent by removal of nonvolatile impurities by zone refining.

The orientation of the single crystals was determined by back Laue photographs. The x-ray analysis of the initial orientation of single crystals of molybdenum, tungsten, and other refractory metals with a bcc lattice, produced by the technique described above, showed that the growth of single crystals in electron-beam zone refining proceeds spontaneously in a

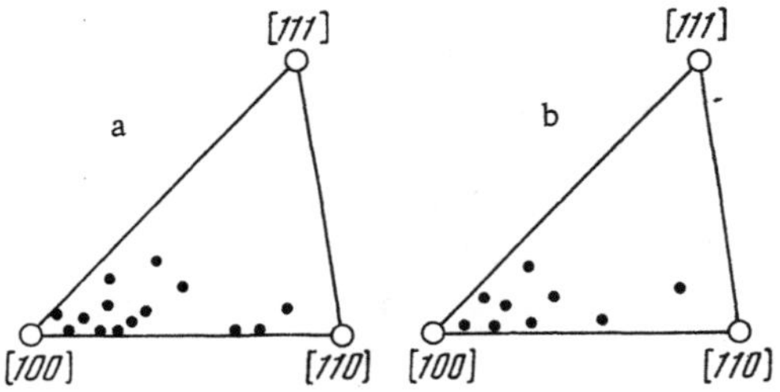


Fig. 99. Axial orientation of single crystals of molybdenum (a) and tungsten (b) on a triangular standard stereographic projection.

direction close to the [100] direction with a deviation 6-18 and sometimes only 1-3° (Fig. 99). Crystals having a [111] or [110] orientation are very rarely produced. In some x-ray photographs, subdivision of the interference spots is observed, showing the presence of disorientation blocks. The disorientation angles do not exceed 0.5°. The existence of a preferred direction of growth of metal crystals has been confirmed by many investigators [42-44]. It is evident that the growing conditions exercise an influence on the preferred growth direction. Goss [42] assumes that the preferred orientation of crystals grown at a high rate differs from the orientation of crystals grown at low rates; this is due to the reduction in the influence of temperature factors in growth at high rates when the characteristics of the crystal structure of the metal play a major part. For crystals having a body-centered cubic lattice and also face-centered metals (both refractory and nonrefractory), the preferred orientation is [100], while for metals having a hexagonal close-packed lattice, the preferred direction is [210]. The view exists that the cause of preferred orientation in single crystals lies in the liquation action of the impurities and above all tungsten carbide [45]; however, this has not been definitely established, since the growth mechanism of crystals has not been finally explained.

## Substructure of Refractory Metal Single Crystals

Compared with ordinary metallographical analysis of cermet and cast refractory metals in the polycrystalline state, the analysis of the microstructure of single crystals of refractory metals has its specific character. For producing polished sections having a definite crystallographic orientation, the specimens are cut by means of thin abrasive or diamond wheels with powerful water cooling on a special apparatus equipped with a goniometric head for fixing the single crystal. The slight surface work hardening is removed by subsequent electropolishing. Published data are available relating to the application of chemical, electrochemical, and electric-spark erosion of metal single crystals for the production of an undeformed surface [46, 47]. Electropolishing and electrolytic etching of single-crystal polished specimens may be conveniently carried out on apparatus of the "Elipovist" type (Zeiss, German Democratic Republic), permitting the observation of these processes under the microscope.

The most effective and universal electrolyte is a 2-5% aqueous NaOH solution.

For revealing the substructure of molybdenum single crystals, an ordinary chemical etch reagent is also used, consisting of concentrated sulfuric and nitric acid (in the ratio of 1:1).

We have revealed the structure of single crystals of tungsten, molybdenum, and rhenium, the specimens of which were cut along definite crystallographic planes. The surface of the specimens was subjected to electropolishing in a 5% aqueous NaOH solution using a voltage of 30-40 V and a current density of 1.7 A/cm$^2$. The duration of polishing depended on the depth of the work-hardened layer to be removed. Electrolytic etching was carried out at a lower current density.

The shape of the etch figures can be a reliable criterion in determining the crystallographic orientation of the crystal planes. Figure 100 shows the substructure of molybdenum and tungsten single crystals in a plane parallel to (100); square etch figures are to be seen, representing the sites of emergence of dislocations in the plane of the polished specimen. The substructure of the (111) plane of molybdenum and tungsten single crystals is characterized by

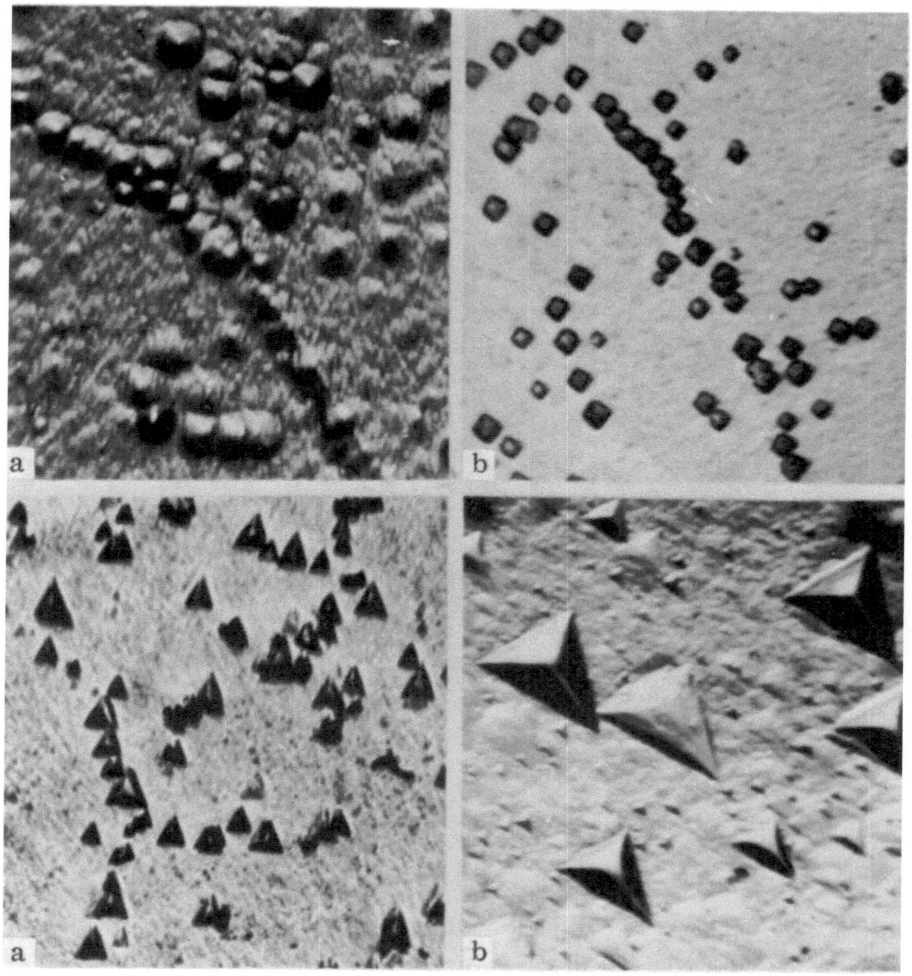


Fig. 100. Substructure of single crystals of molybdenum (a) ×600 and tungsten (b) ×2000. Top row) in the (100) plane; bottom row) in the (111) plane.

etch figures of triangular form. The substructure of a plane parallel to (110) could not be revealed by the given etching method. In addition to single etch figures, distributed over the entire polished surface, disorientation block boundaries, consisting of etch figures, were revealed. The disorientation angles were low, of the order of a few minutes, as determined by the well-known formula $\theta = b/h$ (h being the distance between the dislocations in the subboundary, and b the Burgers vector). The substructure of niobium single crystals consists solely of etch figures distributed statistically over the crystal plane; it has not been possible to reveal a block structure in niobium single crystals.

The etch figures revealed are evidently the sites of impurity concentration and emergence of dislocations on the surface. It should be noted that electrolytic etching reveals only growth dislocations. Dislocations produced in plastic deformation may be revealed by etching with Wolf's reagent (2 parts 25% aqueous $CuSO_4$ solution + 1 part conc. $NH_4OH$). The effect of this reagent is not due to decoration of the dislocation by impurities but it reveals both linear and mixed dislocations, having screw and edge components [48].

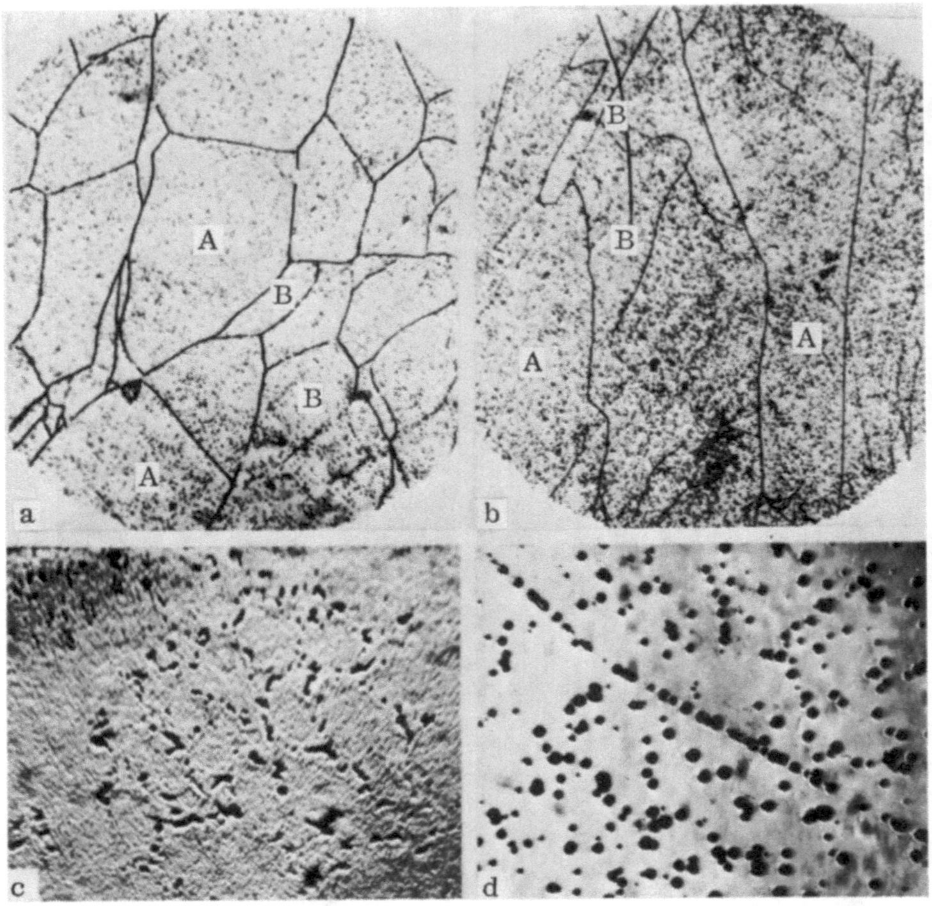


Fig. 101. Substructure of tungsten single crystals in longitudinal and cross sections with respect to the growth direction. a) Cross section (×50); b) longitudinal section (×50); c) third-order subgrains (×1000); d) dislocations in a second-order subgrain and second-order subgrain boundary (×800).

The large number of etch figures in the substructure of a tungsten single crystal shows that tungsten has a less perfect structure and is more contaminated by impurities than molybdenum and niobium. For molybdenum single crystals, grown in a vacuum produced by oil diffusion pumps, the dislocation density (as determined by metallographic and x-ray methods) is about $10^4$-$10^6$ $cm^{-2}$ and for tungsten, it is $10^5$-$10^7$ $cm^{-2}$. As in the case of other bcc metals, we have not succeeded in revealing the structure of tungsten on the (110) rhombic dodecahedron plane. The different etchability of the crystal faces of single crystals is the manifestation of the anisotropy of the properties of metals having a bcc lattice.

Figure 101a,b, according to results of E. M. Savitskii and G. L. Tsarev, shows the substructure of a tungsten single crystal in longitudinal and cross sections with respect to the direction of growth of the crystal. The substructure consists of subgrains of first order of magnitude (A) which in turn contain second-order subgrains (B). The subgrains are strongly elongated in the growth direction of the crystal. In the cross section, the first-order subgrains have a size of 0.2-0.5 mm, while that of the second-order subgrains is 0.05-0.1 mm. In the longitudinal section, the size of first-order subgrains attains 1.5-2 mm, while that of second-order subgrains is 0.5-1 mm. Under higher magnification (×1000), third-order subgrains (cellular growth structure) are also visible (Fig. 101c). A cellular growth structure

TABLE 36. Subgrain Dimensions and Their Disorientation Angles in Tungsten Single Crystals
(Carbon Content 0.025%; $K = \rho_{300°K} / \rho_{4.2°K} = 1100$)

| Order of subgrains | Dimension in transverse direction | Dimension in longitudinal direction | Disorientation angle | Dislocation density, $cm^{-1}$ |
|---|---|---|---|---|
| 1 | 200—500 | 1500—2000 | 17'30" | $1.2 \cdot 10^7$ |
| 2 | 50—100 | 500—1000 | 2'45" | $1 \cdot 10^7$ |
| 3 | 10—18 | 30—75 | — | $4 \cdot 10^8$ |
| 4 | 0.2—0.7 | — | — | $4 \cdot 10^8$ |

was also found in molybdenum single crystals. The existence of a cellular structure was first found in single crystals of fusible metals (tin, copper) [49, 50]. Figure 101d shows second-order subgrains.

Rutter and Chalmers (49) attribute the occurrence of the cellular structure to supercooling close to the interface resulting in redistribution of impurities occurring at the melt–crystal interface.

The size of the cells in tungsten single crystals (in cross section) is 10-18 $\mu$. Electron microscope investigations by E. M. Savitskii and G. L. Tsarev revealed still smaller growth cells 0.2-0.7 $\mu$ in size [51, 89].

Subgrain sizes and values of the disorientation angles are shown in Table 36 [51].

The transmission electron microscope has opened up considerable possibilities in the investigation of the fine structure of single crystals of refractory metals. In the Refractory and Rare Metal Alloys Laboratory, a method has been developed for preparing films of tungsten single crystals for examination by transmission [51]. The initial specimens in the form of tablets are cut from a single-crystal tungsten rod by a thin grinding wheel (0.25 mm). Subsequently, the specimens are subjected to a reduction in thickness by jet electropolishing (Fig. 102). A 2-5% aqueous NaOH solution may be used as electrolyte. Polishing is carried out until an etch pit approximately equal to half the thickness of the specimen is formed. The specimen is then turned over and is polished until an orifice having a diameter of 4-8 $\mu$ appears at the center of the specimen. For uniform reduction in thickness of the specimen the rate of polishing should be a minimum. The structure of the tungsten single crystal was examined in a microscope with an accelerating voltage of 400 kV [52]. The instrument enabled three photographs to be obtained on one negative. An image of a portion of the object, the selector stop, and a microdiffraction pattern of the portion of the object bounded by the selector stop. The image of the selector stop also served as magnification scale, since its diameter was known exactly (1 $\mu$). The investigation showed that tungsten single crystals grown by electron-beam zone refining in a vacuum, produced by an oil-vapor pump, were still not sufficiently perfect or free from interstitial impurities, above all carbon. Figure 103a shows an electron diffraction pattern of a thin film (thickness 1000 Å), etched in a tungsten single crystal (carbon content 0.025%). Under these conditions, inclusions of tungsten carbide of hexagonal form are well visible. The cross-sectional dimension of the inclusions fluctuated in a range of 0.5-2.5 $\mu$. Due to the high accelerating voltage of the instrument, it was possible to obtain clear microdiffraction patterns of the inclusions, in which in addition to tungsten reflections, reflections of a second phase were visible. The character of the symmetry of the reciprocal lattice of the second phase and the calculated interplanar distances corresponded to $W_2C$ crystals [52].

Thus, electron-microscope examination by transmission has shown that tungsten single crystals are actually two-phase systems (the second phase is tungsten carbide $W_2C$). It is

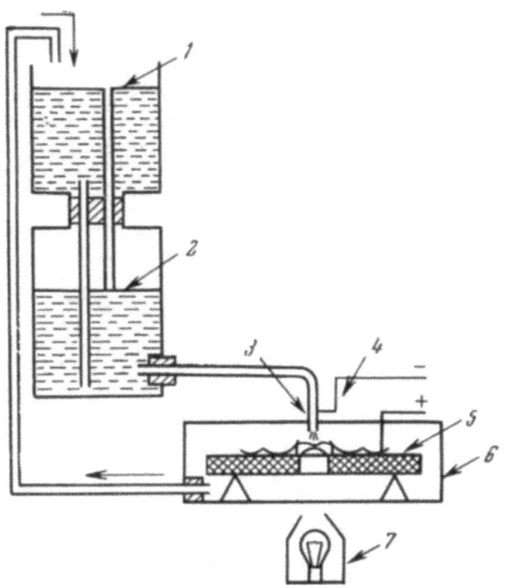


Fig. 102. Diagram of jet electropolishing apparatus for the production of thin films of refractory metals. 1) Stock of electrolyte; 2) constant level of liquid; 3) capillary tube; 4) observation position; 5) nontransparent plate; 6) glass; 7) point source of light.

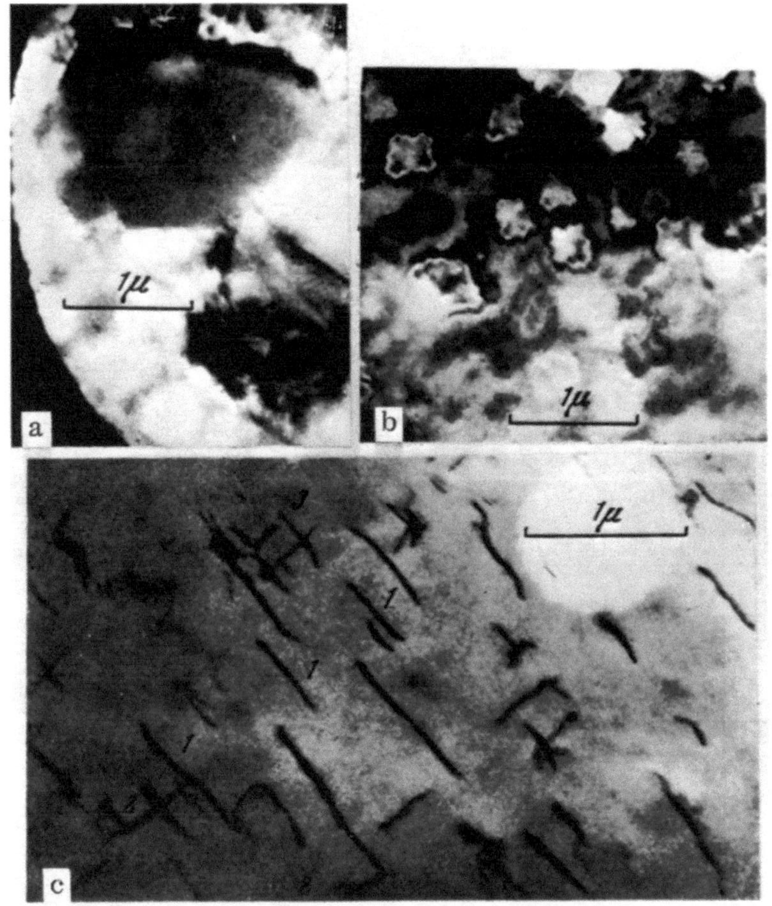


Fig. 103. Electron microscope photographs of a tungsten single crystal. a) Tungsten carbide inclusion of hexagonal form; b) cellular structure of tungsten single crystals in the form of crosses; c) interaction of dislocations with impurities in a bent single crystal [1) screw dislocations, 2) edge dislocations, 3) pile-up of screw and edge dislocations].

curious that no oxides or nitrides were found in the structure of tungsten single crystals, this being evidently due to the low thermodynamic stability of these compounds. The cellular structure of tungsten single crystals in the form of crosses is shown in Fig. 103b. Figure 103c shows the interaction of dislocations with impurities in a slightly bent tungsten single crystal. The dislocation structure of deformed tungsten single crystals is similar to the dislocation structure of other body-centered cubic metals having low solubility for interstitial impurities; the screw dislocations contain a large number of jogs and superjogs, edge dipoles, and individual closed loops [51, 89].

Investigations made in the Refractory Metal Alloys Laboratory of the A. A. Baikov Institute of Metallurgy show that carbon is the most dangerous impurity for tungsten and molybdenum, leading to a sharp reduction in the ductile properties of these metals. E. M. Savitskii and G. L. Tsarev [51] carried out special experiments on the saturation of tungsten single crystals by carbon, using the method of cementation in graphite in a vacuum furnace at 1300°C for 6 hr, followed by a homogenizing anneal at 2000°C for 8 hr. Under these conditions, the carbon content rapidly increased (to 0.054% compared with 0.001%) in a single crystal, previously specially purified from carbon. Microstructure examination showed a sharp increase in disperse carbide inclusions. Dislocation density increased approximately from $10^{-5}$ to $10^{-7}$ $cm^2$. The critical cleavage stress and yield point at 25°C increased from 15.7 to 34.1 kg/$mm^2$ and from 34 to 73 kg/$mm^2$, respectively. The reduction in area under tension fell from 100 to 40%, and what is most important the ductile–brittle transition temperature increased from −196°C (liquid nitrogen) to room temperature. Consequently, work is currently proceeding on the production of tungsten and molybdenum having a low carbon content by modifying a number of technical operations in the chemical technological treatment stage and by taking a number of precautions to prevent the entry of carbon into the metal during zone refining.

## Growing of Refractory Metal Single Crystals of Given Crystallographic Orientation

Frequently, it is necessary to produce single crystals of uniform, predetermined orientation. This is done by using seeds [53]. If a seed with the needed orientation is available, it is a fairly simple matter to grow single crystals with this orientation. The seed is secured in a strictly vertical position in the bottom clamp of the apparatus, and the stock is secured to the top clamp also strictly vertical and in the same axis as the seed. A small gap (up to 1 mm) is left between the stock and the seed. Then, at the joint, equal parts of seed and stock are melted simultaneously, and after fusion, the zone is moved forward in the direction of the stock. On the principle of corresponding orientation, the growing single crystal assumes the crystallographic orientation of the seed. After the production of the single crystal, the seed is cut off and may be used for producing subsequent single crystals with a similar orientation. The deviation of the orientation of the grown crystal from the orientation of the seed does not usually exceed 2°.

The following procedure is adopted when a seed with the necessary orientation is not available. From crystals produced with arbitrary orientation, a crystal is selected which has the least deviation from the necessary orientation and has a regular geometrical shape. A Laue pattern is repeatedly obtained with strictly vertical mounting of the crystal in the x-ray camera, and a stereoscopic projection is made, from which the necessary angles of rotation are found for deducing the necessary crystallographic direction in the vertical position. By rotation of the graduated circle around the axis of the crystal and along one of the arcs of the chamber, the crystal is brought into a position in which the necessary direction is in the vertical position. If the next exposure confirms that the rotation has been correct, a mark is made of the crystal (in order later to recognize the plane of the angle of slope), or an indicator in the form of a flat rule is set up (respectively parallel or perpendicular to the plane

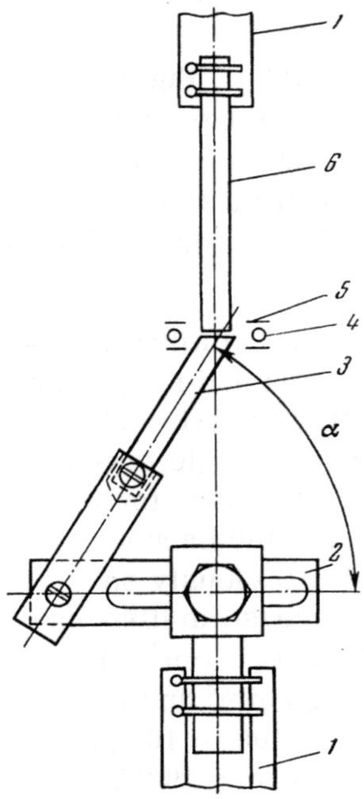


Fig. 104. Holder for seed and initial stock in the electron-beam apparatus. 1) Top and bottom clamps; 2) seed holder; 3) seed; 4) cathode; 5) focusing plates; 6) stock.

of the film) and the angle of slope is noted. The crystal is then removed from the chamber and is secured in the seed-holder (Fig. 104) so that the plane of the slope of the base of the holder coincides with the plane of the slope of the seed. After rotation of the base of the holder to a certain angle of slope $\alpha$, the holder is mounted in the bottom clamp, the starting stock is moved up to the seed with a gap of about 1 mm, and the movement of the zone is commenced. With such a method, the deviation of the orientation obtained depends primarily on the accuracy with which the seed is secured in the holder, and the accuracy with which the latter is fixed in the clamp. If all these operations are carefully carried out, the deviation of the orientation from the required value may be reduced to 2-3° by means of very simple devices.

As shown by the results of investigations on single crystals of a number of semiconductor materials, in addition to impurities, various kinds of crystallographic defects exercise a considerable influence on the semiconductors. It is therefore important to grow single crystals having a perfect structure.

Usually, the degree of perfection of crystals is subdivided on a macroscopic and microscopic scale. Macroimperfections of a crystal include, for example, twinning boundaries or block boundaries having a large disorientation angle, a banded structure, etc., microimperfections and dislocations and various point defects (vacancies, interstitial atoms). High purity of the metal is the first and essential condition for the production of a perfect single crystal. Other significant factors are the absence of vibration, growth rate, temperature conditions of pulling, degree of vacuum, perfection of the seed structure, form of the liquid–solid interface, etc. For developing a method of producing perfect metal crystals from a melt, their growth mechanism must be known. As yet, owing to method difficulties, the study of growth mechanism is confined to low-melting metals only. The growth mechanism of crystals of refractory metals and the thermal physics of the process have not yet been studied. An effective method of growing dislocation-free crystals of metals (even low-melting metals) has not yet been discovered. The development of methods of growing large single crystals of refractory metals having a perfect crystal structure is not only of theoretical interest, but is also of considerable practical interest.

An example of metal crystals possessing a perfect structure is provided by filamentary crystals or what are called "whiskers," produced by the chemical reduction of halides, electrolytic deposition, vapor condensation, etc. Iron "whiskers," for example, have been produced in this way [114]. The investigation of the mechanical properties of the "whiskers" of a number of metals has shown that their strength approaches the theoretical yield point of interatomic bonding [115, 116].

Several theories have been advanced to explain the high strength of "whiskers." Most investigators see the cause of this strength to lie in the perfection of the crystal structure of metal "whiskers." It is supposed that "whiskers" represent perfect single crystals having only screw dislocations along the growth axis; linear dislocations are almost entirely absent. The limited size of the "whiskers" is the principal obstacle to their use on a wide scale as a

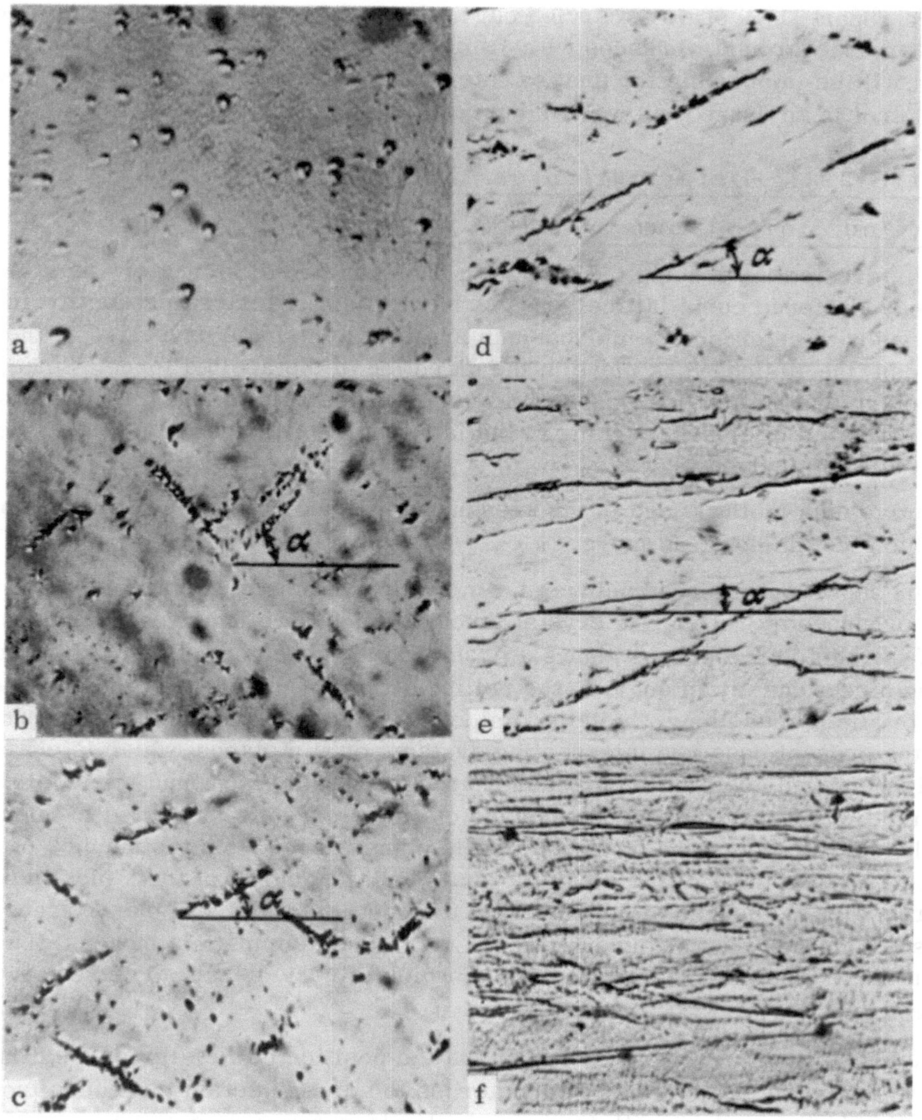


Fig. 105. Change in substructure of a molybdenum single crystal on deformation by rolling between grooved rolls at room temperature (×200). a) Original substructure of single crystal, [113] crystal axis; b) deformation by 30%, $\alpha = 55°$; c) by 40%, $\alpha = 40°$; d) by 50%, $\alpha = 30°$; e) by 75%, $\alpha = 10°$; f) by 85%, $\alpha = 0°$.

high-strength constructional material. However, constructional materials known as "composite" materials, strengthened by metallic inclusions in different combinations, have already found industrial applications [117].

Single-crystal materials are being used on an ever-increasing scale in the new technology. Further work is necessary in the perfection of methods of producing single crystals and semifinished products made from them (production of tungsten and molybdenum single crystals of large diameters having a predetermined orientation, more thorough removal of interstitial impurities, production of single crystals of alloys of refractory metals).

There is a report [118] of work carried out in the USA by Semi-Elements Inc. on the construction of an underground workshop with a microclimate for growing single crystals of rare metals. It will be impossible for dust to enter the workshop from the outside. A rigid temperature control is envisaged, as well as humidity and lighting control.

## Influence of Plastic Deformation on the Structure of Tungsten and Molybdenum Single Crystals

We were interested in studying the change in substructure of refractory metal single crystals having a body-centered cubic lattice, and the mechanism of texture production in the deformation process with reference to molybdenum and tungsten single crystals. In the deformation process, the dislocations move in the direction of the applied force. Since, however, an elementary dislocation possesses minimum energy when its Burgers vector is parallel to the direction of closest packing in the lattice, the preferred slip directions will be the direction of closest packing [3, 40].

The selective etch method used in the present investigation reveals the traces of slip planes along which the dislocations move.

Figure 105 shows the microstructure of a molybdenum single crystal in different stages of deformation produced by rolling the crystal through grooved rolls at room temperature [56]. Examination of the photomicrographs shows a preferred arrangement of the etch figures along two intersecting directions, at an angle to the rolling direction. Evidently, these directions are the traces of slip planes along which the dislocations move in the deformation process, there being in the present case two intersecting systems of slip planes, which is characteristic of the body-centered cubic lattice. Maddin and Cheng [55] also found mined slip in molybdenum on examination of slip lines on the surface of single crystals subjected to tensile stress. The extent of the revealed portions of traces of slip planes depends on the number of piled-up dislocations, i.e., on the purity of the single crystals and the degree of their plastic deformation. With increase in the degree of deformation, the position of the slip plane traces varies; they continuously approach the rolling direction, i.e., in the deformation process, the slip planes rotate in the direction of the applied force, which is very clearly to be seen in the microstructures.

The angle $\alpha$ between the rolling direction and the position of the slip planes diminishes from 55 to 10° when the degree of deformation across the cross section increases from 30 to 75%. The variation in the angle $\alpha$ with increase in the degree of deformation may be represented by the empirical formula $\alpha = (1/k)(c - a)$, where $\alpha$ is the angle between the rolling direction and the position of the slip planes for a given degree of deformation a; c is the degree of deformation, for which the slip planes are arranged along the rolling direction (i.e., when $\alpha = 0$); k is an empirical coefficient ($deg^{-1}$), dependent on the initial orientation of the single crystal. In our case, the formula has the form $\alpha = (85 - a)$ deg [56]. Thus, in the deformation process of molybdenum single crystals, the latter are split into disorientation blocks, the boundaries of which are located in the active slip planes. With increase in the degree of deformation, the blocks are rotated in the direction of the active force, the disorientation of the blocks increases and ultimately leads to the formation of a fibrous structure (texture) at a high degree of deformation (above 80%).

Figure 106 shows the microstructures of a tungsten single crystal deformed by rolling through grooved rolls with various degrees of reduction. In contrast to molybdenum single crystals, high reductions (more than 20%) in tungsten single crystals are possible only if they are heated to 500°C. Thin rods (diameter less than 0.1 mm), were subjected to tensile stress at room temperature. On examination of the microstructure of tungsten single crystals subjected to deformation with a small reduction (6-10%), etch pits were visible at the sites of

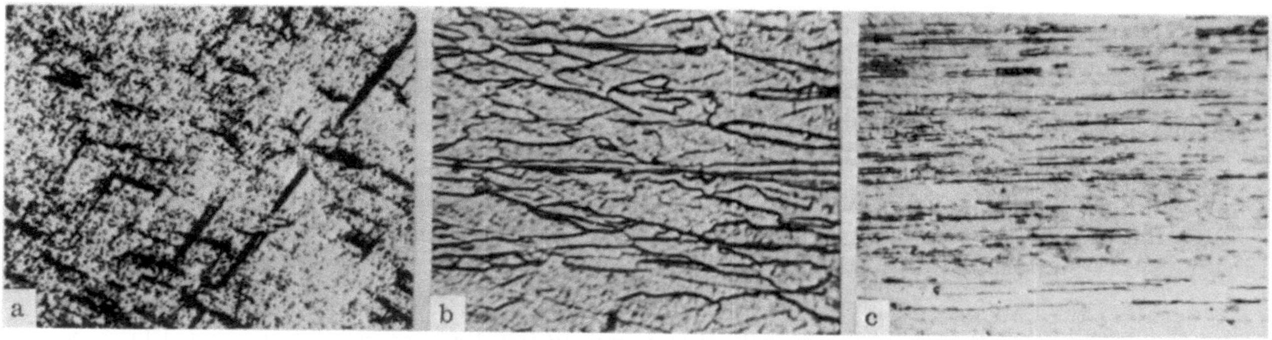


Fig. 106. Microstructure of tungsten single crystals deformed by rolling through grooved rolls at different stages of treatment (×200). Degree of deformation, %: a) 6; b) 47; c) 90.

emergence of dislocations on the slip lines (Fig. 106a). Texture was formed on further deformation. After deformation by more than 90%, the specimens assumed a texture with a small angular scatter. According to [55], plastic deformation of metal single crystals having a body-centered cubic lattice occurs by slip along three families of planes of the [111] zone: (110), (112), and (123). The etch pits are arranged along type (011) and (112) planes. Schadler [57] showed that slip along (011) and (112) planes occurred in tungsten at room temperature; at temperatures of 77 and 20°K, deformation of tungsten proceeds along a type (011) plane, and twinning along the (112) plane. Tungsten single crystals subjected to tension in a narrow temperature range (from −95 to −178°C) showed the formation of deformation twins [58]. An increase in the purity of tungsten single crystals evidently facilitates the twinning process.

For tungsten, therefore, the deformation mechanism is largely determined by the temperature. Such dependence has also been found in other metals [39, 40, 59]. The formation of deformation texture is observed in tungsten single crystals subjected to a high degree of deformation (more than 60%).

## Deformation of Rhenium Single Crystals

Exhaustive data have already been reviewed regarding the deformation of the low-melting hexagonal metals cadmium, zinc, and magnesium in the monograph by E. Schmidt and W. Boas [3] published in 1935. In particular, the existence of a strongly pronounced yield-point anisotropy was shown by the example of magnesium, as well as a considerable increase in the critical shear stress in the alloying of single crystals.

Of the postwar work on hexagonal single crystals, interesting investigations of the deformation mechanism and fine structure of rhenium single crystals have been carried out in the USSR by E. M. Savitskii and G. E. Chuprikov [60, 61] and in England by G. Gich and coworkers [62]. It was found that the relationships discovered for low-melting single crystals of zinc, cadmium, and magnesium are not found in refractory metals. This, for example, relates to the confirmation of lower cold hardenability (lower strain-hardening coefficient) of hexagonal metals compared with the cubic metals. It was found that rhenium single crystals hardened very quickly during plastic deformation. The cause of this is the high value of the elasticity constant of rhenium. The mechanism of the influence of interstitial impurities, especially oxygen, on the strength and ductile properties of rhenium and titanium was found to be very interesting. The deformation of rhenium single crystals as a function of initial orientation occurs by slip, typical of the hexagonal lattice, along the (0001) and (1010) planes of the base and prism in the [1120] direction, and also by twinning on the (0001) [1120] system. Single-crystal rhenium (containing less than 0.001 wt.% of oxygen) with orientation of the axis of tensile stress between the [1010] and [1120] directions is deformed according to the (1010) [1120] system.

According to [60], an increase in oxygen content of more than 0.001 wt.% even with a deformation of 2% causes "wedging" of the usual type (1010) slip planes and the occurrence of the additional (0001)[1120] slip system less favorable for the given orientation and demanding higher stresses. "Wedging" of the slip planes leads to an increase in the resistance to deformation and a reduction in the ductility of the material.

A similar influence of oxygen on the mechanism of deformation of titanium (wedging of some slip planes) was found by the authors of [63] in the elongation of single crystals by 2-5% under room-temperature conditions.

One of the causes of the variation in the mechanism of the plastic deformation of rhenium with increase in its oxygen content is the change in the dimensions of the crystal lattice when oxygen atoms are introduced into it. X-ray structural analyses of rhenium containing oxygen shows that the dissolution of oxygen in rhenium produces an anisotropic distortion of the crystal lattice along the c axis, the lattice parameter a being unchanged (the ratio c/a is increased from 1.615 to 1.633). In view of this, the other planes (with minimum interatomic distances) become more favorable for plastic deformation. Evidently, a no less important cause of the change in the deformation mechanism is the interaction of interstitial impurities with the stress fields in the lattice of the metal.

## Influence of Impurities on the Substructure of Deformed Refractory Metal Single Crystals

The influence of impurities on the character of the distortion of the crystal structure after deformation of molybdenum and rhenium single crystals has been studied by the x-ray method [59, 65]. For estimating the distortion of the structure, use was made of the effect of the increase in width of the x-ray interference maxima of the metals on deformation, due to the reduction in size of the coherent scattering regions and the presence of a residual microdeformation ($\Delta a/a$). For comparison, the investigations were carried out on single crystals and metals of different degrees of purity. The interference spectra were obtained from filtered $K_{\alpha,\alpha_2}$ radiation by automatic recording, using a URS-50I spectrometer. The size of the coherent scattering regions D and the relative microdeformation $\Delta a/a$ were calculated. The investigations showed that due to the insignificant plastic deformation, the size of the coherent scattering regions and the microdeformation for molybdenum and rhenium of commercial purity are much greater than in single crystals of these metals. Dislocations form block boundaries and produce internal stresses in the lattice. Thus, the dislocation density may be determined from the dimensions of the blocks and the microdeformation value. It has been shown that the lower the purity and ductility of molybdenum, the higher is the number of dislocations in pile-ups. In a deformed single-crystal molybdenum specimen, a higher dislocation density is found compared with molybdenum and rhenium of commercial purity, due to the higher plastic deformation occurring up to fracture in the case of single crystals. Increase in the purity and plasticity of molybdenum is accompanied by a decrease in the size of the coherent scattering regions and in microdeformation. It has been found that the dimensions of the coherent scattering regions of commercial molybdenum are approximately three times those of single-crystal molybdenum and that the microdeformation is 1.5 times greater. In refined single-crystal molybdenum, random dislocation distribution was found, while in commercial molybdenum the distribution was in the form of pile-ups. Such a character of dislocation distribution is associated with high ductility in molybdenum single crystals. The high interstitial impurity content of commercial molybdenum prevents the movement of dislocations, as a result of which large dislocation pile-ups are formed, accompanied by stress concentration. This stress concentration assists fracture of the metal at low degrees of plastic deformation.

TABLE 37. Coherent Scattering Regions D and Relative Microdeformation $\Delta a/a$ for Molybdenum and Rhenium with Different Oxygen Contents [61, 65]

| Metal | $O_2$, % | $D \cdot 10^6$, cm | $\Delta a/a \cdot 10^3$ |
|---|---|---|---|
| Polycrystalline vacuum-melted molybdenum | 0.023 | 7.8 | 3.8 |
| The same, deoxidized with lanthanum | 0.002 | 4.7 | 3.6 |
| Single-crystal molybdenum | < 0.001 | 2.5 | 2.6 |
| Polycrystalline vacuum-melted rhenium | 0.020 | – | 6.4 |
| Single-crystal rhenium | < 0.001 | 3.3 | 4.5 |

It is shown in [61] that, as in the case of metals having a body-centered cubic lattice (tungsten, molybdenum, niobium, etc.), the purity of hexagonal rhenium to a considerable degree determines the magnitude of the crystal structure distortions after deformation. An increase in the impurity content of rhenium (particularly oxygen) produces a lesser degree of break-up of the crystal structure and a greater accumulation of residual stresses. The regions of coherent scattering after deformation of a rhenium single crystal are small in size, amounting to $3.3 \cdot 10^{-6}$ cm, comparable with the value of D for single-crystal molybdenum. The microdeformations of the lattice of commercially pure rhenium are approximately 1.5 times greater than in single-crystal rhenium. Table 37 shows the values of D and $\Delta a/a$ for molybdenum in the [110] direction and for rhenium in the [002] direction with different oxygen contents.

Thus, impurities in molybdenum, rhenium, and other refractory metals, substantially affect the mechanism of their deformation and fracture. Interstitial impurities are responsible for the brittle fracture of refractory metals.

## Mechanical and Other Properties of Metal Single Crystals

The principal results of an investigation of the mechanical properties of single crystals of low and medium melting point, acquired by prewar metallurgy, were reported in [3]. Work carried out after the Second World War will be mainly discussed in the following.

### 1. Single Crystals Having a Face-Centered Cubic Lattice

Investigations have been carried out mainly on fcc metals of medium refractoriness. The high ductility of these metals was discovered long ago and is well known. What is new has been the establishment of the fact of the sudden reduction in the recrystallization temperature of zone-refined aluminum to temperatures below room temperature. This aluminum had a purity of 99.998% with a high residual electrical resistance of 7000. When rolled with a high degree of reduction (96%) the metal recrystallized spontaneously even at −60°C [20, 68], compared with 250°C for aluminum of commercial purity (99.7%). The critical cleavage stress of a single crystal of zone-refined aluminum is only 28 g/cm$^2$.

Very much work has been done on the mechanism of slip and dislocation distribution in copper single crystals and bicrystals [69]. In particular, it has been clearly shown that the stresses set up under tension are proportional to the number of systems participating in the slip [70]. It has been shown [20] that the purest copper after zone refining for ten times has a ratio of electrical resistance at room temperature and at liquid hydrogen temperature (20.4°K) of $12.9 \cdot 10^4$. Zone-refined copper polygonizes at 650-700°C, while copper of commercial purity does so at 825-850°C [71]. According to data of Professor B. M. Gokhberg, who has examined the copper single crystals produced in our laboratory, their electrical strength under high-power discharge is much higher than in polycrystalline commercial copper. The investigations showed that inclusions serve as electrical breakdown "centers." According to our results obtained in collaboration with N. I. Novokhatskaya in tensile tests on copper and brass single crystals, the typical brittleness zone (ductility drop) for commercial copper and brass disappears at 400-500°C. This fact is of fundamental significance, since it clearly shows that red shortness of copper and brass is not inherent in the pure metals but is caused by physicochemical actions of the components of the air on the boundaries of the heated metal [40, 72].

Data on the properties of nickel single crystals are contained in [73]. Zone-refined nickel was obtained with a ratio of electrical resistance at room temperature and at liquid helium temperature equal to 3500. In this respect, a considerable influence is exerted by the carbon and iron contents, which by existing methods are determined to an accuracy of greater than $10^{-4}$%. It is pointed out that the presence of some impurities in concentrations of even less than $10^{-4}$% completely vitiate such phenomena as magnetoresistance and cyclotron resonance, which are used for the construction of the Fermi surface.

Of the refractory metals having a face-centered cubic lattice, considerable interest is presented by an investigation of single crystals of group VIII noble metals (platinum, palladium, and especially iridium). The characteristics of the deformation of iridum single crystals produced by electron-beam zone refining have been investigated [74]. The character of the stress–strain curves for iridium single crystals and also for copper, silver, and nickel single crystals has been found to be similar. The critical shear stress (as a function of temperature) and the rate of strain hardening were determined from the stress–strain curves of iridium single crystals. There is information [75] that single crystals of iridium and especially of rhodium withstand a considerable amount of cold plastic deformation (iridium 25% and rhodium 95%), while these metals produced in a polycrystalline state by electric-arc melting or sintering can be deformed only on heating to 1500°C.

## 2. Single Crystals Having a Body-Centered Cubic Lattice

The most significant results have been obtained in the investigations of single-crystal metals having a body-centered cubic lattice. The most important has been the experimental evidence of the absence of cold brittleness in bcc metals of exceptional purity.

For example, in zone-refined iron (less than $15 \cdot 10^{-4}$% carbon and oxygen) the following new properties have been found [5, 20, 71, 75]:

a) absence of ductile to brittle transition on cooling even in liquid helium (4.2°K); under tensile stress, the reduction in area is 90%, whereas commercial iron becomes brittle even at −40°C;

b) presence of a horizontal portion on the elongation curve in the region of the yield point, instead of the well-known "tooth" in Armco iron;

c) low solubility of interstitial impurities: oxygen ($7 \cdot 10^{-4}$ wt.%), hydrogen, carbon, sulfur, phosphorus;

d) reduction in the polygonization temperature to 200°C (as against 850°C) in Armco iron;

e) different form of dilatometer curves at the $\alpha \rightleftharpoons \gamma$ transition temperature.

The walls of magnetic domains are arranged with perfect regularity in crystals of zone-refined iron, whereas they are arranged in disordered fashion in less pure iron. The mobil-

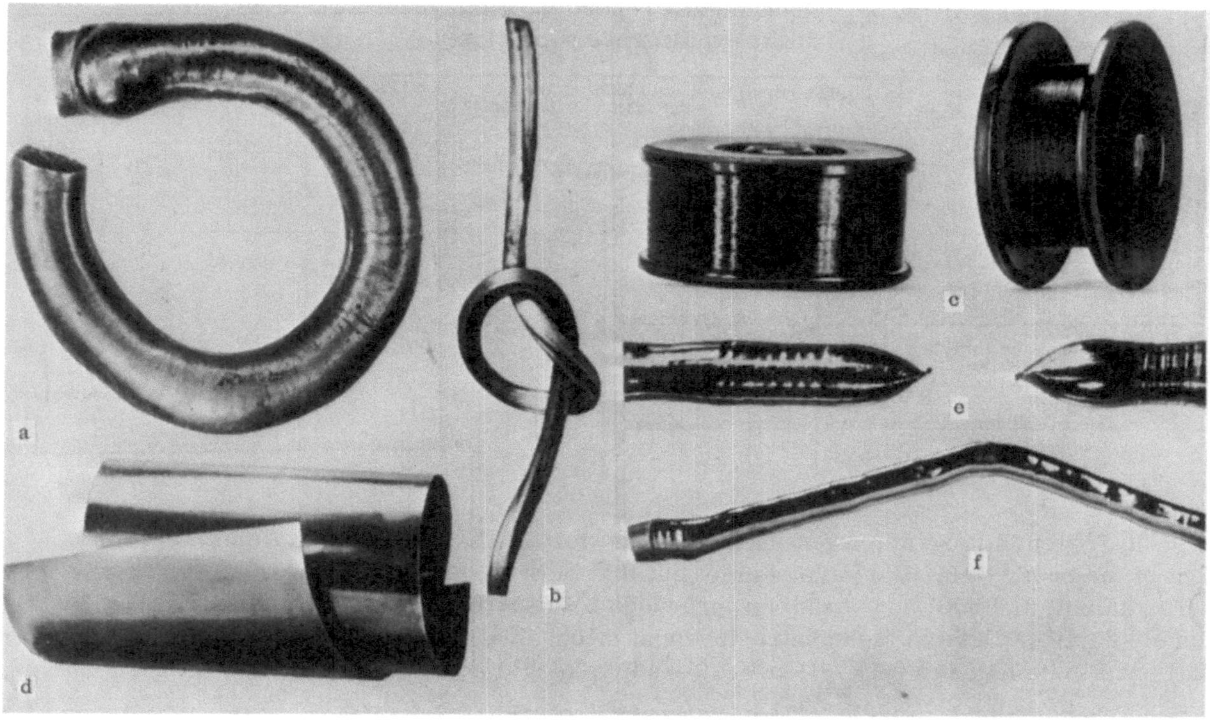


Fig. 107. Illustrations showing the high ductility properties at 20°C of single-crystal molybdenum produced by electron-beam zone refining.

ity of the domain boundaries in zone-refined iron is exceptionally high, and evidently the domains have the maximum dimensions (in thin films, the magnetic domains can be observed in the transmission electron microscope).

Further research on very pure iron will indubitably lead to an extension of the list of its new properties. Thus the foundations have already been laid for an absolutely new metallurgy of high-purity iron, which will undoubtedly be very soon reflected in different branches of science and technology.

What has been said also refers in many respects to the group Va bcc metals – vanadium, niobium, and tantalum. In the production of important components and devices in atomic and electronic engineering and rocket construction, increased demands are being made on metals and alloys with regard to ductility at ordinary and low temperatures, weldability, absence of liberation of gas on heating, stability of properties over lengthy service periods, presence of a predetermined combination of physical properties and compatibility with other materials. Many of these requirements are met by refractory rare metals of groups IVa, Va, and VIa in a state of high purity, and by alloys based on them.

Group Va metals in the solid state have a rather high solubility for interstitial impurities of the order of fractions of one percent. High ductility may be obtained simply by remelting (without zone refinement) in an electron-beam vacuum furnace, principally as the result of the evaporation of the lower oxides of higher vapor tension. A 100-mm-thick slab of electron-beam-melted niobium may be cold-rolled, without annealing, into foil having a thickness of one-tenth of a millimeter. Vanadium, niobium, and tantalum in the form of electron-beam polycrystalline ingots (semifinished products) or in the form of zone-refined single crystals [76] do not become brittle under any cooling conditions. Interstitial impurities may be arranged in the following order of their influence on tendency to brittle fracture ("order of embrittle-

TABLE 38. Mechanical Properties of Molybdenum of Different Degrees of Purity

| Material | Gaseous impurities, % | | | Mechanical properties under tensile stress | | | Hardness, kg/mm² | Temperature at commencement of recrystallization | Ductile – brittle transition temperature |
|---|---|---|---|---|---|---|---|---|---|
| | oxygen | hydrogen | nitrogen | tensile strength, kg/mm² | elongation, % | reduction in area, % | | | |
| Cermet molybdenum after welding | 0.02 | — | 0.004 | 50—100 | — | — | 150—200 | 1260—1390 | 300 |
| Electric arc-melted molybdenum (remelted twice) | 0.003 | 0.0003 | 0.001 | 57.1 | 28.96 | 0.1 | 170—200 | 950—1050 | 25 |
| Single crystal of molybdenum (four zone refinement passes) | 0.0008 | 0.0004 | 0.0006 | 31 | 10—15 | 100 | 160—170 | 650 | —196 |

ment"): hydrogen, nitrogen, oxygen, carbon. Thus, for the ductility of vanadium, a carbon content of the order of 0.1% is permissible, but that of hydrogen is only 0.001%. Single crystals of tantalum, niobium, and vanadium, produced by electron-beam zone refining [76], are extremely ductile. Data on the substructure and mechanical properties of niobium single crystals are to be found in [71, 78], and of vanadium single crystals in [79].

More surprising results have been obtained in research on the properties of single crystals of the group VIa metals, molybdenum and tungsten, purified by electron-beam vacuum melting. This is due to the extremely low solubility of the interstitial impurities (less than 0.0001%). Only by the use of such a powerful method of purification has it been possible to produce these metals without embrittling boundary inclusions of nonmetallic compounds, and to reveal the true, inherent properties of molybdenum and tungsten. The foregoing may be demonstrated, for example, by the molybdenum single crystals, whose fine structure is shown in Fig. 100.

It may be recalled that commercial cermet molybdenum has a ductile–brittle transition temperature of about 300°C. The first property characteristic of highly pure molybdenum is its phenomenal ductility down to the temperature of liquid helium [76-81]. At room temperature, a molybdenum single crystal may be bent without fracture through 180° (Fig. 107a) or beaten into a sheet under the hammer (the degree of deformation attaining 90% or more). Rolled wire produced from a single crystal may be tied in a knot (Fig. 107b). It is possible to produce from a single crystal 12 mm in diameter, by drawing without annealing, a wire 80 μ in diameter or less in a length of several thousand metres (Fig. 107c), or foil with a thickness of 50 μ (Fig. 107d) which in turn may be cold drawn, this being very important for the production of a number of important components of electrovacuum instruments and devices.

Increase in the freedom of zone-refined molybdenum from interstitial impurities substantially reduces its resistance to plastic deformation. Thus, the tensile strength of a molybdenum single crystal is 31 kg/mm² compared with 50-60 kg/mm² for arc-melted molybdenum; under tensile stress, a molybdenum single crystal forms a 100% neck (Fig. 107e) with an elongation of 10-15%. The mechanical properties of molybdenum of different degrees of purity and data on the ductile–brittle transition temperature are given in Table 38. Impact toughness tests on unnotched specimens of molybdenum single crystals showed that at room temperature it is more than 15 kgm/cm². Figure 107f shows a single crystal of molybdenum bent between two supports by the pendulum hammer of an impact testing machine.

TABLE 39. Mechanical Properties of Single Crystals of Refractory Metals Having a Body-Centered Cubic Lattice, Tension, and Bending

| Metal | Tensile strength, $kg/mm^2$ | Elongation, % | Reduction in area, % | Angle of bend at −196°C, deg |
|---|---|---|---|---|
| Chromium . . . | — | — | 40 | — |
| Tungsten . . . | 100 | 10—15 | 100 | 180 |
| Vanadium . . . | — | — | 100 | — |
| Molybdenum. . . | 31 | 10—15 | 100 | 180 |
| Niobium . . . | 15 | 12 | 100 | 180 |
| Tantalum . . . | 20.8 | — | 100 | 180 |

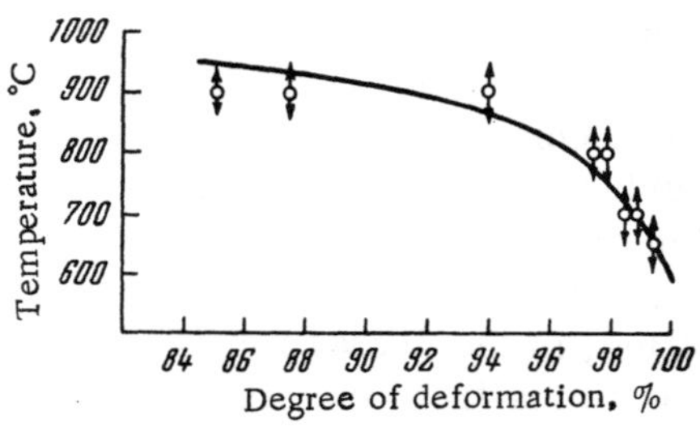


Fig. 108. Dependence of initial recrystallization temperature on the degree of deformation of single-crystal molybdenum.

The data presented in Table 39 show that extremely pure refractory metals having a bcc lattice and characterized by high ductility with a 100% reduction in area, have relatively low elongation values (10-15%). The rupture has the knife-edge form characteristic of single crystals, and due to anisotropy. The scale factor and work hardening of the metal may have a strong influence on the tensile strength of molybdenum. Thus, in the peened condition, the tensile strength of a wire of zone-refined molybdenum 0.1 mm in diameter is 105 $kg/mm^2$, and for wire 0.1 $\mu$ in diameter it is 260 $kg/mm^2$, which opens up a future for the use of molybdenum and its alloys for the production of the resilient components of instruments [64] and for reinforcing dies.

Extreme purification of single-crystal molybdenum leads to a considerable reduction in the initial recrystallization temperature. The recrystallization temperature of single-crystal molybdenum for 98-99% deformation is 650°C as against 935°C in arc-melted polycrystalline molybdenum [80]. At a deformation of 70-80%, evidently only collective recrystallization occurs. A similar process was observed for magnesium and its alloys at low degrees of deformation [83]. Figure 108 shows how the initial recrystallization temperature depends on the degree of deformation of single crystal molybdenum. Recrystallization data were obtained on molybdenum single-crystal specimens with crystallographic orientations in the middle of the stereographic triangle. According to the data of [85], the recrystallization temperature of specimens obtained from single-crystal molybdenum after reduction by rolling of 90% depends to a considerable extent on the initial orientation of the crystal with respect to the rolling direction and rolling plane, and varies within the range of 750-1200°C. A more pronounced rolling texture is accompanied by a higher initial recrystallization temperature. These investigations were recently continued in our laboratory. It was found that by growing molybdenum single crystals with an orientation of the axis along the [110] and [100] crystallographic directions, using initial material with small additions of group IVa metals (for producing the necessary substructure), the recrystallization temperature of wire produced from the single crystals could be increased to 1600-1700°C. Such wire is necessary for a number of instruments of high reliability.

It is particularly important to emphasize the increase in ductility of peened wire and foil made from molybdenum single crystals after annealing in a vacuum (of not more than $10^{-5}$ mm Hg) at 700-800°C (recrystallization treatment). Thus foil 0.1 mm thick in the peened con-

TABLE 40. Thermal Conductigity and Integral Degree of Blackness of Molybdenum Single Crystals

| $T$, °K | λ, W/m·deg | ε | $T$, °K | λ, W/m·deg | ε |
|---|---|---|---|---|---|
| 1300 | — | 0.140 | 1900 | 99.3 | 0.218 |
| 1400 | 111 | 0.154 | 2000 | 98.0 | 0.230 |
| 1500 | 108 | 0.168 | 2100 | 96.4 | 0.241 |
| 1600 | 105 | 0.181 | 2200 | — | 0.252 |
| 1700 | 103 | 0.194 | 2300 | — | 0.263 |
| 1800 | 101 | 0.206 | | | |

dition withstands without fracture four bends through 90° (around a mandrel 1 mm in radius), and in the annealed condition 25-30 bends [80]. This fact is of fundamental importance. As is well known, group VIa metals (molybdenum, chromium, tungsten of commercial purity, especially in the form of cermets), in contrast to other metals, become brittle in the recrystallized condition. This is due to precipitation of brittle nonmetallic compounds along the grain boundaries during annealing, and the decrease in intergranular surface during recrystallization grain growth, as the result of which the specific concentration of interstitial impurities increases. The physical essence of the recrystallization process, consisting in the atoms of the deformed metal overcoming the atomic bonds in the deformed lattice, in the nucleation of new undistorted grains and their subsequent growth by the migration of atoms from the distorted to the undistorted crystals cannot be the cause of embrittlement. On the contrary, in recrystallization, the internal stresses are relieved and the distortion produced in the crystal structure on deformation is eliminated which ought to result in an increase in ductility, as was indeed found in the recrystallization of deformed single-crystal molybdenum.

Table 40 presents data obtained in the High-Temperature Institute (by V. E. Peletskii and V. Yu. Voskresenskii) on the thermal conductivity and integral degree of blackness of molybdenum single crystals.

Unfortunately, progress in the purification of tungsten is not satisfactory as in the purification of molybdenum. This is evidently due mainly to the lower purity of the starting material and stock used for the production of single crystals, although their $\rho_{300°K} / \rho_{4.2°K}$ ratio is 1000-10,000 as against 10-60 for the commercial metal. As already stated, the use of oil-vapor pumps in electron-beam zone refining increases the difficulty of removing carbon impurity from the tungsten. Tungsten single crystals without additional purification currently have a higher ductile—brittle transition temperature as compared with all the other metals having a bcc lattice, and have to be heated in protective shells for plastic deformation. However, the production of parts made from high-purity single-crystal tungsten is evidently the only way of processing cast tungsten to give semifinished products of thin section for electronic engineering (foil, wire, rolled products, strip, capillary tubes, etc.).

Tungsten single crystals having a high degree of purity and high ductility have been grown under laboratory conditions from stock obtained from highly pure tungsten raw material or vacuum-melted tungsten (with additional purification) [51].

Extensive purification sharply reduces the ductile—brittle transition temperature of tungsten; a tungsten single crystal 4 mm in diameter may be bent through 180° at the temperature of liquid nitrogen.

As already pointed out, carbon has a considerable influence on the ductility of tungsten. An increase in the carbon content, accompanied by coarsening of the carbides, leads to a sharp reduction in the ductility of tungsten and an increase in the ductile—brittle transition temperature. In view of the high thermodynamic stability of tungsten carbide compared with other nonmetallic impurities, special attention has been paid to the production of tungsten having a low carbon content, a considerable reduction in the latter being ensured by the production of highly pure tungsten anhydride and its subsequent reduction in quartz boats by hydrogen of ultrahigh purity. Extrusion and sintering of the rods was carried out without the use of carbon-containing binder. In zone refining the sintered rods, entry of oil vapor into the working chamber was prevented by the use of nitrogen traps. An original method of purifying tungsten from carbon was proposed by E. M. Savitskii and G. L. Tsarev [89]. Purification from carbon was carried out by cyclic heat treatment of sintered tungsten rods in an inert gas atmosphere having a low carbon content (0.02%). Long-continued heating of a tungsten rod to 2700-2900°C by the passage of current through it in a weakly oxidizing atmosphere results in the decarburization of the tungsten due to the interaction of the oxygen with the tungsten carbide and the formation of volatile CO and $CO_2$. The efficiency of the process is greatly increased by cyclic heat treatment: heating to 2700-2900°C, cooling to 300-400°C (10 cycles), holding at 2700-2900°C for 15-20 min, and then repeated heating–cooling cycles. In cyclic heat treatment, the rate of diffusion of the carbon sharply increases, since under conditions of thermodynamic instability, and treatment leading to the "swinging" of the metal structure (mechanical treatment, thermal "swing," etc.) increases very considerably the rate of all diffusion processes, and in particular the mobility of impurities [40, 87]. During cyclic heating and cooling, an excessive vacancy concentration is produced in the tungsten rod. On subsequent holding at the high temperature, heterogeneous carbon-vacancy complexes are formed, accompanied by their movement to the grain boundaries acting as sinks for the complexes [51]. Having reached the grain boundaries, the carbon reacts with the oxygen to form volatile CO. As the result of cyclic treatment, the carbon content of the stock fell from 0.04 to 0.001%. The tungsten single crystals produced from such stock showed exceptionally high ductile properties [51, 88, 89]. The mechanical tensile properties and ductile–brittle transition temperature of tungsten single crystals with a carbon content of 0.001% after two phases of the liquid zone, are given below (rate of deformation 0.05 mm/min):

| | |
|---|---|
| Yield point, $kg/mm^2$ | 33.8 |
| Yield point in torsion, $kg/mm^2$ | 15.7 |
| Tensile strength, $kg/mm^2$ | 102 |
| Elongation, % | 10 |
| Reduction area, % | 100 |
| $H_{micro}$, $kg/mm^2$ | 310 |
| $T_{trans}$, °C | Below 196 |

Reduction in the carbon content leads to the production of single crystals having a greater degree of perfection with large subgrains and low disorientation angles. The dislocation density falls from $10^7$-$10^8$ $cm^{-2}$ (for single crystals with a carbon content of 0.025%) to $10^5$-$10^3$ $cm^{-2}$ (for single crystals with a carbon content of 0.001% and less). The low-carbon single crystals have $\rho_{300°K}/\rho_{4.2°K} = 11,200$.

The initial recrystallization temperature of deformed single-crystal tungsten (after 90% deformation) is 850°C. For cermet tungsten and electric-arc vacuum-melted tungsten, the commencement of recrystallization is in the limits 1100-1400°C [90], depending on the degree of purity. Recrystallization annealing leads to a slight increase in the ductility of deformed single-crystal tungsten, while in the case of tungsten of commercial purity, embrittlement is observed after recrystallization annealing [91, 92].

Evidently, recrystallization annealing of high-purity tungsten will necessarily result in an increase in ductility, as shown by the example of single-crystal molybdenum.

TABLE 41. Physicomechanical Properties of Rhenium of Different Degree of Purity

| Material | Gaseous impurities, % | | | Mechanical properties | | Hardness, kg/mm² | Initial recrys-tallization temp., °C |
|---|---|---|---|---|---|---|---|
| | oxy-gen | hydro-gen | nitro-gen | tensile strength, kg/mm² | elonga-tion, % | | |
| Cermet rhenium | 0.07 | 0.008 | 0.006 | 115 | 12 | 250—280 | 1500—1800 |
| Electric-arc-melted rhenium | 0.01 | 0.001 | 0.002 | 75.2 | 19 | 200—250 | 1250—1750 |
| Single-crystal rhenium | 0.001 | 0.0001 | 0.0001 | 42.5 | 41 | 120—100 | 1000 |

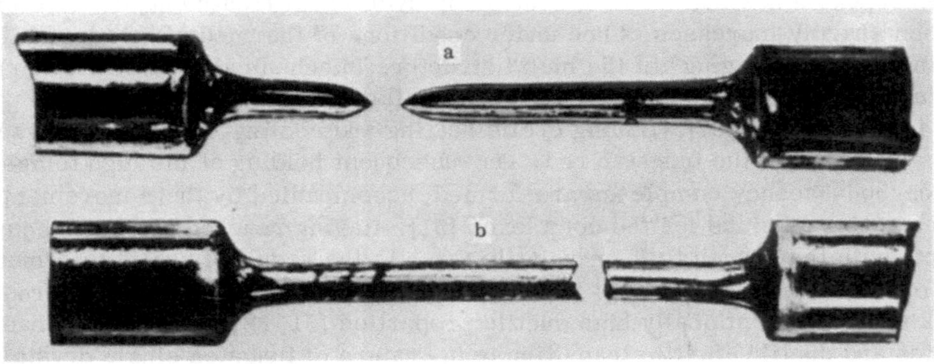


Fig. 109. Tungsten single crystal after tensile test. a) Axis of specimen in the [110] crystallographic direction; b) the same, in the [100] direction.

## 3. Single Crystals of Metals with a Hexagonal Lattice

A reduction in the interstitial impurity content, primarily of oxygen, sharply increases the ductile properties of rhenium. The considerable reduction in oxygen content of single-crystal rhenium reduces the hardness to 100 kg/mm² on the basal plane compared with 200-250 kg/mm² in the initial material. The tensile strength and elongation under room-temperature conditions of single-crystal rhenium with an orientation of the stress axis between the $[10\bar{1}0]$ and $[11\bar{2}0]$ directions of the stereographic triangle are 42.5 kg/mm² and 41%, respectively [58, 60, 93]. Table 41 gives the mechanical properties of single-crystal rhenium with the above-mentioned orientation compared with cermet rhenium and vacuum-melted rhenium.

Considerable anisotropy has been found in the mechanical properties of rhenium, titanium, scandium, and also ruthenium and osmium. Thus, for example, the difference in hardness attains 5:1 for rhenium, 3:1 for $\alpha$-titanium, and 2:1 for scandium and yttrium [86].

## Anisotropy of the Properties of Single Crystals of Metals Having a Body-Centered Cubic Lattice

Anisotropy in the arrangement of the atoms, and the resulting difference in the physical and chemical properties along different planes and directions, as is well known, constitute one of the most characteristic features of crystals. The anisotropy of the mechanical properties

TABLE 42. Mechanical Properties of Tungsten and Molybdenum Single Crystals as a Function of Crystallographic Orientation

| Material | Orientation | Tensile strength, kg/mm² | Elongation, % | Reduction in area, % | Microhardness, kg/mm² |
|---|---|---|---|---|---|
| Tungsten single crystal | [100] | 100 | 2 | 0 | 300—320 |
| | [110] | 98 | 15 | 100 | 300—320 |
| Molybdenum single crystal | [100] | 46 | 23 | 47 | 157 |
| | [110] | 29 | 31 | 100 | 157 |

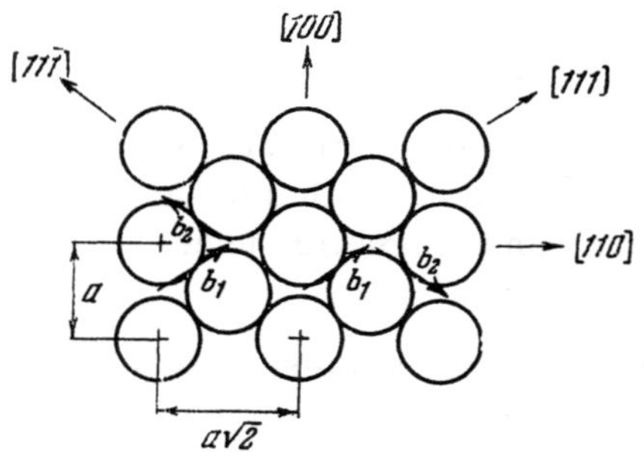


Fig. 110. Arrangement of atoms in a (110) plane of a body-centered cubic lattice.

of metal crystals with crystallographic positions is examined in [3, 55, 71, 94]. With regard to refractory metals having a face-centered cubic lattice, however, the question of anisotropy has not yet been completely clarified owing to lack of experimental material.

We have studied the anisotropy of the mechanical properties of single crystals of tungsten (carbon 0.005%, oxygen 0.001%, nitrogen 0.0001%). Tensile tests were made at room temperature on microspecimens produced by electropolishing in a 5% NaOH solution. Tungsten single crystals with the [100] crystallographic direction showed the lowest ductility (elongation 2%, practically no reduction in area was found). Specimens with the [110] crystallographic direction had maximum ductility. In these specimens with an elongation of 10-15%, a reduction in area of 100% was found. The fracture had a knife-edge form, as in single crystals of molybdenum and other refractory metals (Fig. 109).

In other crystallographic directions, deformation occurred without preliminary necking. The observed anisotropy of ductile properties may evidently be explained by the dependence of the deformation mechanism of body-centered cubic metals on the initial crystallographic orientation. The tensile strength (90-100 kg/mm²) and hardness (300-320 kg/mm²) of tungsten single crystals of the given purity are practically independent of the crystallographic orientation.

Considerable anisotropy of mechanical properties is found in molybdenum single crystals (impurity content: carbon 0.003, nitrogen 0.001, oxygen less than 0.001%).

The influence of the crystallographic orientation on the mechanical properties of tungsten and molybdenum single crystals is shown in Table 42.

When a molybdenum single crystal is subjected to tensile stress in the [110] direction, clearly distinct upper and lower yield points are found; under tensile stress in other directions, the existence of an upper and lower yield point is not so apparent. In [95] this difference in the behavior of molybdenum single crystals is attributed to the anisotropy of the space packing of the atoms. Figure 110 [95] shows the arrangement of the atoms in a [110] plane of a bcc lattice. The directions of the closest packing of the atoms are [111] and [11$\bar{1}$]. The packing of the atoms in the [110] and [100] directions is different. It is assumed that the impurity atoms are arranged in the voids between the metal atoms. Under tensile stress in the [100] direction, the metal atoms move away from each other in this direction and approach the perpendicular. The impurity atoms are provided with a greater volume and exercise less influence in the course of deformation under tensile stress. Under tensile stress in the [110] direction, the impurity atoms are as it were squeezed by the metal atoms, and consequently have a considerable influence on their movement. Although in tungsten and molybdenum single crystals the impurity atoms are arranged mainly along the grain boundaries, and also around dislocations and at point defects, according to R. Palme [95], the mechanism described above is also true of these cases.

As already pointed out, the formation of deformation texture when single crystals of bcc metals are rolled depends on the crystallographic orientation of the crystal in relation to the rolling direction and plane. It is precisely in this way that one can explain the dependence of hardening on orientation in the rolling of molybdenum single crystals [95]. According to Rexer [96], a specimen with the initial orientation {100}[110] after rolling (77% reduction) showed a slight increase in hardness. The rolled specimens also had an increased recrystallization temperature. According to [97], a specimen having the initial orientation {100}[110] after rolling (77% reduction) remained a single crystal and retained the same orientation.

Crystals of the cubic system [31, 94], in contrast to crystals of lower symmetry, show isotropy with respect to specific electrical conductivity. Evidently, this also applies to their thermal conductivity, thermal expansion, thermoelectromotive force and magnetic susceptibility.

Considerable interest attaches to the comparatively recently discovered anisotropy of thermionic emission on different faces of a tungsten single crystal. The lowest values of the work function (4.39 eV) is obtained from the (111) octahedron plane and the maximum value from the (110) plane [98]. The maximum value of the work function has not yet been determined accurately. Various investigations give values of 4.69 to 6 eV. According to Smith [99], the work function for the (110) face is 5.26 eV, while according to Shuppe [98] it is 5.14 eV.

The study of electron emission of single-crystal tungsten in a spherical electron projector is of great interest [100]. Emission patterns have been obtained and the work functions have been measured on three faces by the Richardson method. The results of these tests are $\varphi_{(110)} = 5.33 \pm 0.03$ eV; $\varphi_{(100)} = 4.40 \pm 0.03$ eV; $\varphi_{(111)} = 4.3 \pm 0.03$ eV.

S. Smoluchovskii has made a theoretical calculation of the work function of the emission of electrons from different faces of a tungsten single crystal [101] and gives the following arrangement of the planes in descending order of work function: (110), (100), (112), and (111). This is in agreement with the experimental results; on the introduction of a correction for nonuniform distribution of the charge, the descending order of work function would be (110), (112), (100), and (111). In our view, generally, the maximum work function in a vacuum evidently ought to be from planes having the closest atoms: (110) for bcc, (111) for fcc, and (1000) for hexagonal lattices. In cesium plasma, on the contrary, the work function for these faces of single crystals will be a minimum and the thermionic current a maximum. We may thus speak of a correlation between the phenomenon of thermionic emission and plastic deformation of metal crystals, since the latter also occurs along planes most closely packed with atoms. Experimental results show a much higher anisotropy than do theoretical calculations. According to recent data, the work function anisotropy for different faces of a tungsten crystal attains values of 1.2-1.6 eV.

The work function anisotropy of single-crystal molybdenum was studied by E. M. Savitskii and co-workers [102]. Single crystals with the dimensions 2 × 2 × 0.15 mm and definite crystallographic orientation were prepared for measuring the work function. The starting stock for the preparation of such specimens was a zone-refined single-crystal molybdenum rod 10-12 mm in diameter and 150-200 mm in length with varying crystallographic orientation. The single crystals contained less than 0.001% oxygen, less than 0.0001% hydrogen, less than 0.0001% nitrogen, and about 0.01% carbon; the content of metallic impurities (iron, tungsten, silicon) was less than 0.01%.

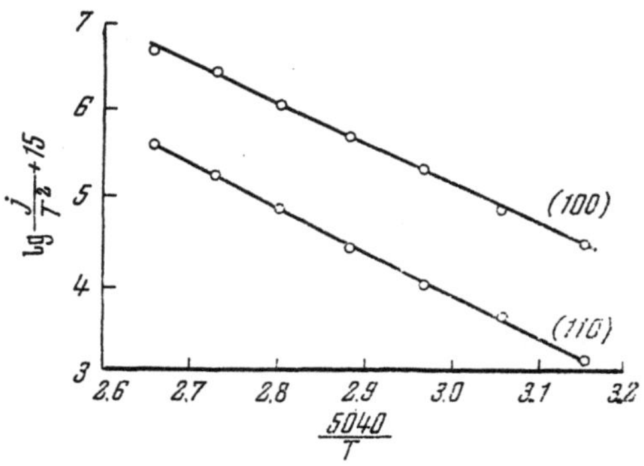


Fig. 111. Richardson straight lines for the (110) and (100) faces of a molybdenum single crystal.

The specimens were prepared in several stages. First of all, specimens of the dimensions $2 \times 20 \times 1$ mm were cut off by means of grinding wheels. The thickness of the specimens was then reduced to 0.2 mm by mechanical grinding and polishing. Finally, the specimens were electropolished for removing the work-hardened surface. Laue photographs were made for checking the crystallographic orientation. The measurements were made in sealed-off tubes (see Fig. 41) in a vacuum with a residual pressure of less than $1 \cdot 10^{-7}$ mm Hg, measured by an LM-2 tube. The initial evacuation of the tubes was carried out by means of oil-vapor pumps. After lengthy heating of the glass (400°C, 20 h), cathode (1900-2000°K, 10 h), and other metallic parts, sealing-off of the measuring tube together with the LM-2 tube and the gettering system of the tubes was checked. In the evacuated apparatus, further reduction of the pressure of the residual gases was carried out by means of Feba getter, titanium iodide, and was measured by means of the LM-2 tube.

The cathode temperature was measured by means of a tungsten–iridium thermocouple. Before measurement of the emission current, the cathode was flashed at 2200°K.

The emission-current densities were measured in the range 1600-1900°K (every 50°K). The work function was determined from the Richardson straight lines. Figure 111 shows the temperature dependence of the electron currents determined on the basis of a series of experiments. The effective work function $\varphi$ for the (110) and (100) faces, as found from the slope of the straight lines in the graph, were $4.9 \pm 0.076$ and $4.35 \pm 0.07$ eV, respectively.

The work function for polycrystalline molybdenum is $4.26 \pm 0.05$ eV [103]. The values obtained for the work function from pure surfaces of molybdenum single crystals are evidence of the considerable anisotropy of the emission properties in different crystallographic directions.

## Production and Properties of Refractory Metal Alloy Single Crystals

As already pointed out, a considerable obstacle to the production of alloys based on refractory metals is the reduction in technological ductility on alloying refractory metals with group Va and especially group VIa elements, which form substitutional solid solutions. As shown by scientific and industrial experience with group IVa refractory metals, especially titanium and zirconium, every possible means of producing alloys may be utilized, provided only that the metal base and also the alloying additions are obtained in a state of high purity, particularly as regards interstitial impurities. In this connection, the production of single-crystal refractory metal alloys free from interstitial impurities is of paramount importance. By means of electron-beam zone refining, we have been able for the first time to grow single crystals of some refractory metal alloys in the solid-solution region of the molybdenum–niobium, molybdenum–tungsten, tungsten–niobium, tungsten–rhenium, and tungsten–tantalum systems [1, 77, 81, 104, 105].

The starting alloy was first prepared by arc melting with overflow into a copper mold in a vacuum or inert atmosphere. If, however, considerable difficulties arise in the mechanical treatment of the cast alloys for producing stock for single crystals, melting of the components may be carried out during actual zone refining. In this case, the starting materials are rods of the high-purity components of suitable cross sections, corresponding to the mean composition of the alloy. For example, to produce a single-crystal alloy of tungsten with 10% Mo, a tungsten single crystal is first grown and this is joined along the entire length by a calculated addition of molybdenum in rod or wire form. During fusion, the molybdenum mixes with the tungsten and for a certain rate of advance of the zone a single crystal of the alloy is grown. To obtain a uniform distribution of the components, the method of zone refining with several passes of the fusion zone in opposite direction is used (zone leveling). The end portions of the resulting single-crystal rod are discarded. Chemical analysis of the single-crystal rods confirmed the uniform distribution of the components over the length. Analysis of Laue photographs showed that the preferred growth direction of single crystals of alloys based on refractory body-centered cubic metals was along the [100] axis, just as for the corresponding pure metals grown without seeding.

In growing single-crystal alloys, the seeds used were crystals of the pure metal base. In such a case, a more dilute alloy is obtained close to the seed, but even at a distance of 15-20 mm from the seed the composition of the alloy evens out. In other respects, the growing of a single crystal remains the same as already described.

As is well known, the molybdenum – niobium system represents a continuous series of solid solutions. The two components do not differ much with regard to melting point and have a bcc lattice with approximately the same parameters and atomic dimensions.

In an electron-beam zone-refining apparatus using a nitrogen trap in a vacuum of $10^{-5}$ mm Hg, we have grown single-crystal solid solutions of the molybdenum – niobium system with given crystallographic orientation over the entire concentration range [105]. We used rods of vacuum-melted molybdenum and niobium. Melting of the components was carried out directly in the electron-beam zone-refining apparatus. For the production of single-crystal alloys of uniform composition, the last two passes of the fusion zone were made in opposite directions, the last pass for producing the single crystal being made from the seed. The rate of movement of the fusion zone was 0.5-3 mm/min depending on the composition of the alloy.

As shown by the experiments made, the production of single crystals of high concentration was found to be impossible without seed, this being evidently due to the considerable distortions of the crystal lattice and the resulting stresses set up in the alloy. It was only by the use of a seed of a single crystal of pure molybdenum or niobium (or of their alloys) as substrate that it was possible to obtain single crystals of alloys having a high content of the second component.

All the alloys of the investigated system were grown with [100] and [110] crystallographic orientations. The deviation of the orientation from the given direction in growing from a seed was 1-5°.

Figure 112 shows the arrangement of etch figures of single crystals of molybdenum – niobium alloys on the (100) plane. Individual etch figures in the form of squares and also interrupted mosaic block boundaries, consisting of etch figures, are to be seen. The fine structure of single-crystal alloys on the molybdenum side is characterized by a certain increase in dislocation density, a reduction in size of the block structure, and an increase in the disorientation of the blocks with increase in the alloying element content. In Laue x-ray photographs of single-crystal alloys, splitting of the spots is observed, thus confirming the existence of a mosaic structure and an increase in the degree of disorientation of the blocks in single-crystal alloys.

The hardness and specific electrical resistance of single-crystal molybdenum – niobium alloys vary with the composition, as in polycrystalline specimens, in accordance with the

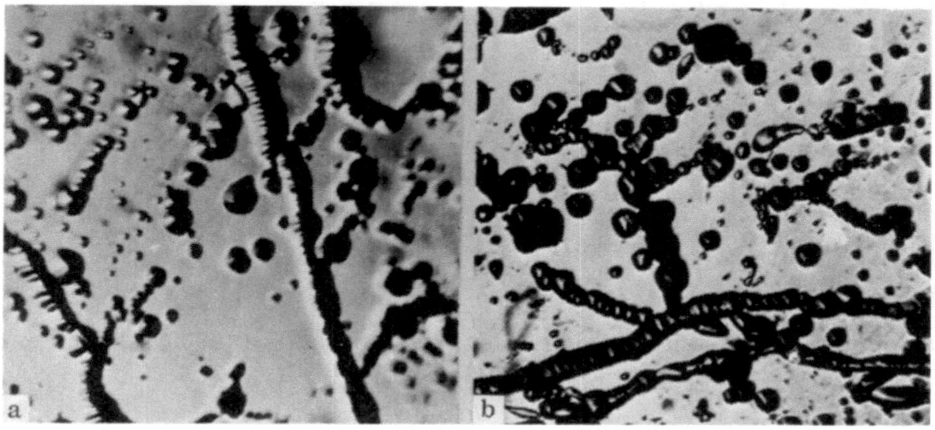


Fig. 112. Substructure of single crystals of molybdenum–niobium solid solutions on the molybdenum side (×500). Niobium content, %: a) 4; b) 10.

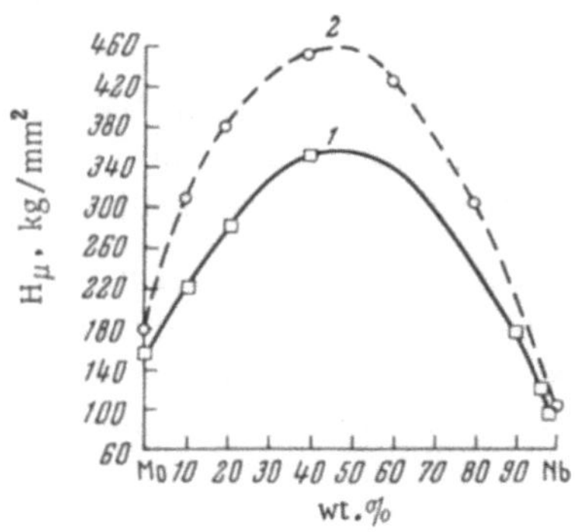


Fig. 113. Variation in hardness of molybdenum–niobium alloys in the single-crystal state (1) and polycrystalline state (2) as a function of the niobium content.

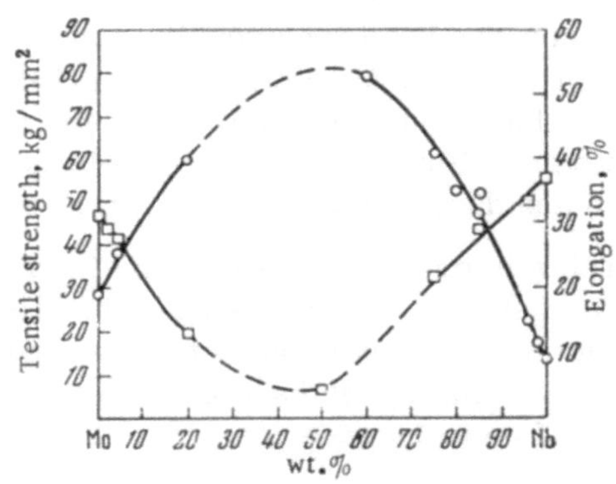


Fig. 114. Mechanical properties of single-crystal molybdenum–niobium alloys under tension. Orientation of crystals [110].

Kurnakov rule. In Fig. 113, the bottom curve shows the variation in hardness of single-crystal alloys of the molybdenum–niobium system, while the top curve is that of the same alloys but vacuum arc-melted. Due to the high purity with regard to interstitial impurities, the single-crystal specimens of the alloys have a lower hardness value and are characterized by better machinability than the corresponding arc vacuum-melted alloys.

The anisotropy of the mechanical properties of single crystals of molybdenum and molybdenum–niobium alloys is of considerable interest. Mechanical tests were made at room temperature using a Gagarin machine with recording of the elongation diagram. The test specimens were prepared by electrolytic polishing in a 10% NaOH solution using the apparatus described in [72], the work-hardened layer being completely removed. The surface of microspecimens prepared in this way was smooth and lustrous; the specimens retained their single-crystal properties.

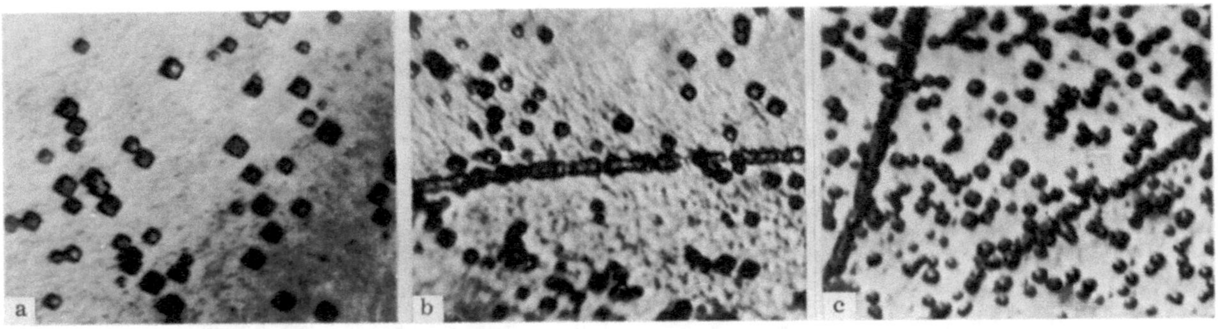

Fig. 115. Substructure of tungsten–rhenium alloy single crystals (×2000). Rhenium content, %: a) 0; b) 0.5; c) 1.5.

The results of tensile tests of molybdenum–niobium alloys (over the entire concentration range) at room temperature are shown in Fig. 114. Substantial anisotropy of the mechanical properties of single-crystal molybdenum–niobium alloys was found. As in the case of the pure metals, alloys with [110] orientation had the maximum ductility.

With increase in niobium content, the tensile strength of single-crystal molybdenum–niobium alloys increased and the ductility diminished, while remaining at a fairly high level. Single crystals of the alloy Mo–20%Nb with [110] orientation showed a reduction in area of 40% on tension and an elongation of 13%; for the [100] crystallographic direction, the reduction in area for the same alloy was 19%, for an elongation of 6%. The difference in the values of tensile strength and ductility of single crystals with the orientations [100] and [110] was 50%.

Considerable interest is presented by the investigation of single crystals of alloys of the tungsten–rhenium system as promising materials for cathodes of thermionic converters and also in connection with the existence of superconducting properties in these alloys [106]. Using the method of zone refining, we have produced single crystals of alloys of molybdenum with a rhenium content of up to 10 wt.%.

Alloying a tungsten single crystal with even slight additions of rhenium (0.5%) leads to a considerable increase in dislocation density. Figure 115 shows the substructure of single crystals of tungsten–rhenium alloys. A considerable increase in etch pit number may be observed with increase in the rhenium content of the alloy, the disorientation of the crystal block increasing at the same time. The increase in the number of dislocations facilitates the process of plastic deformation of tungsten–rhenium single crystals, deformation occurring by both slip and twinning. Single-crystal tungsten–rhenium alloys (containing small additions of rhenium) may be deformed at room temperature or with slight heating (to 200-400°C). This is not found in the case of polycrystalline alloys of the same composition with a higher content of interstitial impurities. Small additions of rhenium produce an appreciable reduction in the hardness of tungsten single crystals, which is also a measure of the resistance of a material to plastic deformation. Research is in progress on the thermionic emission properties of tungsten–rhenium alloy single crystals on different faces of the crystal.

In the Refractory and Rare Metal Alloys Laboratory of the A. A. Baikov Institute of Metallurgy investigations have been started on the production of single crystals of metallic compounds (NiAl, etc.) [104]. Information is available on single crystals of the compounds TiC, WC, $W_2C$, and TaC [108, 109]. It is interesting to note that the tungsten carbides, found in tungsten single crystals, were in single-crystal form [51].

TABLE 43. Influence of Degree of Deformation on the Mechanical Properties at 25°C of Single-Crystal Molybdenum Wire [64, 80]

| Diameter of wire, mm | Tensile strength, $kg/mm^2$ | Elongation, % | Remarks |
|---|---|---|---|
| 1.5 | 55,0 | 4—5 | Unannealed |
| 0.8 | 90—92 | 3—4 | |
| 0.3 | 104—105 | 2—3 | |
| 0.2 | 105 | 2 | |
| 0.8 | 49—51 | 10—15 | Annealed at 900°C 1 h |
| 0.5 | 48—50 | 10—12 | |

## Production of Semifinished Products of Tungsten and Molybdenum Single Crystals and Their Properties

The high freedom from interstitial impurities, minimum liberation of gases during high-temperature treatment, satisfactory technological properties, stability of shape, and vacuum tightness is far from being a complete list of the requirements which materials of the devices and constructions of the new technology, particularly electronics, have to meet.

Tungsten and molybdenum cermets currently employed in electronics technology possess adequate technical properties for the production of semifinished products from them, but they fail to meet the requirements regarding purity, gas liberation, stability of properties, and so forth. Vacuum arc-melted tungsten and molybdenum have lower interstitial impurity contents. However, due to the high ductile–brittle transition temperature, these metals have to be worked at temperatures of up to 1500°C, during which contamination of the metals by different impurities is possible.

For selecting the optimum conditions of working the single crystals into various semifinished products (foil, wire), a study was made of the influence of the degree of deformation and temperature of vacuum annealing on the mechanical properties of tungsten and molybdenum single crystals. The optimum variations of the technological systems were selected [80]. The production of wire from molybdenum single crystals comprised the following basic operations: 1) forging of single-crystal rods in a rotary forging machine or rolling between grooved rolls; 2) drawing in a chain-type wire-drawing machine through Pobedit or diamond dies. For the production of foil and strip, molybdenum single crystals were rolled between smooth rolls. The high ductility of molybdenum single crystals enabled them to be worked at room temperature and without intermediate annealing; slight heating (200-300°C) was permitted in the initial stages of reduction. It was found that with increase in the degree of deformation of molybdenum single crystals, the strength increased considerably and the ductility decreased compared with the initial specimens (Table 43).

In the final working stage, the wire and foil were subjected to heat treatment – vacuum annealing for recrystallizing treatment (700-800°C). The conditions of treatment mentioned ensure the conservation of purity and the production of ductile semifinished products from single-crystal molybdenum (strip, foil, wire, sheet).

Tungsten single crystals not subjected to preliminary purification from carbon do not withstand high degrees of deformation at room temperature. A more thorough purification from carbon and other interstitial impurities is necessary for tungsten than for molybdenum. When tungsten single crystals are worked into foil and wire, heating in the first working stages

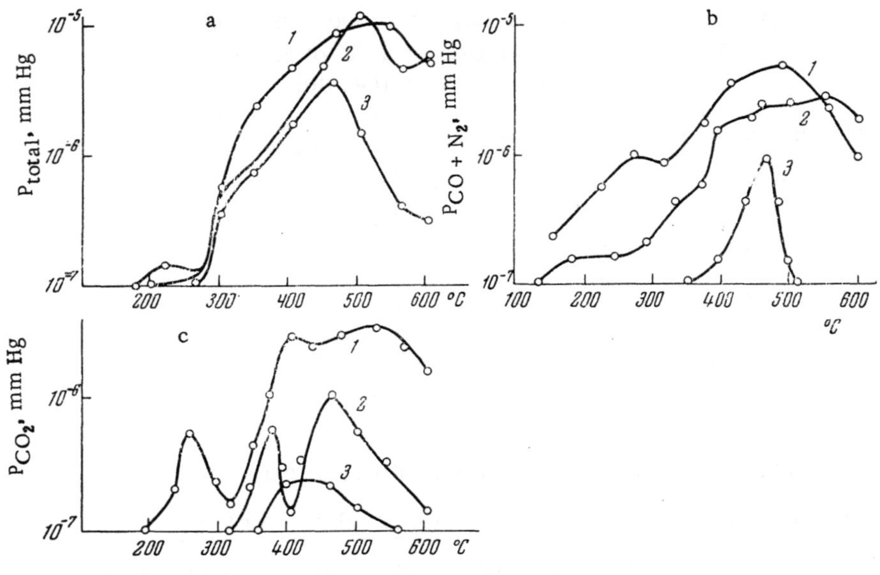


Fig. 116. Liberation of gas from different specimens of molybdenum foil. a) Total liberation of gas; b) liberation of CO + $N_2$; c) liberation of $CO_2$.

is essential. To preclude the possibility of contaminating the material by interstitial impurities, all pressure treatment operations of tungsten single crystals are carried out in a steel sheath at a temperature not exceeding 600°C. With increase in the overall reduction the working temperature is gradually reduced; wire 0.2 mm in diameter (total reduction 99.5%) withstands drawing at room temperature.

It is thus possible to use single-crystal molybdenum strip, foil, and wire as cathodes, anodes, plates, grids, heaters, and other components of electrovacuum apparatus [80]. In addition to high ductility, deep-drawing properties, welding and soldering properties, and the absence of the liberation of gases in vacuum work and constancy of shape under high-temperature conditions are attractive features in the use of molybdenum of single-crystal purity in electrovacuum apparatus. It is therefore now being used on an ever-increasing scale as material for components and individual units of electrovacuum apparatus.

Figure 116 presents data of Professor V. A. Afanas'ev and E. V. Chibisova on the liberation of gas from foil made in our laboratory from single-crystal molybdenum. The liberation of gas during heating was determined by means of an omegatron type of mass spectrometer. The total amount of gas liberated from foil made from single-crystal molybdenum by the "cold" process described above (curve 3) is several times less than from molybdenum cermet foil (curve 2) and vacuum arc-melted molybdenum (curve 1). It is interesting to note that for a given degree of vacuum, the liberation of (CO, $N_2$, $H_2$, $H_2O$, $CO_2$) from single-crystal foil falls sharply at 450-500°C, while for the other sorts of molybdenum, the liberation of gas continues to increase with further increase in temperature.

Single crystals of molybdenum, tungsten, and semifinished products made from them may be used for components of devices and apparatus of the electronic and other fields of engineering where high demands on purity and stability of properties have to be met. The technical properties of these materials (especially molybdenum) ensure the possibility of making from them components of complex shape and their considerable freedom from interstitial impurities ensures long and reliable service in electrovacuum apparatus and devices. In a number of cases, an improvement in the electrical parameters of the devices is even observed. The value of the coefficient of diffusion of potassium in polycrystalline molybdenum is 2.5-5 times higher than in the single-crystal material [119].

Ductility, considerable freedom from interstitial impurities, vacuum tightness and structural stability as far as the melting point, stability to the action of plasma of alkali and alkaline earth metals are the principal advantages of these materials over the ordinary polycrystalline materials.

The principal fields contemplated for the use of tungsten single crystals are cathode heaters for certain electrovacuum devices at 2000-2500°C, current pick-up rods in magnetohydrodynamic electric power generators, and as material for the cathode tubes of thermionic converters. In all such apparatus, use is made of the stability of single-crystal tungsten to recrystallization, even at very high temperatures, and also stability in cesium plasma and other media. For long service in thermionic converters, stability of the single crystals in cesium plasma (above 1000°C, the cesium "seats" through the boundary of the single crystals) and the property of anisotropy of thermionic emission from different faces of a tungsten single crystal are important. The use of turbine blades of single crystals has been reported [121]. According to [122], the turbine blades of a jet engine were made of single crystals of a nickel alloy and showed advantages in service over blades of ordinary polycrystalline material. The resistance to creep and shock was increased by a factor of two.

Single crystals are now no longer a laboratory curiosity. The requirements of certain branches of engineering having been taken into consideration, the industrial production of single crystals of refractory metals (tungsten, molybdenum, niobium) and their semifinished products (strip, foil, tubes, wire) has been launched in the Soviet Union. The introduction of the industrial production of single crystals of refractory metal alloys has been commenced. A single-crystal industry has also been created in the USA, England, Japan, and France where dozens of firms are engaged on the production of single crystals [34, 71].

## LITERATURE CITED

1. E. M. Savitskii, Izv. Akad. Nauk SSSR, Metally, No. 5, p. 3 (1965).
2. U. Czochralski, Z. Chem., 92:219 (1917).
3. E. Schmidt and W. Boas, Plasticity of Crystals, Especially Metallic [Russian translation], GONTI (1938).
4. In: Growth Processes and the Growing of Single Crystals [Russian translation], IL (1963).
5. W. G. Pfann, Zone Melting, Wiley, New York (1958).
6. A. Calverley, M. Davis, and R. E. Leker, J. Scient. Instrum., 34(4):142 (1957).
7. E. M. Savitskii et al., Zavod. Lab., No. 8, p. 879 (1961).
8. D. P. Schadler, Trans. Met. Soc. AIME, 218(4):649 (1960).
9. Van Arkel, Reine Metalle, Berlin (1939).
10. V. S. Emel'yanov, A. I. Evstyukhin, G. A. Leont'ev, and A. N. Semenikhin, Metallurgy and Physical Metallurgy of Pure Metals [in Russian], No. 4, Atomizdat (1963).
11. Express Information, Science of Metals and Heat Treatment of Metals [in Russian], VINITI (Abstract 104) (1964).
12. V. D. Kuznetsov, Crystals and Crystallization [in Russian], Moscow, Gosteoretizdat (1953).
13. J. Lindmayr, J. J. Casey, and R. R. Garnache, J. Electrochem. Soc., No. 7, 153C (1964).
14. A. V. Stepanov, Zh. Tekh. Fiz., 29:381 (1959).
15. A. A. Shakh-Budagov and A. V. Stepanov, Zh. Tekhn. Fiz., 29:394 (1959).
16. B. M. Gol'tsman and A. V. Stepanov, Izv. Akad. Nauk SSSR, OTN, Metallurgiya i Toplivo, No. 5, p. 49 (1959).

17. B. M. Gol'tsman, in: Growth of Crystals [in Russian], No. III, Izd. Akad. Nauk SSSR, p. 408 (1961). [English translation: Consultants Bureau, New York (1962), p. 289.]
18. P. I. Antonov and A. V. Stepanov, Izv. Akad. Nauk SSSR, Neorganicheskie Materialy, 2(5):950 (1966).
19. N. G. Sushkin, Teplofiz. Vys. Temp., No. 2, p. 313 (1963).
20. In: Physical and Chemical Properties of High Purity Metals [in Russian], Izd. "Metallurgiya" (1964).
21. A. N. Holden, Physical Metallurgy of Uranium, Addison-Wesley, Mass. (1958).
22. E. M. Savitskii, V. F. Terekhova, I. V. Burov, and O. P. Naumkin, Tsvetn. Metally, No. 5, p. 51 (1963).
23. A. A. Bochvar and E. M. Savitskii, Tsvetn. Metally, Nos. 5–6, p. 99 (1937).
24. N. L. Parr, Zone Refining and Allied Techniques, St. Martin's Press, New York (1960).
25. J. W. Rutter and B. Chalmers, Canad. J. Phys., 31:15 (1953).
26. B. Chalmers, Impurities and Defects [Russian translation], Metallurgizdat (1960).
27. J. A. Burton, J. Chem. Phys., 21:1991 (1953).
28. W. G. Pfann and D. W. Hagelberger, J. Appl. Phys., 27:12 (1956).
29. G. F. Barber and R. Bekin, Introduction into the Technology of Electron-Beam Processes [Russian translation], Izd. "Metallurgiya," p. 111 (1965).
30. H. Sell and W. Grimes, Rev. Sci. Instr., 35(1):64 (1964).
31. G. F. Ivanovskii and T. N. Zagorskaya, Izv. Akad. Nauk SSSR, Metally, No. 3, p. 65 (1965).
32. A. D. Donskoi and V. V. Bashenko, Application of Electron-Beam Heating in Industry [in Russian], Izd. "Znanie," Leningrad (1962).
33. P. H. Brau, A. Cochardt, and G. Comentz, Rev. Sci. Instr., 26:303 (1955).
34. D. P. Schadler, in: The Art and Science of Growing Crystals, New York, p. 343 (1963).
35. G. T. Hahn, A. Gilbert, and R. I. Jaffee, Refractory Metals and Alloys, Inter. Publ., New York (1963).
36. E. M. Savitskii and G. E. Chuprikov, Izv. Akad. Nauk SSSR, OTN, Metallurgiya i Toplivo, No. 4, p. 137 (1962).
37. H. Seel, Ultrapure Metals and Alloys, Westinghouse Engr., 24(3):94 (1964).
38. K. H. Berthel, Phys. Stat. Sol., 5:399 (1964).
39. K. H. Berthel, Phys. Stat. Sol., 5:159 (1964).
40. E. M. Savitskii, Influence of Temperature on the Mechanical Properties of Metals and Alloys [in Russian], Izd. Akad. Nauk SSSR (1957).
41. N. M. Popov, Vestnik Akad. Nauk SSSR, No. 1, p. 39 (1964).
42. A. Goss, Proc. Phys. Soc. B, 66:525 (1953).
43. G. Tammann, Z. Metallkunde, 21:277 (1929).
44. D. T. Herl, Mechanism of the growth of metal single crystals from the melt, in: Growth Processes and the Growing of Single Crystals [Russian translation], IL (1963).
45. D. Walton and B. Chalmers, Trans AIME, 215:447 (1959).
46. R. Palme and H. Wagner, Metall, No. 9, p. 752 (1955).
47. G. V. Indenbaum and Yu. D. Chistyakov, Zavod. Lab., 29(2):189 (1963).
48. S. Weissmann, B. S. Lement, and M. Copen, Refractory Metals and Alloys, 17:159 (1962).
49. J. W. Rutter and B. Chalmers, Canad. J. Phys., 31:15 (1953).
50. F. D. Rosi, J. Metals, 5:1661 (1953).
51. E. M. Savitskii and G. L. Tsarev, Izv. Akad. Nauk BSSR, Seriya Fiz.-Tekhn. Nauk, No. 2, p. 84 (1965).
52. N. M. Popov, E. M. Savitskii, and G. L. Tsarev, Dokl. Akad. Nauk SSSR, Tekhnicheskaya Fiz., 162(1):64 (1965).

53. E. M. Savitskii, G. S. Burkhanov, G. L. Tsarev, and N. N. Bokareva, Pribory i Tekhnika Eksperimenta, No. 4, p. 248 (1965).
54. J. W. Pugh, Trans. AIME, 50:1072 (1955).
55. R. Maddin and I. N. Cheng, Uspekhi Fiz. Met. [in Russian], Vol. 2, Metallurgizdat (1958).
56. E. M. Savitskii and N. N. Bokareva, Dokl. Akad. Nauk SSSR, 159(5):1137 (1964).
57. D. P. Schadler, Trans. AIME, 218(4):649 (1960).
58. R. S. Koo, Acta Metallurgica, 11(9):1083 (1963).
59. E. N. Andrade and Y. S. Chow, Proc. Roy. Soc. A, 175:290 (1940).
60. E. M. Savitskii and G. E. Chuprikov, Izv. Akad. Nauk SSSR, OTN, Metallurgiya i Toplivo, No. 6, p. 167 (1962).
61. E. M. Savitskii, G. E. Chuprikov, and A. A. Babaréko, Izv. Akad. Nauk SSSR, OTN, Metallurgiya i Gornoe Delo, No. 3, p. 166 (1963).
62. R. A. Jeffery, G. A. Gich, and E. Smith, in: Problems of Modern Metallurgy [Russian translation], IL, No. 5, p. 139 (1960).
63. A. T. Chyrchman, Proc. Roy. Soc., 226(1165):216 (1954).
64. E. M. Savitskii, Z. A. Timofeeva, and M. A. Tylkina, Obmen Opytom v Radioelektronnoi Promyshlennosti, No. 10, p. 5 (1964).
65. E. M. Savitskii, A. A. Bobaréko, and Tao Tsu-ts'un, Izv. Akad. Nauk SSSR, OTN, Metallurgiya i Gornoe Delo, No. 1, p. 176 (1964).
66. M. V. Klassen-Neklyudova, Mechanical Twinning of Crystals [in Russian], Izd. Akad. Nauk SSSR (1960).
67. D. W. Rhys, J. Less-Common Metals, 1(2):269 (1959).
68. Mantariol, R. Reich, P. Albert, and G. Chandron, Compt. Rend., 238(7):815 (1954).
69. R. V. Honeycombe, Effect of Temperature and Alloying Additions on the Deformation of Metal Crystals, Pergamon, New York (1961).
70. H. Wiedersich, J. Metals, 16(3):425 (1964).
71. G. Chandron, in: Ultrapure Metals [Russian translation], Izd. "Metallurgiya," p. 134 (1966).
72. E. M. Savitskii and N. N. Novokhatskaya, in: Growth and Imperfections of Metal Crystals [in Russian], Izd. "Naukova Dumka," p. 144 (1966).
73. V. J. Albano and R. R. Soden, J. Electrochem. Soc., 113:115 (1966).
74. H. Hieber, B. L. Mordike, and P. Haasen, Platinum Metals Rev., 8(3):102 (1964).
75. A. Calverley and D. Rhys, Nature, 183:599 (1959).
76. E. M. Savitskii, Ch. V. Kopetskii, A. I. Pekarev, and M. I. No'osadov, Izv. Akad. Nauk SSSR, OTN, Metallurgiya i Toplivo, No. 6, p. 74 (1961).
77. E. M. Savitskii, G. S. Burkhanov, Ch. V. Kopetskii, and G. E. Chuprikov, in: Transaction of a Conference on the Growth of Crystals [in Russian], Izd. "Nauka," Vol. 6, p. 308 (1965).
78. A. Bergheran, Metalle für die Raumfahrt, Metallwerk Plansee, A. G. Reutte/Tyrol, p. 460 (1965).
79. G. W. Edington and R. E. Smallman, Acta Metallurgica, 13:765 (1965).
80. E. M. Savitskii, G. E. Chuprikov, and A. S. Grokhochinskii, Obmen Opytom v Radioelektronnoi Promyshlennosti, No. 10, p. 1 (1964).
81. E. M. Savitskii, G. S. Burkhanov, Ch. V. Kopetskii, N. N. Bokareva, and V. G. Kardashevskaya, Production and properties of single crystals of refractory metals and alloys, in: Investigations on Heat-Resistant Alloys [in Russian], Izd. Akad. Nauk SSSR (1966).
82. S. I. Gubkin, Working of Metals by Pressure [in Russian], Metallurgizdat (1949).
83. E. M. Savitskii, Plastic Properties of Magnesium and Its Alloys [in Russian], Izd. Akad. Nauk SSSR (1941).
84. J. H. Bechtold, J. Metals, 5:1469 (1953).
85. E. M. Savitskii and Ta'o Tsu-ts'un, Izv. Akad. Nauk SSSR, OTN, Metallurgiya i Toplivo, No. 4, p. 133 (1962).

86. E. M. Savitskii, V. F. Terekhova, and V. E. Kolesnichenko, in: Growth and Imperfections of Metal Crystals [in Russian], Izd. "Naukova Dumka," p. 149 (1966).
87. K. A. Osipov, Some Questions of Heat-Resistance Theory [in Russian], Izd. Akad. Nauk SSSR (1959).
88. E. M. Savitskii and G. L. Tsarev, in: Research on Heat-Resistant Alloys [in Russian], Izd. "Nauka," p. 290 (1966).
89. E. M. Savitskii and G. L. Tsarev, in: Growth and Imperfections of Metal Crystals [in Russian], Izd. "Naukova Dumka," p. 290 (1966).
90. E. M. Savitskii, M. A. Tylkina, A. I. Pekarev, M. A. Gavrilyuk, and A. V. Zabanova, Dokl. Akad. Nauk SSSR, 140(6):1301 (1961).
91. K. Agte and I. Vatsek, Tungsten and Molybdenum [in Russian], Izd. "Energiya" (1964).
92. C. D. Smithells, Tungsten, Chemical Publishing Co., New York (1953).
93. E. M. Savitskii and G. E. Chuprikov, in: Rhenium [in Russian], Izd. "Nauka," p. 110 (1964).
94. C. S. Barrett, Structure of Metals, McGraw-Hill, New York (2nd ed.) (1952).
95. R. Palme, Plansee Berichte für Pulvermetallurgie, Metallwerk Plansee A. G. Reutte/Tyrol, 13:30 (1965).
96. E. Rexer, Metalle für die Raumfahrt, Metallwerk Plansee A. G. Reutte/Tyrol, 394 (1965).
97. Express Information, Metals and Heat Treatment [in Russian], VINITI, No. 34, Abstract 170 (1964).
98. G. N. Shuppe, Electron Emission of Metal Crystals [in Russian], Tashkent, Izd. Sredne-Aziatskogo State University (1959).
99. G. Smith, Phys. Rev., 94:295 (1954).
100. H. F. Webster and P. L. Read, Report N 63-RL-3347 E. General Electric Research Lab., Schenectady, New York (1963).
101. S. Smoluchovskii, Phys. Rev., 60:661 (1941).
102. E. M. Savitskii, I. V. Burov, L. N. Litvak, G. S. Burkhanov, and N. N. Bokareva, Zh. Tekhn. Fiz., No. 7, p. 1310 (1966).
103. M. A. Filyand and E. I. Semenova, Properties of Rare Elements (Reference Book) [in Russian], Izd. "Metallurgiya" (1964).
104. E. M. Savitskii, Ch. V. Kopetskii, and E. P. Arskaya, Izv. Akad. Nauk SSSR, OTN, Metallurgiya i Gornoe Delo, No. 6, p. 129 (1964).
105. E. M. Savitskii, G. S. Burkhanov, and N. N. Bokareva, in: Growth and Imperfections of Metal Crystals [in Russian], Izd. "Naukova Dumka," p. 297 (1966).
106. E. M. Savitskii, N. A. Tylkina, and K. B. Povarova, Rhenium Alloys [in Russian], Izd. "Nauka" (1965).
107. G. I. Federer and R. M. Steele, Nature, 205(4971):587 (1965).
108. G. A. Geach and F. O. Jones, J. Less-Common Metals 1:56 (1959).
109. H. Hillman, Metall, No. 2, p. 102 (1961).
110. E. M. Savitskii, G. S. Burkhanov, and Ch. V. Kopetskii, Izv. Akad. Nauk SSSR, OTN, Metallurgiya i Gornoe Delo, No. 6, p. 12 (1963).
111. V. E. Ivanov, I. I. Papirov, G. F. Tikhvinskii, and V. M. Amonenko, Pure and Superpure Metals [in Russian], Izd. "Metallurgiya" (1965).
112. G. P. Benediktova, G. N. Dubinin, M. G. Karpman, and G. V. Shcherbedinkii, Metallevedenie i Termicheskaya Obrabotka Metallov, No. 5, p. 55 (1966).
113. Okkerse, Philips Techn. Rev., 21:340 (1959–60).
114. F. Blaha, Problemy Sovremennoi Metallurgiya, No. 4, p. 108 (1959).
115. S. S. Brenner, J. Appl. Phys., 27:1484 (1956).
116. J. J. Gilman, Mechanical Engineering, 83:55 (1961).
117. Express Information, Metallovedenie i Termoobrabotka [in Russian], VINITI, No. 4 (Abstract 12) (1965).

118. Am. Metal Market, 71(168):11 (1964).
119. D. Dimpel, Metall, No. 3, p. 211 (1962).
120. E. M. Savitskii, G. S. Burkhanov, A. S. Grokhochinskii et al., Dokl. Akad. Nauk SSSR, Teknicheskaya Fizika, No. 6 (1966).
121. New Scientist, 29(489):835 (1966).
122. J. J. Gilman, Transaction Quarterly, No. 5, p. 59 (1966).

CHAPTER VI

# ALLOYS OF REFRACTORY METALS

Modern theory of metals has not yet attained a level of development such that previous calculation of the composition of an alloy with given properties is possible. The choice of materials for the new technology has to be made experimentally by taking into account already known relationships in the variation of the mechanical, electrical, and physicochemical properties of metal alloys as a function of the variation in their composition, structure, temperature, system of active forces, rate of strain, action of corrosive media, penetrating radiations, and so forth (see Chapter III).

The prospect of previous calculation of the composition of alloys of given structure and properties, starting from the binding forces between the atoms or any other properties, is very attractive [1]. For the present, however, despite all the achievements of modern physics and chemistry and the existence of some interesting hypotheses, the optimum compositions and alloys are found experimentally, taking into account the already known relationships in the variation of the properties as a function of composition, structure, service conditions, defects, and so forth.

A change in the properties of metals by melting them with other elements, as is well known, may be effected by making use of the following four mechanisms: 1) formation of substitutional solid solution on melting with metals; 2) formation of interstitial solid solutions on melting with nonmetals; 3) formation of second phases on melting with either metals or nonmetals; 4) by heat treatment of the alloys obtained for the purpose (most often successively) of stress relief, polygonization, recrystallization, homogenizing the solid solution, formation of Kurnakov compounds in the case of ordered solutions, redistribution of second phases in substitutional or subtractional solid solutions.

With regard to group IVa refractory metals (especially titanium and zirconium), all the mechanisms enumerated are used. With regard to group Va metals, chiefly pure niobium and tantalum, the first mechanism is widely used (although with some limitations due to inadequate ductility) and a start has been made with the use of the fourth mechanism. For group VIa metals (chromium, molybdenum, and tungsten of commercial purity) so far even the first mechanism has practically not yet been used.

Metals which form substitutional solid solutions, harden the metal base and increase the ductile–brittle transition temperature. As already discussed, group Va and VIa metals are soluble in each other almost unlimitedly in the liquid and solid states. For binary continuous solid solutions according to Kurnakov's rule, maximum strengthening and hardness as a function of composition is obtained at 40-60% [2]. Figure 117 shows the hardness of alloys in the niobium–molybdenum and niobium–tungsten systems at 20 and 1000°C [4]. The results of research on the influence of tungsten, molybdenum, iron (up to 0.1%), and interstitial impurities on the physicomechanical properties of rhenium show that interstitial impurities, par-

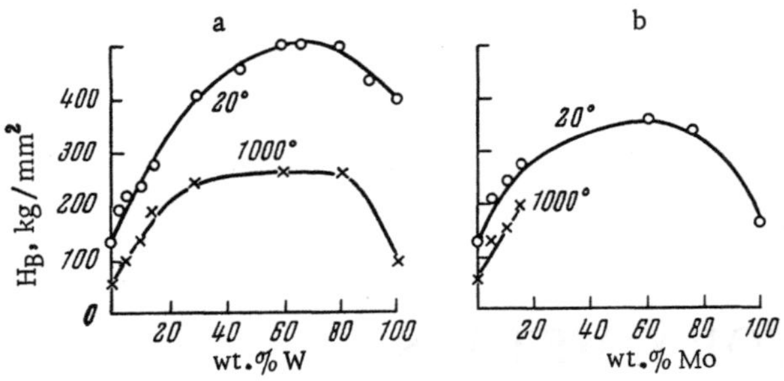


Fig. 117. Hardness of niobium–tungsten (a) and niobium–molybdenum (b) alloys.

ticularly oxygen, act more powerfully than the enumerated substitutional metals [5, 6]. Alloying of group Va and VIa metals with substitutional elements leads to an increase in the strength properties at elevated temperatures. Thus, for example, when niobium is alloyed with certain other refractory metals the result is an increase in strength by 500% [1]. Such a material is quite promising in various apparatuses and devices operating at a temperature about 1000°C. As will be shown below, however, for technological reasons it has so far not been possible to employ to the full the mechanism of strengthening by alloying as applied to refractory metals.

The strengthening effect of a solid solution is directly proportional to the difference in the atomic radii of the solvent and the soluble addition [2]. This relationship may be expressed in the form of strengthening as a function of the variation in the lattice parameter. By way of example, Fig. 118 shows the relationship $\Delta HB = f(\Delta a)$ at room temperature ($\Delta a$ is the difference in atomic radii of niobium and the soluble addition [7]).

As shown by E. M. Savitskii and A. M. Zakharov [7], at very high temperatures, the influence of the alloying addition on the melting point of the metal base becomes of primary importance.

The principal obstacle to the use in technology of exceptionally strong, highly alloyed substitutional solid solutions based on group Va metals (vanadium, niobium, tantalum) is the reduction in commercial ductility. Therefore, for example, no more than 25% W, 10% Mo, 7% V, 5% Zr, and 3-5% Cr is introduced into niobium [8, 9]. For this reason, the method of strengthening by alloying with group VIa metals (chromium, molybdenum, and tungsten) has practically not been used so far. Existing alloys based on these metals are low-alloy solid solutions, produced by treating the melts with deoxidizers, such as carbon, titanium, zirconium, vanadium, hafnium, and rare-earth metals. The considerable strengthening of the alloys observed in this case is mainly the result of the precipitation of disperse inclusions of a second phase formed as the result of the reaction of the refractory metals with interstitial impurities. Usually, the second phase is precipitated in the form of nitrides, oxides, carbides, or complex compounds (oxycarbides, oxynitrides). The carbide disperse phase has a more effective action on increasing the heat resistance than oxides and nitrides. Perhaps the only exception in the production of highly-alloyed alloys of industrial significance on the basis of VIa group metals are alloys of the tungsten–rhenium, molybdenum–rhenium, and tungsten–molybdenum systems (the last mentioned is most often prepared by the cermet method). As shown by scientific and industrial experience acquired in group VIa metals (titanium, zirconium, hafnium), every possible means of alloying to any considerable extent may be used, provided only that the metal base and the alloying addition are produced in a state of high purity, especially with regard to interstitial impurities.

At the present time, the bulk of the refractory metals which can be produced – tungsten, molybdenum, vanadium, niobium – is required by ferrous metallurgy (for alloying iron and steel) and the hard alloy industry.

In this chapter we shall examine data on the structure and properties of some alloys of refractory metals which have been utilized or are quite promising for industrial use. The discussion commences with alloys based on group VIa and VIIa metals, then alloys based on

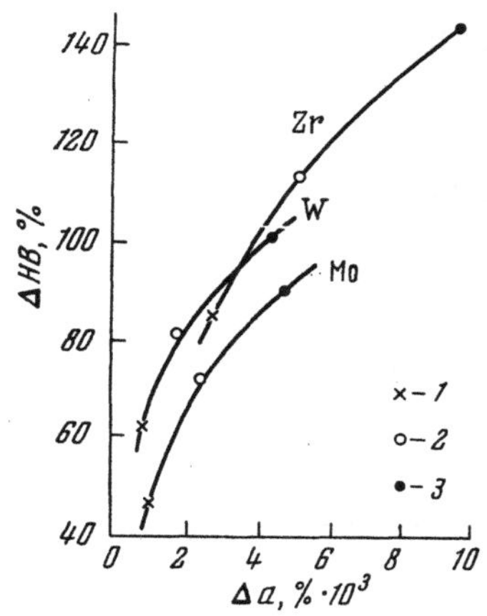


Fig. 118. Relative increase in hardness of niobium at 20°C as a function of the degree of distortion of its crystal lattice on alloying with tungsten, molybdenum, and zirconium. Content of alloying element, at.%: 1) 2.5; 2) 5; 3) 10.

group Va and IVa metals are examined in the order of refractoriness of the metals in the limits of each group. The chapter ends with a brief description of refractory alloys based on the platinum metals, and also materials of SAP type and refractory metallic compounds.

## Chromium – Based Alloys

High refractoriness (melting point 1875°C), good resistance to oxidation, and relatively extensive raw-material resources make chromium a very attractive base for the production of constructional alloys employed at high temperatures. The brittleness of chromium alloys, however, considerably restricts their possible uses in engineering (hazard of fracture of parts and the difficulty of machining chromium alloys).

The strength of aluminothermic chromium is doubled when the latter is heated to 700°C [2].

Parke and Bens [10] found that alloys of the chromium–iron–molybdenum system, especially the alloy 60% Cr, 15-20% Fe, 15-25% Mo, possessed a good combination of properties (ductility and strength). They considered that for optimum combination of ductility and strength, the carbon in these alloys should not exceed 0.05% and silicon 0.2%, with oxygen and nitrogen as low as possible. However, the alloys had low technological ductility and were difficult to machine.

Salli [11] cites data on the investigation of chromium–iron–tantalum, chromium–iron–molybdenum, chromium–iron–niobium, chromium–iron–tungsten, and chromium–iron–vanadium alloys. Chromium–iron alloys, being solid solutions, show the highest resistance to deformation, and therefore in many cases they have been used as base for developing alloys of more complex composition. Niobium, tantalum, tungsten, molybdenum, and vanadium have been introduced in Cr–10%Fe and Cr–20%Fe alloys in amounts of up to 30% by weight. All the alloys enumerated were solid solutions in structure. Attempts to produce high-grade semifinished products from these alloys ended in failure.

According to data of E. M. Savitskii and V. F. Terekhova, additions of yttrium, lanthanum, cerium, and rare-earth metals have a substantial influence on the structure and mechanical properties of chromium; they refine the grain, increase the ductility and tensile and compressive strength, lower the ductile–brittle transition temperature by 100-150°C and increase the resistance to oxidation and raise the recrystallization temperature [14, 15].

The good creep resistance of chromium alloys containing tungsten and titanium is mentioned in [12]. The production of high-grade semifinished products from these alloys was accompanied by considerable technological difficulties.

The structure, strength properties, and corrosion resistance of chromium alloys was studied in [13]. An increase in strength of chromium was obtained by the production of a carbide phase, formed by the interaction of titanium and zirconium additions with carbon. Yttrium, rhenium, and manganese were also used as alloying additions. Yttrium effectively reduced the oxygen and nitrogen content, and also considerably increased the resistance of chromium to oxidation and prevented the absorption of nitrogen [13, 14]. The microstructure of an extruded alloy of chromium with titanium was characterized by the presence of small elongated

TABLE 44. Mechanical Properties of Experimental Chromium-Based Alloys at Room Temperature and on Heating to 315°C [13]

| Chemical composition of alloy, % | Test temperature, °C | Tensile strength, kg/mm² | Proof stress (0,2%) kg/mm² | Elongation, % |
|---|---|---|---|---|
| Cr + 1Y | 24 | 22.3 | — | 0 |
| | 66 | 31.6 | 31,5 | 1,2 |
| | 149 | 33.0 | 28.1 | 33,0 |
| Cr + 1Y + 0,3Hf + 0,1Ti + 0,2Zr | 24 | 36.4 | 35,4 | 1,0 |
| Cr + 0,3Y | 24 | 37.7 | — | 0 |
| | 149 | 37.9 | — | 0 |
| Cr + 0,3Y + 2,4Ti + 0,3Zr + 0,3C | 24 | 28.8 | — | 0 |
| | 149 | 30.5 | — | 0 |
| | 204 | 40.5 | — | 0,2 |
| Cr + 0.3Y + 2.4Ti + 0,5Zr + 0.5C | 24 | — | — | — |
| | 149 | 43.5 | — | 0 |
| | 204 | 44.2 | 44.1 | 0.5 |
| | 315 | 50.0 | — | 0 |

grains containing numerous precipitates of a disperse phase consisting of a complex carbide (Ti, Zr)C having a face-centered cubic lattice. The possibility of the solution of nitrogen and oxygen in the monocarbide was also noted. The introduction of manganese was found to be irrational. High tensile strength was shown by a chromium alloy with 2.4% Ti, 0.5% Zr, 0.5% C, and 0.3% Y (Table 44).

Alloys of the chromium–yttrium system had the lowest ductile–brittle transition temperature (about 65°C). The mechanical tests at elevated temperature showed that carbide-hardenable alloys had high stress-rupture strength values (higher than heat-resistant nickel alloys). The basic possibilities of hardening chromium alloys by heat treatment was also shown in [13].

Comparatively recently, data were published on the properties of cermet dispersion-resistant "plastic chromium compositions" containing about 6% magnesium. It was shown that the elongation of these materials at 20°C was 15-25% (tensile strength 28 kg/mm$^2$), and at 1095°C it was 40% [16].

The principal spheres of application of alloys of cermet compositions based on chromium are parts of magnetohydrodynamic generators, heat-resistant bottoms and other parts of annealing furnaces, and molds for stoneware and metal casting.

## Molybdenum-Based Alloys

The industrially interesting properties of molybdenum are high melting point, high modulus of elasticity, low coefficient of linear thermal expansion, good heat resistance, high electrical conductivity and thermal conductivity, comparatively small cross section of capture of thermal neutrons, almost half the specific gravity of tungsten, etc. Creep and stress-rupture strength of molybdenum appreciably exceed these properties in many metals employed as base for heat-resistant materials. The strength of molybdenum may be increased considerably by alloying.

The influence of alloying additions on the hardness of molybdenum at room and elevated temperatures was investigated in [17]. A pronounced increase in hardness resulted from al-

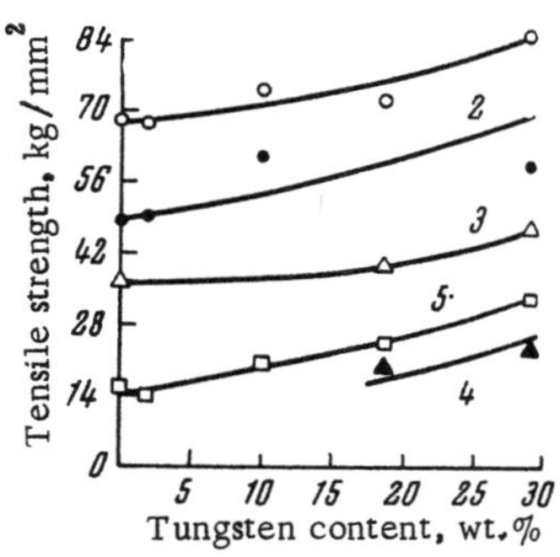


Fig. 119. Tensile strength of molybdenum–tungsten alloys. 1) Short-duration test at 20°C, after stress-relief anneal; 2) the same, after recrystallization; 3) short-duration test at 980°C after stress-relief anneal; 4) the same, after recrystallization; 5) stress-rupture strength (at 980°C after 110 h), after stress-relief annealing.

loying molybdenum with boron and silicon. Some reduction in hardness with a low content of titanium, zirconium, niobium, tantalum, or vanadium was attributed to the deoxidizing action of these elements on molybdenum. In agreement with the variation in hardness, the strength properties of molybdenum also vary on alloying [18]. The comparatively high ductile–brittle transition temperature of commercial molybdenum, as well as the increase in the transition temperature and the impairment of the technological ductility of molybdenum on alloying retard its application and make it difficult to develop alloys based on it. Other shortcomings of molybdenum are its poor weldability and strong oxidizability at high temperatures.

During the past 15 years, in different countries a large number of molybdenum alloys have been developed for use at high temperatures. Usually, in the arc-melted metal, the content of interstitial impurities exceeds by about one order (and for carbon even by three orders) their limit solubility in molybdenum at room temperature. Thus, the bulk of the impurities must be in the form of a second phase of variable composition (evidently oxynitrocarbides of molybdenum and some alloying metals) [19]. Thermodynamically active alloying elements in the presence of small amounts of carbon effectively increase the tensile strength, mainly due to the hardening effect associated with the precipitation of the above-mentioned fine carbide inclusions in the aging process. According to [91], complex carbides of variable composition are formed in alloys of the transition metals containing carbon.

According to [8], at 1095°C, the tensile strength of a work-hardened alloy of molybdenum with 0.5% Ti and 0.02% C is 150% higher, and the stress-rupture strength is 300% higher than in molybdenum (42.2 and 23.9 kg/mm², compared with 28 and 8.5 kg/mm², respectively). A molybdenum alloy with 1.4% Nb and 0.05% C at 980°C has a tensile strength of approximately 60 kg/mm² (as also has the molybdenum alloy with 0.05% Zr and 0.02% C). The effective increase in short-time and stress-rupture strength at high temperatures in the case of complex alloying of molybdenum with titanium and zirconium is also noted [8]. According to the data of this article, a molybdenum alloy with 0.46% Ti and 0.07% Zr in the peened condition at 980° and 1315°C has a tensile strength of 60.5 and 38 kg/mm², respectively. An increase in the titanium, zirconium, and carbon content increase the heat resistance of the molybdenum alloy. High stress-rupture strength at 1315°C is shown by a molybdenum alloy with 1.27% Ti, 0.29% Zr, and 0.30% C (23.2 kg/mm²); the tests were made in the peened condition).

Comparative data on the heat resistance of molybdenum alloys based on nickel, chromium, cobalt, and iron are given in [20]. A detailed review of the mechanical properties of molybdenum alloys is made in [21]. There is evidently a connection between the ratio of the diameters of the atoms of the alloying metal and molybdenum on the one hand, and the technological ductility of the alloys on the other. However, besides tungsten, rhenium, and tantalum for preserving minimum technological ductility, the permissible addition of other metals is, as a rule, about 1 wt.% [21]. For arc-melted alloys, it is less than in alloys produced by the cermet method.

There is considerable interest in the development of alloys having a melting point higher than that of molybdenum. As follows from the phase diagrams, the melting point of molybdenum

TABLE 45. Mechanical Properties of Sheets of Alloy TZM [26]

| Sheet thickness, mm | Minimum tensile strength, kg/mm² | | Minimum yield point, kg/mm² | | Minimum elongation (on a length of 50.8 mm), % | |
|---|---|---|---|---|---|---|
| | 20° | 1200° | 20° | 1200° | 20° | 1200° |
| 0.25—0.64 | 98.4 | 35.2 | 80.9 | 17.6 | 4 | 5 |
| 0.64—1.15 | 94.9 | 35.2 | 77.3 | 17.6 | 5 | 6 |
| 1.15—2.28 | 91.4 | 35.2 | 74.8 | 17.6 | 7 | 8 |
| 2.28—4.75 | 91.4 | 35.2 | 74.8 | 17.6 | 8 | 10 |

is increased by tungsten and tantalum. Alloys of the molybdenum–tungsten system made by vacuum arc melting with a consumable electrode have been investigated over the entire concentration range. If melting of the initial molybdenum and tungsten resulted in the formation of a coarse columnar structure, an appreciable grain refinement was found in molybdenum–tungsten binary alloys [22].

Figure 119 shows the tensile strengths in short-duration testing and the stress-rupture strengths of these alloys as a function of their composition. Up to a content of 30%, tungsten exercises a comparatively slight influence on the strength of molybdenum, which is evidently due to the small difference in the atomic radii of these components and their weak chemical interaction in the formation of solid solutions. More exact data on the properties of alloys of this system could be obtained by using tungsten and molybdenum of high purity.

Communications are to be found in the literature on attempts to produce alloys of more complex composition based on the molybdenum–tungsten system, and on the development of an alloy of molybdenum with 25% W, 0.1% Zr, and 0.05% C. At 1315°C this alloy has maximum strength (in work-hardened condition 52 $kg/mm^2$); in stress-rupture strength, it is inferior to low-alloyed molybdenum alloys [8].

Research on the mechanical properties of alloys based on the molybdenum–tantalum system is also of definite interest, since the alloying of molybdenum with tantalum leads to an increase in the melting point and the recrystallization temperature of the alloy. It is to be expected that alloying molybdenum with tantalum ought to result in an increase in the heat resistance of molybdenum. There is no published detailed information on the mechanical properties of alloys of this system.

It is stated [23] that molybdenum having a carbon content amounting to 0.06 wt.% still retains its ductility when forged in the hot condition, but has a ductile–brittle transition temperature well above room temperature. Communications have appeared [24, 26, 27, 29] to the effect that in the USA alloys on a molybdenum base are undergoing experimental tests in aviation and space research. They are being used for the nose cones and nozzles of rockets, thrust rings, power units, rudders, leading edges of wings, radiation shields, and reinforcement parts subject to high-service temperatures. In 1963, in the capitalist countries, 35,834 tons of molybdenum concentrate was used. The demand for molybdenum is constantly increasing [25].

Information is available on the development of a molybdenum alloy TZM with a low carbon content [26]. The chemical composition of the alloy is 0.5% Ti, 0.1% Zr, the rest molybdenum. The carbon content of this alloy does not exceed 0.01-0.02%. The alloy has a low cold-brittleness temperature and good technological ductility. Tubes and sheets are made from the alloy. It is stated [26] that sheets of this alloy have been used for sheathing the constructions of cosmic gliding apparatus intended for return to earth from cosmic space. It is also as-

TABLE 46. Mechanical Tensile Properties of Molybdenum Alloys [8, 24]

| Composition of alloy, % | Temperature, °C | Tensile strength, kg/mm² | Elongation, % | Stress-rupture strength (100 h), kg/mm² | | | |
|---|---|---|---|---|---|---|---|
| | | | | 930° C | 1093° C | 1204° C | 1315° C |
| Molybdenum by arc melting (0.001–0.03 C) | 980 | 24.6 | 41 | — | — | — | — |
| Mo + 0.5 Ti (0.01–0.03 C) | 980 | 47.7 | 20 | — | 23.9 | 12.6 | 7.3 |
| | 1315 | 13.1 | 70 | — | — | — | — |
| Mo + 0.5 Ti + 0.08 Zr (0.02–0.08 C) | 980 | 53.5 | 18 | — | 35.2 | — | — |
| | 1093 | 50.2 | 18 | — | — | — | — |
| Mo + 1.27 Ti + 0.29 Zr (0.3 C) | 980 | 65 | 11 | — | — | — | — |
| | 1315 | 42.9 | 26 | — | — | — | 23.2 |
| Mo + 1.4 Nb (0.05 C) | 980 | 61.2 | 17 | 48.5 | — | — | 8.4 |
| | 1315 | 16.9 | 77 | — | — | — | — |
| Mo + 25W + 6.1 Zr (0.05 C) | 980 | 30.2 | 46 | 29.3 | — | — | 8.4 |
| | 1315 | 20.4 | 58 | — | — | — | — |

sumed that sheets of the alloy TZM will be used in the construction of guided missiles, as heating elements and reflectors in vacuum furnaces and other parts used at high temperatures and requiring the production of a vacuum-tight weld. The mechanical properties of sheets of the alloy TZM are given in Table 45.

A number of alloys having good heat resistance and based on molybdenum have been proposed. In the Soviet Union, the industrial production of molybdenum alloys has been mastered [27].

The composition and strength properties of some molybdenum alloys produced in the USA are shown in Table 46. The recrystallization temperature of these alloys is in the range 1400–1700°C. It is assumed that in future it will be possible to produce alloys based on molybdenum for service temperatures of 1400°C.

The influence of alloying elements on the malleability and creep of molybdenum at 1000°C have been studied [28]. A considerable increase in strength of molybdenum was found when it was alloyed with chromium, but the alloys were very brittle and quite unsuitable for machining.

A molybdenum alloy with 0.5% Ti is used extensively in the USA; the alloy is produced industrially in the form of strip, sheet bars, sheets, rods, and tubes. A photograph of turbine blades made of this alloy (after 50 h service) is shown in Fig. 120.

Although alloying elements increase the hardness and strength of molybdenum at room and elevated temperatures, the maximum quantity which (as a rule) may be introduced into molybdenum with retention of its technological ductility, is limited to tenths of a percent. Therefore, the principal methods of varying the mechanical properties of such alloys are work hardening and recrystallization. One of the objects of alloying is to increase the recrystallization temperature of the work-hardened metal for preserving the mechanical properties (imparted by work hardening) at elevated temperatures. Data on the effect of alloying additions on the recrystallization temperature of molybdenum are given in [20, 35] (see also Fig. 32).

Such an approach to the development of molybdenum alloys is, however, correct for molybdenum of commercial purity. The production of molybdenum of high purity with regard

Fig. 120. Turbine blades of molybdenum alloy with 0.5% Ti after testing for 50 h.

to interstitial impurities and the production on its basis of ductile alloys having a high content of alloying components will preclude the possibility of brittleness at temperatures above the recrystallization temperature. Thus, the ceiling of the service temperatures of molybdenum alloys will be raised considerably.

Industrial production in the USA includes molybdenum sheets 915 × 2440 mm in size with a thickness of up to 0.1 mm (and above), and also forged molybdenum rings (diameter 350-380 mm) for jet engine nozzles. The American metallurgical industry produces forgings weighing up to 1135 kg from arc-melted molybdenum alloys (diameter of initial ingot 380 mm).

According to [30], the production of individual parts of gas turbines and rocket engines of heat-resistant alloys based on molybdenum permits the service temperature to be increased by 200-300°C, which, in its turn, has a considerable effect on the increase in power of the engine. The conditions and possibility of using molybdenum alloys, as well as other alloys of refractory metals, in the constructions of space vehicles is examined in [29]. In the opinion of the authors of this book, due to the combination of technological and strength properties, molybdenum alloys, together with some niobium alloys, are the most promising materials for the heat screens of space apparatus in places where high temperatures are to be expected.

As is well known, molybdenum is widely used in electronic vacuum instruments. Its use in this field will grow in future, the tendency being to replace cermet gas-saturated molybdenum by molybdenum and its alloys produced by vacuum melting methods.

Work on the production of new molybdenum alloys is directed toward obtaining alloys having a higher recrystallization temperature and a lower ductile–brittle transition temperature.

An interesting trend is also the production of highly alloyed molybdenum alloys which in addition to satisfactory technological ductility and high recrystallization temperature would also possess a number of other necessary physical properties. As yet, however, there is no possibility of using alloys having a relatively high content of alloying elements. Increased alloying element content in an alloy, despite the refinement of structure, leads to a sharp increase in the resistance to deformation and an increase in the ductile–brittle transition temperature. The destruction of the cast structure requires heating to a high temperature (which will lead to additional contamination by gaseous impurities), as well as special high-power equipment.

The solution of the problem of the production of highly alloyed molybdenum alloys, as in the case of tungsten, is associated with the development of methods of producing molybdenum which is highly ductile and pure (with regard to interstitial impurities), and its protection from contamination by impurities, especially carbon and oxygen, causing embrittlement in the process of the production and treatment of molybdenum-based alloys.

## Tungsten-Based Alloys

Owing to its exceptional refractoriness and high-strength properties, tungsten is the most promising base for the production of high-strength alloys. The program of development of heat-resistant alloys in the USA envisages the production of a tungsten alloy having a tensile strength of 10.6 kg/mm$^2$ at 2200°C [24, 31]. There is information on the use of tungsten for fuel rocket engines, parts of plasma engines, and nose cones of rockets returning into the dense layers of the atmosphere [31, 32]. The low ductility of tungsten and its oxidizability at high temperatures make it difficult to develop and use such alloys.

Currently, the radioelectronic and illuminating engineering industry makes extensive use of tungsten alloys with additions of inert oxides (for increasing the recrystallization temperature), and also alloys of the tungsten–molybdenum system, prepared by powder metallurgy methods. Tungsten alloys with various metal additions produced by vacuum melting (arc- or electron-beam) have failed to find any extensive industrial application owing to their brittleness, and are still in the development stage. An exception is provided by alloys of the tungsten–rhenium and molybdenum–rhenium systems, which will be considered separately. In recent years, however, an ever-increasing demand is to be found for semifinished products of fused tungsten. In the USA 25% of all tungsten products are made of fused tungsten.

Intensive work is being carried out on the study of the influence of alloying on the properties of tungsten, special interest being found in alloying for increasing not only high-temperature strength but also high-temperature oxidation resistance and ductility. The majority of tungsten alloys under development are substitutional solid solutions, strengthened by disperse particles of nonmetallic compounds. Alloying additions used are chromium, nickel, molybdenum, niobium, tantalum, titanium, zirconium, hafnium, and carbon. As a rule, the amounts of these elements added to tungsten do not exceed 5%. The study of tungsten alloys containing large amounts of substitutional elements is difficult owing to their low technological ductility.

It is stated in [8] that the tensile strength of the alloys W–0.6%Nb–0.001%C and W–0.9%Nb–0.018%C, and also the alloy W–0.1%Zr–0.001%C at 1650°C is 200-300% higher than the tensile strength of recrystallized tungsten (34-42 kg/mm$^2$ as against 13.3 kg/mm$^2$ for pure tungsten). Evidently, such a high increase in strength of low-alloy tungsten is mainly obtained as the result of precipitation hardening. The mechanical properties of tungsten alloys containing carbon in combination with additions of niobium, zirconium, and hafnium are examined in [33]. The carbon content has a considerable influence on the strength of tungsten alloys and also on the recrystallization temperature. The recrystallization temperature, however, does not depend on total carbon content alone, but also on the character of the distribution of the structural components. Reference may be made to the high stress-rupture strength of the alloy W–1.92%Nb–0.23%Hf–0.008%C (2.8 kg/mm$^2$ at 1930°C and 1.5 kg/mm$^2$ at 2205°C), despite relatively low carbon content. High values are also reported [34] for the high-temperature tensile strength (54.6 kg/mm$^2$ at 2760°C) of forged rods of tungsten alloys containing thousandths of a percent of boron, tenths of a percent of zirconium and hafnium, and about 1.5% of niobium. The ductile–brittle transition temperature of such alloys, however, is rather high, being about 500°C.

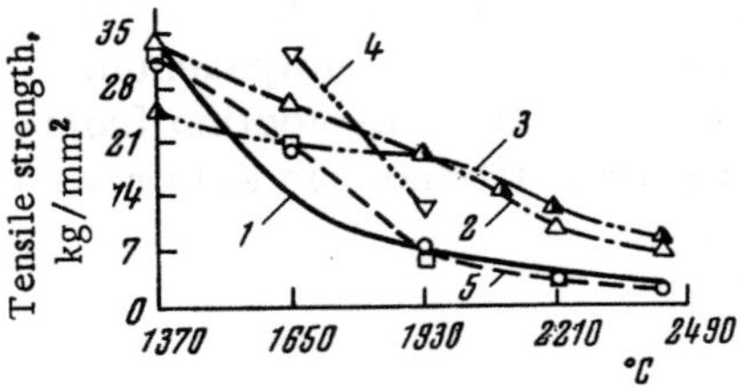


Fig. 121. Dependence of tensile strength of some tungsten alloys on temperature. 1) Unalloyed tungsten; 2) 1% $ThO_2$; 3) 2% $ThO_2$; 4) 0.88% Nb; 5) 25% Mo.

Interest attaches to the alloying of tungsten and its alloys with boron. Although according to [33], boron is a less effective strengthening agent than carbon, evidently in combination with other alloying elements, it will be a useful addition to tungsten alloys owing to its refining and modifying action.

As already mentioned, carbon is an effective strengthening agent for tungsten alloys (in consequence of precipitation hardening), but an increase in its content in the alloy produces an increase in the ductile–brittle transition temperature and a decrease in technological ductility. Plastic deformation of tungsten alloys having a high carbon content is accompanied by considerable difficulties.

The problem of the production of tungsten alloys combining high strength and ductility has not yet been solved. In this connection interest attaches to the development of highly alloyed tungsten alloys in the substitutional solid-solution range with low content of carbon and other interstitial impurities. This problem can be solved only if the initial components of the alloy are produced in a highly pure state as regards interstitial impurities, and if this purity is conserved in the melting and mechanical and heat-treatment processes. One of the methods of increasing the ductility of tungsten alloys which shows considerable promise is the production of alloys having a strength–ductility eutectic structure, in which the hard component will be formed by crystals of tungsten or tungsten-based solid solution, and the soft component will be crystals of a ductile metal. An example of such a system could be provided, for instance, by alloys of tungsten with titanium (see Chapter III), but their production is very complex, owing to the considerable difference in melting point and vapor tension of the two components. Effective strengthening of tungsten is produced by additions of refractory oxides of metals. In contrast to other refractory metals, particularly molybdenum, tungsten is strengthened more effectively by oxides than by carbides. Maximum strength is attained in tungsten alloys hardened by thorium oxide. Figure 121 shows the dependence of tensile strength on temperature for some tungsten alloys [32].

There is considerable interest in alloys of tungsten with the platinum metals. It has been found that additions of osmium to tungsten increase its ductility [3]. The mechanism of the effect of osmium is as yet not clear. Considering the fact that osmium, like rhenium, is a metal having a hexagonal structure, it may be assumed that in the present case the cause of the increase in ductility consists in an increase in twinning as complementary to the slip deformation mechanism. Evidently, ruthenium will have the same effect on tungsten.

The favorable effect of group VII and VIII metals on the deformability of tungsten is also associated with a change in the electronic structure of the substitutional solid solutions formed [35, 36], in agreement with the theory put forward by Robinson [37], according to which a reduction in the hardness of group VIa metals when alloyed with elements of higher valence is due to a reduction in the solubility of interstitial impurities.

A more detailed study of the structure and properties of alloys of the tungsten–manganese system would be of indisputable interest, since manganese belongs to group VII of the periodic system and comes close to rhenium in chemical properties.

However, due to the high vapor tension of manganese, the production of tungsten–manganese alloys is a complex technological problem, and obviously ought to be carried out under considerable pressure of inert gases or by using cermet methods.

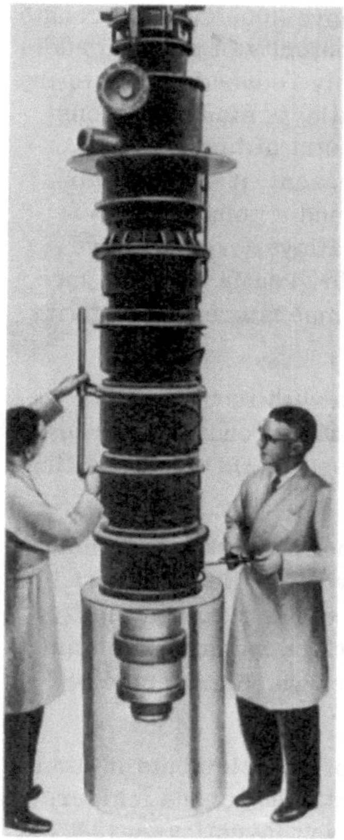

Fig. 122. Pulse amplifying klystron (length 3250 mm, diameter 512 mm).

Considerable interest is also presented by alloys of tungsten with molybdenum. At temperatures of up to 2200°C, a W–15%Mo alloy is superior, in its strength properties, to unalloyed tungsten. In the USA, this alloy is being made in the form of ingots 300 mm in diameter [38]. It is reported that alloys of the tungsten–molybdenum system are being used for the blades of jet engines. The properties of some alloys of the tungsten–molybdenum system are discussed in [3]. A 50%W–50%Mo alloy has maximum electrical resistance and minimum temperature coefficient which enables this alloy to be used for the heaters for electrovacuum apparatus; the alloy is made by cermet methods (as already mentioned, vacuum melting of highly alloyed tungsten alloys has not yet found wide application in industry owing to their poor deformability).

As shown by investigations, in the molybdenum–tungsten system, characterized by complete solubility of the components, the alloys are distinguished by high strength at high temperatures. Thus, for example, an alloy with 75% W, 25% Mo, 0.11% Zn, and 0.05% C at 980° has a tensile strength of 69.1 kg/mm$^2$, and at 1315°C, 51.7 kg/mm$^2$ [39]. According to data in [40], maximum high-temperature strength in the tungsten–molybdenum system is possessed by the alloy W–15%Mo, exceeding in strength unalloyed tungsten up to 2215°C.

There is information on the promising nature of tungsten alloys containing small additions of niobium (up to 1%) and about 3% Re. The yield point of wires of a W–3%Re alloy at 1982°C is about 40 kg/mm$^2$ [39]. These alloys are distinguished for their high recrystallization temperature, about 2000°C [41, 42].

Alloys of the tungsten–molybdenum–rhenium, tungsten–niobium–molybdenum, and tungsten–tantalum–molybdenum systems are being investigated. It is reported that the alloy W–20%Ta–12%Mo has a strength of over 42 kg/mm$^2$ at 1649°C. In the USA heat-resistant deformable ternary alloys based on tungsten with 1-10 wt.% Re and 1-10 wt.% Ta [46], and in the USSR an alloy containing 33-57% W, 15-34% Mo, and 25-40% Re [47] have recently been patented. Alloys of this series possess high mechanical properties at 1800°C with good ductility at room temperature in the recrystallized condition. The influence of temperature on the tensile strength of some tungsten alloys is shown in [43].

Until recently, pure tungsten has been used in the new technology more than alloys based on it. This is due to the limited choice of tungsten alloys and the difficulty of making semi-finished products from them owing to the low technological properties of these alloys. Intensive research has recently been conducted in connection with the development of tungsten-based alloys containing a small amount of molybdenum, niobium, and zirconium. According to [52], small additions of niobium and zirconium considerably increase the strength up to temperatures of 1600-1700°C; when the temperature is increased to 2000°C, a considerable drop in strength is found in these alloys, evidently due to the need for more complex alloying. High heat resistance is possessed by an alloy of tungsten with 30% Re, and by artificial tungsten–thorium oxide and tungsten–tantalum carbide disperse systems produced by the power metallurgy method.

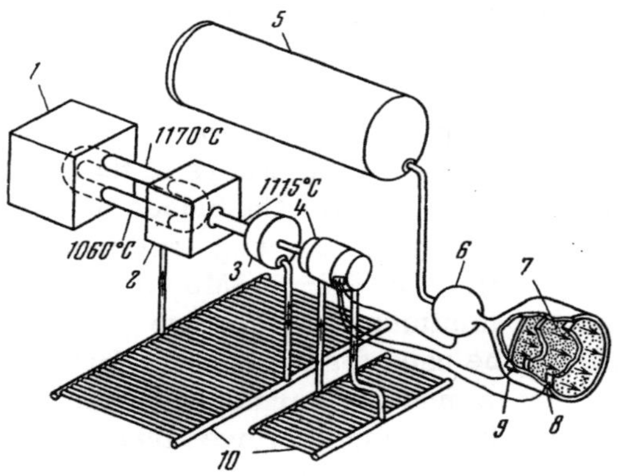


Fig. 123. Diagram of turboelectric nuclear power plant. 1) Nuclear reactor; 2) boiler; 3) turbine; 4) electric generator; 5) liquid cesium; 6) vaporizer chamber; 7) ionic chamber; 8) cathode; 9) heaters; 10) radiator.

At temperatures above 2000°C, the strength-increasing effect of the majority of alloying additions on tungsten is sharply reduced, owing to the low melting point of the alloys, since additions which raise the melting point of tungsten have not yet been found. At present, it does not appear possible to recommend a composition of vacuum-melted tungsten alloys (except alloys with rhenium) which would greatly surpass pure tungsten in strength at temperatures close to its melting point.

Alloying of tungsten with iron, cobalt, nickel, and titanium is difficult owing to the intense vaporization of these metals at the melting point of tungsten.

As already mentioned, tungsten and some of its alloys have found wide application in radio-electronics. Figure 122 shows a pulse amplifying klystron (diameter 512 mm, length 3250 mm) many of the components of which are made of tungsten [50].

The possibilities of using tungsten and its alloys as components of fuel elements in atomic power plants, parts of plasma jet engines, magnetohydrodynamic generators, porous ionizers, ionic engines, cathodes of thermionic emission converters, etc., are being investigated [24, 31, 32, 48].

Tungsten as well as other refractory metals and their alloys may be used in nuclear power plants for the production of liquid-metal containers and also in turbines. The basic requirement for such materials is compatibility with low-melting alkali metals and mechanical strength at high temperatures. One construction of a nuclear power plant is shown in Fig. 123 [49].

Low self-diffusion in tungsten is a valuable property for the ionizers of ionic engines, in which it is necessary to preserve constant porosity for a considerable length of time at high temperatures. The use of tungsten has been reported for the production of parts of a rocket engine using solid fuel [49]. Owing to the fact that the combustion products of solid fuel are reducing agents, the high oxidizability of tungsten and also of other refractory metals is not an obstacle to their use in rocket nozzles. Under such conditions, tungsten is found to have good resistance to erosion up to the melting point [49].

As the result of the use of tungsten in rocket engine nozzles, a need arose for the development of the technology of making large-dimension tungsten components. This problem is solved by extruding and forging vacuum-melted tungsten, and also by the methods of powder metallurgy. The machinability of tungsten is considerably improved by the method of impregnating tungsten with silver or copper [49]. The heat transfer and the resistance to thermal shock are greatly improved compared with tungsten sintered by the usual methods. A new field of application for tungsten and its alloys is as material for the cathodes of thermionic converters. Obviously, in the new technology, particularly in thermionic converters, a use will be found for single crystals of tungsten and its alloys.

In recent years, there has been a growing use of semifinished products of melted tungsten and its alloys. There is information concerning the production of tungsten 2-mm-gage sheets 510 × 915 mm in size. In the USA, attempts are being made to cast jet engine blades of tungsten–molybdenum alloys and work is also being carried out on the extension of components of W–3%Mo alloy.

For the wider application of tungsten alloys in modern technology, a number of very difficult problems have still to be solved. The most important of these problems is the production of highly ductile and weldable tungsten. It is obvious that only under this condition will it be possible to solve radically the question of the production of highly alloyed tungsten alloys, possessing high-temperature strength and good technological ductility. Research on the development of methods of producing high-purity ductile tungsten is of the utmost importance.

The production of precipitation hardening tungsten alloys and the establishment of the optimum conditions for their mechanical and heat treatment also represent a very important line of research on tungsten which will make it possible to produce alloys of high strength at temperatures above 2000°C.

The high chemical activity of tungsten alloys in the heated condition makes it imperative to find protective coatings for them. However, satisfactory protective coatings for the long-term service of tungsten and its alloys at high temperatures in an oxidizing medium have not yet been found.

## Rhenium Alloys

Of the alloys of refractory metals, rhenium alloys occupy a special position, owing to the peculiarities of the interaction of rhenium with refractory body-centered cubic metals and the influence of rhenium on the physical and mechanical properties of these metals, in particular on the ductility of molybdenum and tungsten. The authors of [53] divide all the alloys in the systems of rhenium with transition metals into two groups: group I, alloys in the range of solid solutions of rhenium in the transition metals; group II, alloys representing metallic compounds of the type of $\sigma$- and $\varkappa$-phases; group III, alloys in the range of solid solutions based on rhenium.

The most interesting and promising materials are rhenium alloys of group I, primarily alloys of tungsten–rhenium, molybdenum–rhenium, and tungsten–molybdenum–rhenium systems and also more complex alloys based on these systems. All these alloys deform well and may be produced in the form of different semifinished products (sheet, foil, tube, rod, wire, etc.). As the result of research on systems with the participation of rhenium based on tungsten, molybdenum, tantalum, niobium, vanadium, chromium, cobalt, nickel, palladium, zirconium, titanium, and other metals in the Refractory and Rare Earth Metals Alloys Laboratory of the A. A. Baikov Institute of Metallurgy in association with industry, different compositions of alloys and the technology of producing semifinished products from them have been developed [27, 53-60]. The alloys VR-27VP (W–27%Re) and MR-47VP (Mo–47%Re) and semifinished products made from them are manufactured on an industrial scale [27].

Alloys of rhenium with tungsten, tantalum, molybdenum, and niobium are currently the most heat-resistant [61].

Alloying of pure tantalum and niobium with rhenium leads to a sharp increase in their strength and a reduction in ductility (as is usually observed with the formation of interstitial solid solutions); a similar occurrence is also found on alloying group VIII metals with rhenium. Quite a different effect is obtained when tungsten and molybdenum of commercial purity are alloyed with rhenium; rhenium increases both the strength and ductility of these metals. Figure 124 is a diagram showing the variation in strength and ductility of transition metals of

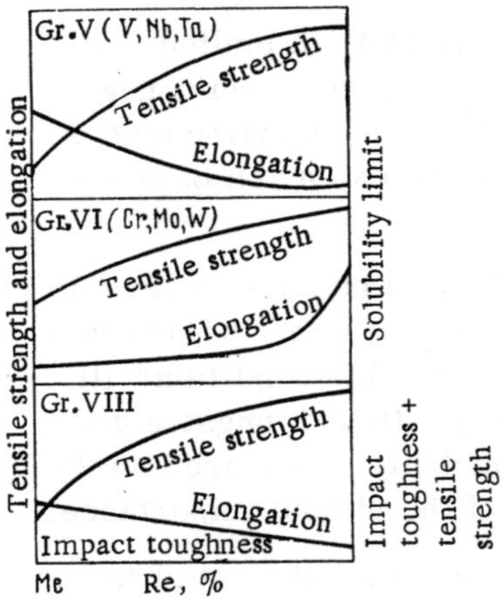


Fig. 124. Diagram showing variation of mechanical properties of group Va, VIa, and VIII transition metals on being alloyed with rhenium.

groups V, VI, and VIII [27]. Such an effect of an alloying addition is exceptional among the various alloy compositions encountered and is evidently typical of the alloying of brittle metals. Research has shown that the optimum compositions of alloys of tungsten and molybdenum with rhenium, possessing good ductility, technology, and weldability and at the same time high-strength properties at room temperature and high temperatures, lie close to the saturation boundary of solid solutions (20-30 wt.% Re in tungsten and 40-50 wt.% Re in molybdenum). These properties of the components apply to metals of commercial purity. It is possible that the use of tungsten and molybdenum of high purity with regard to interstitial impurities will enable optimum ductility and strength properties to be attained with lower rhenium contents. The cause of the increase in ductility of tungsten and molybdenum when alloyed with rhenium cannot yet be regarded as finally elucidated. One of the most important causes of an increase in ductility, as already mentioned in Chapter IV, is the occurrence of an additional deformation mechanism (twinning) [42]. It is shown [62] for the alloy Mo−35%Re that twins are formed along the (112) plane with [111] shear direction. The twinning intensity increases with the rate of strain. The ductile−brittle transition temperature of this alloy is below −196°C. Similar occurrences are also found in alloys of the tungsten−rhenium system.

According to Jaffee [63], one of the possible causes of the high ductility of tungsten−rhenium and molybdenum−rhenium alloys is the redistribution of the oxygen and the formation of complex oxides, which are arranged not in the form of brittle boundary films but in the form of globules (in the actual grains and along the boundaries). In such an arrangement, the oxides do not exercise any embrittling effect. The presence of the complex oxides MoReO and WReO in the form of globules is due to the high value of the surface tension of complex oxides of the tungsten−rhenium and molybdenum−rhenium systems. In these alloys, a higher oxygen content is more acceptable than in pure tungsten and molybdenum. A molybdenum−rhenium alloy is ductile with an oxygen content of 0.03-0.05%, whereas molybdenum retains its ductility only with an oxygen content of less than 0.001%.

In our view, the existence of continuous solubility in the solid state between rhenium and dicarbides of molybdenum or tungsten in the corresponding ternary system is also of substantial importance [29]. The favorable influence of rhenium may possibly be due to the disturbance of the electronic structure of tungsten and molybdenum on the formation of solid solutions, and to the change in character of the interatomic bond forces of alloying. It is pointed out in [64] that the strengthening curve of the alloy Mo−35 at.%Re resembles more the curves of group Va metals than those of the alloy components; the authors of [64] make the assumption that in this alloy, the rhenium acts as an additional source of electrons, filling the free orbitals and forming pairs with the electrons participating in the metallic bond. The result is that the electronic structure of the alloy becomes similar to the electronic structure of group Va metals.

E. M. Savitskii and co-workers [60, 65] made a detailed study of the properties of alloys of the tungsten−molybdenum−rhenium ternary system.

A number of high-strength alloys were developed. The investigations showed that alloying the refractory metals with rhenium increased their initial recrystallization temperature and their mechanical properties at elevated temperatures, due to the increase in the forces of

TABLE 47. Tensile Strength ($kg/mm^2$) of Alloys of Tungsten, Molybdenum, and Tantalum with Rhenium [60]

| Composition of alloy, wt.% | Annealed | Deformed | | |
|---|---|---|---|---|
| | 20° C | 20° C | 1500° C | 1800° C |
| W — 30 Re | 140 | 272 | 34.1 | 14.3 |
| W — 25 Re | 119 | 271 | 30 | 15.9 |
| Mo — 50 Re | 95 | 178 | 15.2 | 6.0 |
| Mo — 40 Re | 84 | 153 | 13.6 | 6.2 |
| 40 W — 20 Mo — 40 Re (1 : 1 : 1)* | 125 | 185 | 21 | 11.8 |
| 57 W — 15 Mo — 29 Re (2 : 1 : 1) | 98 | 129 | 26 | 9.3 |
| 45 W — 24 Mo — 31 Re (3 : 2 : 2) | 90 | 113 | 24 | 8.6 |
| 33 W — 34 Mo — 33 Re (1 : 2 : 1) | 92 | 124 | 21 | 9.3 |
| Ta — 1,5 Re | — | 60 | 16.9 | 8.7 |
| Ta — 5 Re | — | 118 | 23 | 13.8 |

*The proportions of the elements in at.% are shown in brackets.

the interatomic bond on the formation of solid solutions [42, 60]. Results are given in [60] of mechanical tests on alloys of rhenium with tungsten, molybdenum, and tantalum at 1500° and 1800°C (Table 47). A study has been made [66] of the high-temperature strength of rhenium and its alloys with tungsten and molybdenum. Figure 125 gives the stress-rupture strength of a tungsten alloy containing 25% Re at high temperatures in hydrogen [66]. Alloying of tungsten and molybdenum with rhenium produces an appreciable increase in the modulus of elasticity [27, 67]; an increase in the rhenium content of tungsten–rhenium alloys from 3 to 20% increases the modulus of elasticity by 10%. Alloying of tungsten and molybdenum with rhenium appreciably increases their specific electric resistance, but reduces their thermal conductivity. A small addition of rhenium and some stable oxides ($ThO_2$) to tungsten produces a nonsagging wire combining high ductility with increased recrystallization temperature, strength, specific electrical resistance, and resistance to oxidation [68].

Alloys of tungsten and molybdenum with rhenium possess yet another valuable advantage over other alloys based on tungsten and molybdenum, and that is good welding properties. Tough, sound welds were obtained by electron-beam welding of tungsten–rhenium, molybdenum–rhenium, and tungsten–molybdenum–rhenium alloys in a vacuum of $2 \cdot 10^{-5}$ mm Hg [27]. Methods of welding these alloys will be described in Chapter VII.

According to [69], alloying of chromium with rhenium (up to 35.6 at.% Re) seems not to have any effect on the density of dislocations nucleated under conditions of deformation by compression at room temperature. The dislocations have a screw orientation, which is in agreement with the tendency found in the alloys to deform by twinning. However, as the author points out, the ductility of the alloys is not correlated with the observed dislocation structure.

Alloying a nickel–chromium alloy of Nimonic type with rhenium increases the time to rupture at 800°C and a stress of 10 $kg/mm^2$ by a factor of 10 while retaining good deformability [53]. The alloy retains its high strength at 900 and 1000°C [70, 71]. Rhenium considerably increases the heat resistance of titanium alloys (alloys of titanium with chromium and aluminum) [70, 71]. Rhenium-based alloys in the solid-solution range have as yet been little studied. The solid-solution range from the rhenium side is fairly narrow, of the order of a few weight percent [27]. Research commenced in this direction shows that small alloying additions sharply alter the physical properties of rhenium (electrical resistance, hardness, etc.). However, the limited raw material resources of rhenium evidently do not permit the extensive use of these alloys in components of considerable weight. For this reason, the search for available substitutes for rhenium in tungsten and molybdenum alloys is an interesting scientific problem of considerable importance.

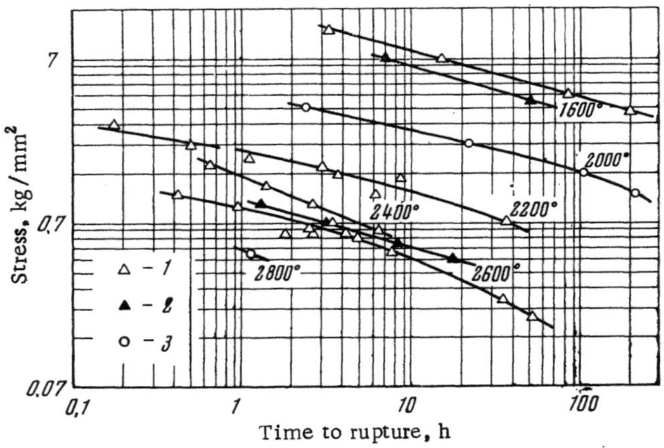


Fig. 125. Stress-rupture strength of a molybdenum alloy with 25% Re (tested in a hydrogen atmosphere). 1) Sintered at 2400°C; 2) the same at 3000°C; 3) arc-melted ingot.

Alloys of tungsten and noble metals with rhenium are well known as materials for high-temperature thermocouples [72-74]. Thermocouples of tungsten alloys have a high thermo-emf and high constancy of electrical properties when used in a vacuum and inert atmospheres.

Rhenium alloys are beginning to be widely used in electronics for different parts of cathode units (heaters, control grids, cathode holders, cathode bases, etc.) [27, 75-77]. The use of rhenium alloys in electrovacuum apparatus considerably increases their reliability and life. The Siemens firm (Federal Republic of Germany) recommends the use of a tungsten–rhenium alloy as anticathode of powerful x-ray tubes. Reference has been made to the possibility of using rhenium and its alloys as thermocathodes in thermionic emission energy converters [27]. Thoriated tungsten–rhenium alloys are used as direct-heated cathodes of electrovacuum devices on account of their high thermionic emission [78].

Reference may be made to the considerable advantage of rhenium alloys over other materials for use as contacts in electrical apparatus [27, 81, 82]. The use of rhenium and its alloys with tungsten, silver, and other metals under severe working conditions (high temperatures, high humidity, corrosive media, etc.) ensures reliable service of electric contacts.

A number of alloys containing rhenium possess superconductivity [27]. An alloy of Mo–50%Re has a superconductivity transition temperature of 10.9°K [79, 80].

The use of alloys of noble metals with rhenium as high-ohmic potentiometer materials is very interesting. Semifinished products and components made of pure rhenium are used in a number of electronic engineering components. It is also reported that high melting point, ductility in the recrystallized state, and resistance of water-cooling heating cycling make rhenium a very promising material for heating elements of electrovacuum devices.

A complete review of alloys containing rhenium, their structure, properties, and fields of application is to be found in the monograph [27].

Despite the fact that interest began to be taken in rhenium only comparatively recently, substantial results determining its fields of application have already been obtained. The principal fields are: 1) electronics (in the form of the pure metal or alloys with tungsten, molybdenum, nickel, etc.); 2) heat-resistant alloys with tungsten and molybdenum; 3) instrument making (details of tension wires and cathode bases, contacts working in corrosive media, high-temperature thermocouples); 4) details of devices and apparatus for the direct conversion of heat energy into electric energy, etc.

Rhenium is used in all these fields, despite its high price and short supply, since it has revealed properties which cannot be obtained on the basis, or with the participation, of other materials.

The potentiality of rhenium production in the USA is estimated to be about 9-13 tons annually, with the present level of production being 0.45-0.9 tons per year [27].

## Vanadium-Based Alloys

The principal consumer of vanadium is ferrous metallurgy (for alloying cast iron and steel). The titanium industry is becoming a new and important consumer of vanadium, since the constructional titanium-based alloys which this industry produces contain 4-6% and even 11-13% V.

One of the principal forms of vanadium raw materials in the USA consists of the tailings from uranium ore concentrates; in 1964, 3224 tons of vanadium pentoxide were produced [25]. A special feature of Soviet metallurgy is the separation of vanadium from low-grade iron ore [85].

The considerable natural resources of vanadium, its high melting point, resistance to oxidation when heated to 600°C, relatively low specific gravity, and a number of other physical properties make it a promising constructional material. The ductility of vanadium achieved by extreme purification from interstitial impurities by vacuum electron-beam melting reveals considerable possibilities for the development of alloys based on this metal. Yu. V. Efimov, E. M. Savitskii, and V. V. Baron have developed an effective method of purifying vanadium from interstitial impurities by means of cerium [86].

Research is being conducted in the USA on the development of vanadium-based constructional alloys capable of being shaped, welded, and worked at temperatures up to 650°C [87]. Sheet, strip, rod, wire, and tube products have been produced from such alloys. One of the most promising alloys contains up to 20% titanium (the remainder is vanadium); at 650°C its tensile strength is 35 $kg/mm^2$, which is higher than the values obtained for titanium alloys. At 1000°C, vanadium has a tensile strength of 6 $kg/mm^2$ and an elongation of 35-40%.

It is expected that semifinished products of vanadium and its alloys will be used in aircraft, rocket, and atomic engineering [87]. Alloys of vanadium with niobium, aluminum, and tantalum may be used as constructional materials. Due to the high resistance to corrosion, tubes, sheet, and rods of vanadium and its alloys with niobium, titanium, and aluminum are quite promising materials for use in the chemical industry. Vanadium is also used for the production of superconducting alloys and compounds (alloys of vanadium with gallium, silicon, titanium, scandium, etc.) [88-90].

## Niobium-Based Alloys

A fortunate combination of high strength, ductility, weldability, low cross section of thermal neutron capture, low ductile–brittle transition temperature, and other properties makes niobium a very promising base for the production of high heat-resistant alloys. Most compositions of multicomponent and highly alloyed alloys of the refractory metals are produced on a basis of niobium. Compared with group VIa metals, niobium is less sensitive to interstitial impurities and retains its ductility even when highly alloyed with other metals [61]. A shortcoming of niobium and its alloys is their high oxidizability in air at elevated temperatures.

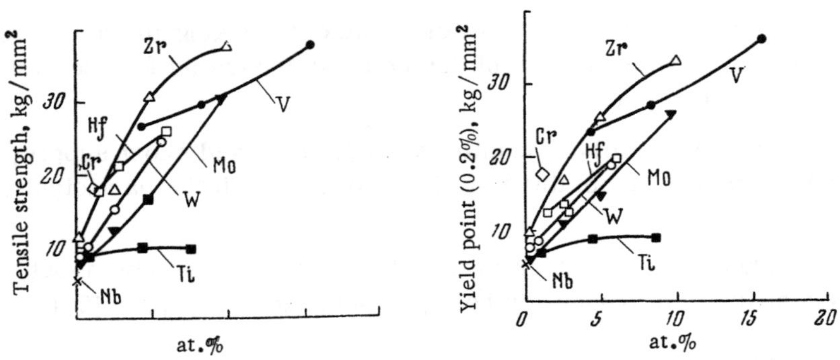


Fig. 126. Strength properties of binary niobium alloys at 1095°C versus different additions of alloying elements.

The influence of titanium, zirconium, molybdenum, vanadium, tungsten, and hafnium on the heat resistance of niobium has been studied [8, 61, 91-98]. Strengthening of niobium alloys is mainly due to the formation of substitutional solid solutions. Strengthening by disperse particles on decomposition of a supersaturated solid solution is as yet utilized very little for niobium alloys. Niobium alloys using high-temperature aging have begun to appear. At 1095°C, the most effective strengthening have been found to be vanadium, zirconium, tungsten, and molybdenum (Fig. 126) [121].

The influence of alloying elements on tensile strength and yield point of niobium alloys is shown in [94]. The most effective alloying element per 1 at.% addition was found to be tungsten. The effectiveness of titanium additions above 2 at.% diminishes considerably in the temperature range 870-1200°C. Hot-hardness measurements showed that niobium–tungsten alloys were the hardest at elevated temperatures (see Fig. 117) [94]. Data on the high-temperature hardness and strength of more complex niobium alloys were given in Chapter III.

It has been shown by E. M. Savitskii and A. M. Zakharov [99] that the hot-hardness measurement method may be used for a preliminary estimate of the high-temperature strength of niobium alloys.

Data on the mechanical properties of niobium alloys containing different amounts of oxygen (from 0.05 to 0.5%) are of interest. An increase in the oxygen content appreciably increases the strength of alloys of niobium with titanium, zirconium, and hafnium at elevated temperatures [100, 101]. This is evidently due to precipitation hardening (precipitation of oxides of the active additions – zirconium or hafnium), and is one of the typical examples of the aging of alloys owing to interstitial impurities, in the present case oxygen. Since the aging effect disappears even at moderate temperatures (500-700°C), "oxygen strengthening" cannot be used in particularly heat-resistant alloys; in addition, oxygen even at low contents sharply increases the hardness and brittleness of refractory metals. Nevertheless, the use of stable disperse phases formed in systems of niobium with group IVa metals (titanium, zirconium, hafnium) and oxygen (or carbon) for strengthening the niobium matrix should always be kept in mind when developing niobium alloys.

The hot hardness and short-term strength of a large group of niobium alloys were investigated by Maykuth and Jaffee [102].

Data are given in [42, 91, 103] on the effect of refractory additions on the recrystallization temperature on niobium alloys. Additions of tungsten, molybdenum, titanium, zirconium, and vanadium appreciably increase the recrystallization temperature of niobium.

The properties of a number of multicomponent niobium-based alloys are investigated in [91-93, 95, 96, 104].

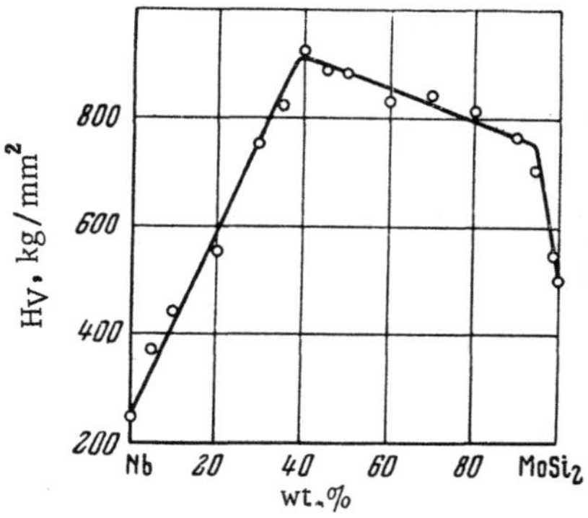


Fig. 127. Hardness of alloys of the Nb−$MoSi_2$ cross section.

Alloys of niobium with molybdenum and silicon were studied in [105, 106]; alloying of niobium with these elements results in a considerable increase in hardness of the alloys (Fig. 127). At the same time, both in the range of substitutional solid solutions and on the appearance of a second phase (in the case of the alloy of niobium with 5% $MoSi_2$, a eutectic is formed, consisting of $Nb_4Si$ and a solid $\alpha$-solution on a niobium base) the alloys retain sufficient plasticity and may be worked at elevated temperatures.

The mechanical properties of alloys of the niobium−tungsten−titanium system were studied in [104, 105]. Figure 128 shows the results of tests on a niobium alloy containing 28% W and 7% Ti (as forged) at different temperatures. The influence of heat treatment on the structure and properties of this alloy were studied. Recrystallization took place at 1500°C. Recrystallization annealing led to an increase in elongation, but no considerable reduction in strength was found. The influence of alloying additions on the properties of the ternary alloy was also studied in [104]. It was found that vanadium and molybdenum did not lead to any increase in the strength of ternary alloys of the niobium−tungsten−titanium system, evidently owing to the reduction in the tungsten content of the alloy containing molybdenum and vanadium.

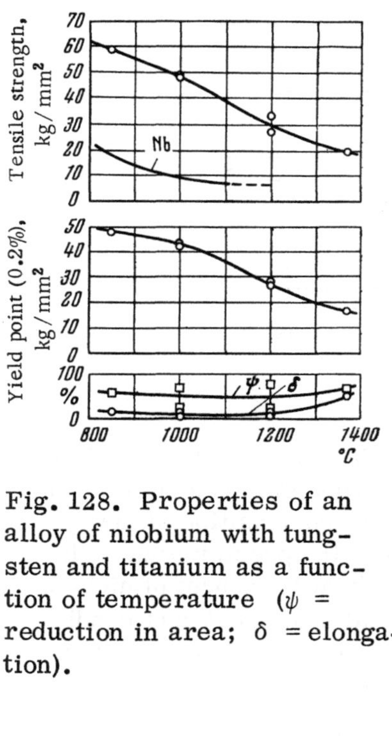


Fig. 128. Properties of an alloy of niobium with tungsten and titanium as a function of temperature ($\psi$ = reduction in area; $\delta$ = elongation).

The strength properties of niobium−tungsten−titanium alloys, and also of other ternary and multicomponent alloys may be increased by the introduction of small additions of zirconium. Molybdenum and especially zirconium inhibit grain growth. Microscopic and x-ray analysis of the alloy Nb−28%W−7%Ti showed the presence in the structure of finely disperse precipitates, similar to TiC [104]. Heating the alloy to 1500°C converted this finely disperse phase into a solid solution, while alloying with zirconium increased the stability of the carbides (heating even to 1700°C failed to convert them into solid solution). The presence of these insoluble inclusions may evidently explain the influence of zirconium on the reduction of grain growth.

According to the combination of properties (heat resistance, ductility, weldability, stability in molten alkali metals), the most promising system for the development of constructional niobium alloys has been found to be the niobium−tungsten−molybdenum−zirconium system. The structure and properties of alloys of this system and constituent systems has been investigated in detail by E. M. Savitskii and A. M. Zakharov [107, 108] and also by E. M. Savitskii and O. B. Dashevskaya [191].

An analysis of the work of Soviet and other investigators on niobium alloys justifies the assumption that some of the effect of hardening of these alloys on the introduction of alloying additions is due to the precipitation of disperse compounds in the form of oxides, carbides, or in a more complex form (oxycarbides, oxynitrides, etc.). Particularly effective hardeners are small additions of zirconium, forming an excess phase with the interstitial impurities.

TABLE 48. Mechanical Properties of Multicomponent Niobium Alloys Developed in the USA [8, 24, 109]

| Alloy sort | Chemical composition, % | Manufacturer | Test temperature, °C | Tensile strength, kg/mm² | Elongation, % | Reduction in area, % | Endurance stress, 100 h, kg/mm² |
|---|---|---|---|---|---|---|---|
| Nb-1Zr | 0.6—1.2 Zr (Remainder nickel) | Westinghouse | — | — | — | — | — |
| D-31 | Nb + 10 Ti + 10 Mo | Dupont | 24<br>980<br>1095<br>1205 | 70<br>—<br>24.6<br>17.6 | 22<br>—<br>12<br>14.0 | 47<br>—<br>9<br>13 | —<br>9.8<br>—<br>— |
| F-48 | Nb + 15 W + 5 Mo + 1 Zr | General Electric | 24<br>1095<br>1205 | 85.7<br>35.0<br>24.6 | 24<br>28.0<br>35.0 | —<br>—<br>— | —<br>14<br>7.7 |
| Cb-74 | Nb + 5 Zr + 10 W | Union Carbide Alloys | 1205 | 31.6 | 23 | — | — |
| Cb-65 | Nb + 7 Ti + 0.8 Zr | Union Carbide Alloys | 24<br>1095 | 50.6<br>16.2 | 35<br>25 | 80<br>94 | —<br>2.8 |
| Fs-85 | Nb + 27 Ta + 11 W + 0.8 Zr | Finesteel | 20<br>1095 | 73.8<br>32.8 | —<br>— | —<br>— | —<br>— |
| C-129 | Nb + 10 W + 10 Hf | Van Cheng | 20<br>1095 | 78.8<br>33.0 | —<br>— | —<br>— | —<br>17.6* |
| Cb-752 | Nb + 10 W + 2.5 Zr | Haynes Stellite | 20<br>1095 | 59.1<br>29.6 | —<br>— | —<br>— | —<br>12.6 |
| B-66 | Nb + 5 Mo + 5 V + 1 Zr | Westinghouse | 20<br>1095 | 80.8<br>— | —<br>— | —<br>— | —<br>18.3* |

*After ten hours.

According to [103] oxides of $ZrO_2$ and $HfO_2$ were found in the alloy Nb–1%Ti–1%Zr–1%Hf. In a number of niobium alloys of the niobium–tungsten–molybdenum–zirconium system, a carbide phase has been found, the size and distribution of which depend on the composition of the phase. The niobium alloy with 1% Zr is strengthened mainly owing to the precipitation of disperse particles of $ZrO_2$ and ZrC. In connection with these data, the heat treatment of niobium alloys assumes considerable importance. The development of optimal heat-treatment conditions will enable the strength of niobium alloys to be considerably increased. Researches on these lines have only just commenced.

Table 48 shows the composition and mechanical properties of some niobium alloys which enjoy semi-industrial or industrial importance outside the USSR. Various semi-finished products are made from these alloys by extrusion and other methods of pressure treatment. The niobium alloy having the highest strength is the alloy F-48, but the prospects of using it are limited owing to inadequate ductility and poor weldability [24]. In the general case, the technological difficulties in working niobium alloys naturally increase with increase in alloying and the accompanying increase in strength at room temperature and elevated temperatures. Of the alloys cited in Table 48, the following alloys according to [109] are easy to work: Nb-1Zr, Cb-65, and Fs-85. After destruction of the cast structure, these alloys may be deformed at 20°C. Pressure working of the other alloys given in Table 48 involves considerable technological difficulties. The alloys F-48 and F-50 are subjected to preliminary hot extrusion at temperatures above 1500°C; however, sheets of large dimensions are obtained from them (although with difficulty).

The above-mentioned niobium alloys have the following physical properties [109]:

| Alloy sort | Nb-1Zr | Fs-82 | Cb-65 | Cb-74 | D-31 | F-48 | F-50 | Unalloyed niobium |
|---|---|---|---|---|---|---|---|---|
| Melting point, °C | 2400 | 2510 | 2260 | 2430 | 2270 | 2480 | 2430 | 2468 |
| Density, g/cm³ | 8.66 | 10.2 | 8.06 | 8.98 | 8.08 | 9.4 | 9.1 | 8.66 |
| Crystal structure | bcc | bcc | bcc | bcc | bcc | bcc | bcc | bcc |
| Lattice constant, Å | – | – | 3.295 | – | 3.229 | – | – | 3.294 |
| Thermal conductivity, cal/cm · sec · deg | 0.10 (25°C) | – | 0.053 (315°C) 0.046 (1200°C) | – | 0.17 (1200°C) | 0.094 (25°C) 0.14 (1095°C) | 0.082 (25°C) | 0.127 (25°C) 0.156 (600°C) |
| Linear coefficient of thermal expansion, $\alpha \cdot 10^6$ | – | – | 77 (25-1200°C) | – | 7.4 (0-1000°C) | 8.8 (25-1150°C) | 9.0 (25-1150°C) | 6.97 (18-1000°C) |
| Specific heat, cal/g | 0.065 (25°C) | – | 0.074 (25-1200°C) | – | 0.074 (20-1100°C) | – | – | 0.065 (0-1400°C) |
| Modulus of elasticity, kg/mm² | | | | | | | | |
| at 20°C | 10,600 | 11,600 | 10,540 | 11,240 | 11,600 | 17,570 | 16,860 | 10,600 |
| at 1095°C | – | – | – | — | 5970 | 12,640 | 11,600 | – |
| Thermal neutron capture cross section, barns/atom | 1.1 | 5.2 | 1.7 | 2.0 | 2.0 | 2.5 | 2.9 | 1.1 |

In the USSR, the heat-resistant deformable alloys VN-1, VN-2, and RN-6 are produced, together with a number of other alloys based on niobium and possessing good technological ductility and good weldability. The mechanical properties of the alloys VN-1 and VN-2 at different temperatures are shown in Fig. 129 [110]. The alloy RN-6 at room temperature has the following mechanical properties:

| | Tensile strength, kg/mm² | Elongation, % |
|---|---|---|
| Deformed condition | 89 | 4 |
| Annealed at 0°C | | |
| 1000 | 76 | 4.8 |
| 1200 | 64 | 15 |
| 1500 | 58.9 | 20 |
| 1700 | 58.2 | 25.7 |

The stress-rupture strength of the alloy RN-6 after 100 h at 1200°C is 11 kg/mm².

The chemical composition of the alloy is: W, 4.5-6%; Mo, 4.5-6%; Zr, 1-1.5%; O, 0.008-0.016%; N, 0.009%; C, 0.022-0.040%; remainder Nb [192].

The alloy was developed in the Refractory and Rare Metal and Alloys of the A. A. Baikov Metallurgical Institute of the Academy of Sciences of the USSR with the collaboration of V. V. Baron and K. N. Ivanova.

The initial (as cast) structure has a considerable influence on the final mechanical properties of the alloy, despite the intervening heat and mechanical treatments [192]. The finer is the grain in the as cast condition, the higher are the values of these properties. The difference in the strength properties of an alloy, produced by different melting methods, attains 20 kg/mm² [192]. As remarked in [192], in the present case, the initial structure is "inherited." Research on the influence of different melting methods (electric-arc, electron-beam, and elec-

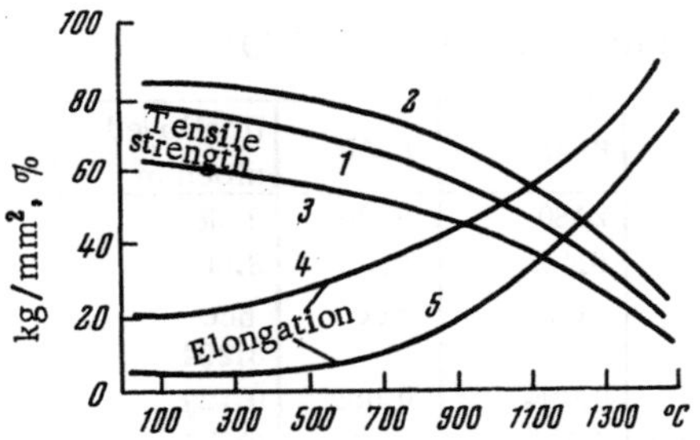


Fig. 129. Mechanical properties of the alloys VN-1 and VN-2 as a function of test temperature. 1) Peened alloy VN-1; 2, 5) the same, alloy VN-2; 3, 4) recrystallized alloy VN-2.

tric-arc with centrifugal casting) on the mechanical properties of the alloy RN-6 showed that the maximum strength properties were found after electric-arc melting with centrifugal casting ensuring the production of a structure having maximum grain fineness [192]. At the same time, the alloy possessed high ductility comparable with the ductility of an alloy produced by electron-beam melting and containing a lower content of gaseous impurities.

The maximum hardening effect in niobium alloys is found to be due to the action of two mechanisms: for mation of substitutional solid solutions and the precipitation hardening. Examples of such compositions are alloys of the niobium–tungsten–molybdenum–zirconium, niobium–tungsten–molybdenum–zirconium–carbon [191], and other systems. The number of compositions of niobium alloys will increase with future research on multicomponent systems based on niobium. Recrystallization annealing of niobium alloys does not reduce their strength at elevated temperatures, but increases the ductile–brittle transition temperature. Evidently, the embrittlement of an alloy is due to the precipitation from the solid solution of excess phases formed by interstitial impurities. Precipitation hardening of niobium alloys will be examined later (Chapter VII). Owing to considerable resources of raw material, its excellent ductility, fairly high strength at high temperatures, weldability, higher melting point of the oxide, and lower volatility, niobium is more suitable as base for high-strength and high-temperature alloys than molybdenum.

The production of and demand for niobium have risen rapidly in recent years. The growth of niobium production in the USA is shown by the following figures: 1958, 30 tons; 1959, 60 tons; 1960, 120 tons. The capacity for niobium production in 1960 was estimated at 200 tons. The productive capacity of the US niobium industry may be doubled or trebled in the not far distant future. The bulk of niobium concentrates is used in the production of ferroniobium, amounting to 454 tons in 1960, 907 tons in 1961, 1270 tons in 1962, and 1406 tons in 1965 [25]. About 100 tons of niobium was used in the USA in 1963 in the production of alloys.

The high cost of semifinished products of niobium heat-resistant alloys is due to the small scale of production, imperfect technology, and lack of certain types of special equipment. According to a forecast made by the General Electric Company, in the course of the next decade it will be possible to produce annually 1000 tons of rolled niobium on the following conditions: fuller utilization of niobium ores, rolling mills operating in a protective atmosphere, and progress in the technology of sheet rolling. Niobium alloys are more promising than Nimonic alloys for use in gas turbines. The application of niobium alloys in turbines increases the efficiency of the latter (according to data of the General Electric Company).

For the use of niobium alloys at high temperatures in an oxidizing atmosphere, reliable protective coatings have been developed (on the basis molybdenum disilicide, chrome-yttrium, etc.) [115] for durations of tens and hundreds of hours. Some high-strength niobium alloys (molybdenum, tungsten, and other alloys) are quite promising for use in space engineering. In particular, it is pointed out [34] that the fields of maximum use of niobium alloys in the next 5 or 10 years will be heat-protective systems (for atmospheric re-entry) and containers for alkali metals in the heat exchangers of nuclear reactors.

Figure 130 shows an example of the use of niobium (together with tungsten and molybdenum) for the production of an aircraft component. Each metal is used in its own temperature range.

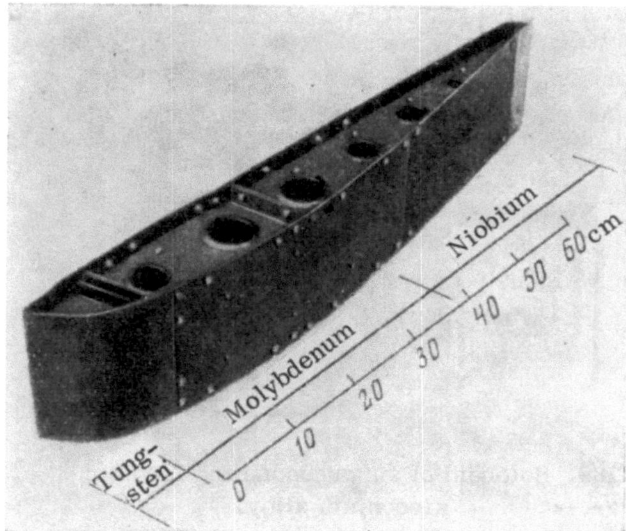


Fig. 130. Complicated component made of three refractory metals: tungsten, molybdenum, and niobium

Fig. 131. Vacuum chamber for testing materials.

A chamber is described in [111] simulating space conditions for testing materials, constructional elements, and units of spacecraft. It is reported that in the USA more than 100 similar chambers having a capacity of 20 to 1000 $m^3$ are in use. One of the largest space chambers is shown in Fig. 131 (chamber diameter about 9 m). The vacuum in the large chambers attains $10^{-8}$ mm Hg. The firm of "Balzers" has developed a chamber for testing ion engines, in which an air lock is provided for the entry of units without disturbing the vacuum.

Recently, heat-resistant niobium alloys have been used in electronic engineering to replace tantalum (screens) which is in short supply, and molybdenum (tubes and other components) which has lower ductility.

The good corrosion resistance and small thermal neutron capture cross section (1.1 barns) make niobium a promising constructional material for nuclear reactors. Research has

Fig. 132. Solenoid of superconducting niobium–zirconium alloy.

shown that niobium is capable of resisting the action of molten uranium and its alloys, and also that sodium and lithium, freed from oxygen, sodium–potassium alloys, and of other low-melting metallic materials, may be used as heat carriers of power systems owing to their high heat-transfer coefficient [34, 49, 121].

In recent years, an entirely new field of application of niobium as base for superconducting materials has been disclosed [90, 119, 154]. Of the metals, niobium has the maximum superconductivity critical temperature (8.9°K). Alloys of the niobium–zirconium system have values for the critical magnetic field of up to 80 kG, critical current density 4-6 $\cdot 10^4$ A/cm$^2$, and critical temperature 11°K. In the USSR, the industrial superconducting alloys RNS-1, RNS-2, and RNS-3 have been developed on the basis of the niobium–zirconium system [117] (Fig. 132). Several years ago, the A. A. Baikov Institute of Metallurgy and the Lebedev Physics Institute of the Academy of Sciences of the USSR produced a small-size solenoid with a magnetic field strength of 57 kG [118]. Considerable interest attaches to alloys of the niobium–tin system, including the compound $Nb_3Sn$ having very high superconducting properties. Superconducting magnets of 100 and 130 kG have already been produced on the basis of this compound [119].

The development of industrial methods of producing articles made of brittle superconductors is of considerable technical value. In addition to the construction of solenoids, a start has been made in the development of magnetic pumps for increasing the magnetic field strength by magnetic pumping. Figure 133 shows a magnetic pump made of the compound $Nb_3Sn$ [120]. The pump is a cylinder having a cavity of figure-eight shape, one loop being considerably larger than the other. A powerful magnetic field is produced inside the hollow cylinder and a cylinder made of a superconductor is then inserted in the large opening. Since the magnetic flux does not pass through the superconductor, the magnetic lines of force are "repelled" from the large opening to the small opening, producing a concentrated field in the latter. By means of this method, a number of brittle superconducting materials may be used for producing magnetic fields of high strength without first having to make the materials into wire. According to [120], it has currently been possible by means of a magnetic pump to produce superpowerful magnetic fields of a strength of up to 22,000 Oe. It is to be expected that the use of superpowerful magnetic fields, produced by means of superconducting magnetic materials consisting of niobium alloys and compounds, will lead to enormous new advances in space engineering, atomic engineering, power engineering, and will open up wide prospects for solid state physics [90].

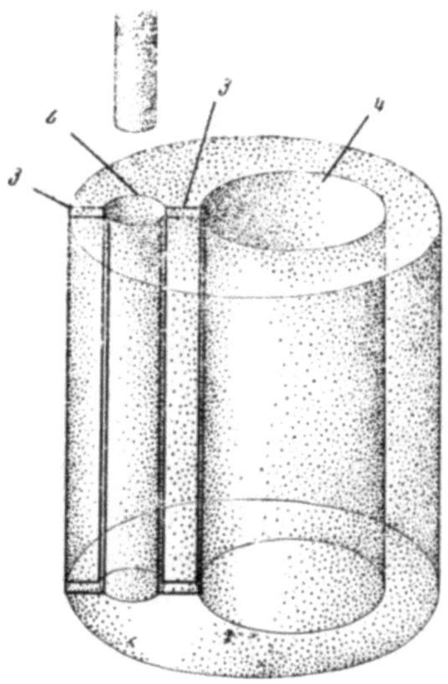

Fig. 133. Magnetic pump made of the compound $Nb_3Sn$. 1) Piston of superconducting material; 2) compressor cavity; 3) heating coil; 4) working chamber.

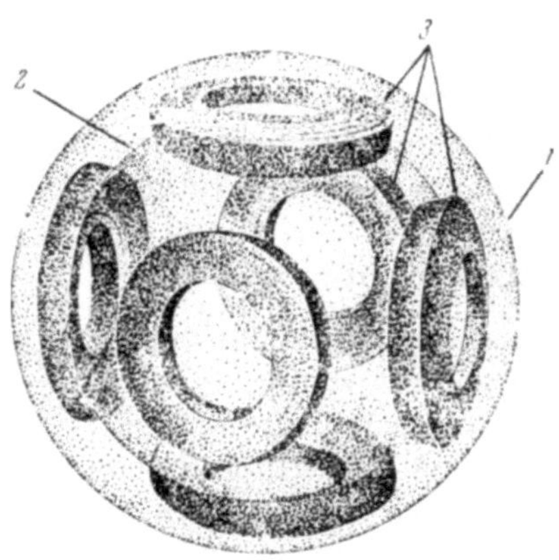

Fig. 134. Diagram of cryogenic gyroscope. 1) Superconducting cell; 2) rotor; 3) coil support.

Some isolated examples will now be given to illustrate the use of the phenomenon of superconductivity in technology. Figure 134 shows diagrammatically a cryogenic gyroscope [120]. A superconducting sphere, made of niobium or other superconducting material, is suspended in a magnetic field. The rotating sphere is a peculiar type of gyroscope rotor. The application of magnetic suspension disperses with bearings. The complete absence of friction increases the reliability and life of the gyroscopes.

Considerable interest attaches to the use of superconductors in the production of magnetohydrodynamic generators. It is evident that magnetohydrodynamic generators with superconducting magnets will find extensive application as sources of high-power electrical energy in spacecraft and terrestrial electric power stations. It is proposed to construct an electric motor, the rotor and stator of which will be made of superconducting alloys. The efficiency of such an electric motor ought to be 99.9% [120].

Evidently, superconductivity will make it possible to solve the problem of the use of thermonuclear reactions for producing new, improved computor devices, as well as electrical devices for the reception and transmission of signals at ultrahigh frequencies. Problems of the transmission of power to considerable distances, the protection of spacecraft from radiation in interplanetary flights, etc., will be solved in a new way. It is at present difficult to imagine all the domains in which the application of superpowerful magnetic fields, produced by means of superconducting solenoids, will result in revolutionary progress.

Niobium alloys, because of their resistance to corrosive media, ought also to find extensive application in the chemical industry, where they may be used in the form of sheet, rod, and tube, the technical production of which has already been developed [114].

## Tantalum-Based Alloys

Pure tantalum is highly ductile down to the temperature of liquid helium, owing to the relatively high solubility of interstitial impurities, especially nitrogen and oxygen. The limit solubility of nitrogen and oxygen in tantalum at room temperature is 0.44% and 0.26%, respectively. Because of the high ductility of tantalum, it is possible to carry out without any particular difficulty technological operations (rolling, stamping, welding, etc.) for producing from tantalum semifinished products and finished articles. Figure 135 shows welded tubes 3 m long produced from tantalum sheet 1.2 mm thick [121].

Judging by the combination of physical and mechanical properties (melting point 3000°C, excellent workability, and high ductility), tantalum may be a promising material for exception-

Fig. 135. Welded tubes 3 m long made from tantalum sheet 1.2 mm thick.

ally heat-resistant alloys. Unfortunately, the natural resources of tantalum are very small, being estimated to be 1/70 of the niobium reserves.

Alloying of tantalum with tungsten and molybdenum considerably increases its modulus of elasticity and strength at elevated temperatures, but at the same time sharply reduces its ductility. Tungsten and molybdenum have approximately the same effect on the strength of a solid solution of tantalum. In the tantalum–tungsten system, the alloy Ta–10%W has maximum mechanical properties. While retaining sufficiently high technological ductility this alloy in the recrystallized condition has a tensile strength at 1200°C of about 30 kg/mm$^2$; alloys having a higher tungsten content are not very amenable to treatment by pressure.

Tantalum–niobium alloys may be well worked throughout the entire concentration range but up to the present they have not found any extensive industrial application, evidently because of the low strength of these alloys at elevated temperatures and the short supply of tantalum.

An increase in strength of binary tantalum alloys may be obtained by the introduction of further additions of various components. Alloys of the tantalum–niobium–tungsten system, as well as alloys of tantalum with rhenium, hafnium, and molybdenum, were investigated in [122-124]. Tantalum–niobium–tungsten alloys have high strength at room temperature and elevated temperatures, and a high recrystallization temperature (1400°C). Hafnium, rhenium, vanadium, and molybdenum also effectively increase the strength of tantalum.

Additions of hafnium, molybdenum, rhenium, and tungsten are most effective in increasing the recrystallization temperature of tantalum, i.e., those elements which, as already pointed out, also show an appreciable influence on the high-temperature strength of tantalum [103, 123]. Of the tantalum alloys investigated, the alloy Ta–20%W had the highest recrystallization temperature (1600°C). The recrystallization temperature of the alloys Ta–15%W and Ta–10%W–0.063%C is 1500°C. The addition of titanium to tantalum (up to 30%) evidently reduces the recrystallization temperature of pure tantalum. A reduction in the short-term high-temperature strength of these alloys is found at the same time. The alloys of the tantalum–niobium–vanadium system were investigated in [103]. The best ternary alloy was found to be the alloy Ta–30%Nb–50%V, which in the recrystallized condition has a tensile strength at room temperature of 67.3 kg/mm$^2$. The alloy Ta–10%Hf has a good combination of strength and ductility at room temperature. According to [40], this alloy also has the highest strength of the investigated tantalum alloys at 1200°C. Alloys of tantalum with tungsten and rhenium have maximum high-temperature strength [125].

Alloying of the solid solution with relatively large amounts of tungsten, niobium, hafnium, vanadium, and rhenium is evidently the most effective means of increasing the strength of tantalum, the permissible alloying limits being set by the technological properties of the alloys. Data [38] are given below on the limit content of alloying elements for which it is still possible to obtain high-quality strip at the indicated rolling temperatures of tantalum alloy (wt.%):

| | 24° C | 982° C | 1600° C |
|---|---|---|---|
| Titanium ....... | >80 | — | — |
| Niobium ....... | >50 | — | — |
| Zirconium ...... | <5 | >40 | — |
| Hafnium ....... | >5 | >35<40 | — |
| Vanadium ....... | <10 | <30 | >30 |
| Tungsten ....... | >5 | 15 | <30 |
| Molybdenum ....... | <5 | 7.5 | <20 |

An increase in the purity of the initial alloying components with regard to interstitial impurities is accompanied by an increase in the attainable alloying limits in the attainable strength-increasing effect. Electron-beam melting enables the limits of alloying tantalum with tungsten to be increased to 15-20% (as regards workability of the alloy).

According to the data of [24], the alloys Ta–10%W, Ta–12.5%W, and Ta–8%W–2%Hf are being produced industrially in the form of various semifinished products. These alloys are produced by arc melting and also by electron-beam melting. Ingots of the alloy Ta–10%W may be rolled to sheet bar at 1100-1370°C [38]. After scarfing and annealing, the 25-mm-thick bar is rolled at room temperature to sheet of a thickness down to 0.26 mm without intermediate annealing. The sheets retain good ductility after cold rolling, and also show high ductility in welded joints when welded by electron-beam in a vacuum or with a tungsten electrode in an inert gas atmosphere.

Some properties of industrial tantalum alloys, according to [24], are given below:

| Alloy .............. | Ta — 10% W | Ta — 12,5% W | Ta —8% W — 2% Hf |
|---|---|---|---|
| Density, g/cm³........ | 16.78 | 16.82 | 16.65 |
| Strength (0.2% proof stress/ tensile strength), kg/mm² | | | |
| 20° C ....... | 71.0/76.5 | 59.8/71.7 | 67.5/78.7 |
| 1093° C ....... | 38.7/42.9 | — | — |
| 1316° C ....... | 24.6/28.1 | — | 23.2/28.1 |
| 1649° C ....... | 11.2/13.4 | — | 12.6/14.1 |
| 1927° C ....... | — | — | 7.7/7.7 |
| 2705° C ....... | — | 0.98/1.3 | — |
| 2927° C ....... | — | 0.61/0.65 | — |
| Stress-rupture strength (100 h), kg/mm² | | | |
| 1093° C ....... | 35.1 | — | — |
| 1316° C ....... | 19.7 | — | 21 |

The principal advantage of tantalum-based alloys is the combination of high strength at high temperatures (up to 2000°C), high ductility, good workability, and good welding properties. Alloys of tantalum and niobium considerably surpass other materials based on refractory metals as regards low-temperature ductility. The high ductility of tantalum and some of its alloys in the recrystallized state is retained down to the temperature of liquid helium (provided these alloys are made in a state of high purity with regard to interstitial impurities).

Pure tantalum is finding wide application in capacitors (main sphere of application), chemical apparatus (Fig. 136) [121] and for the production of constructional elements of electrical lamps [75, 109].

Work is being done in the USA to ascertain the possibility of using tantalum alloys in rocket engine nozzles, construction of ballistic rockets, and returning spacecraft. According to Jaffee, alloys of the tantalum–tungsten–hafnium system, if reliable antioxidation protective coatings are developed for them, may be used for the heated envelope of spacecraft [49]. Kieffer [121] has reported the production of rocket nozzle components made of the alloy Ta–10%W (Fig. 137). In small amounts, this alloy is also used as spring material in electronic devices.

Fig. 136. Heat exchanger with tantalum tubes.

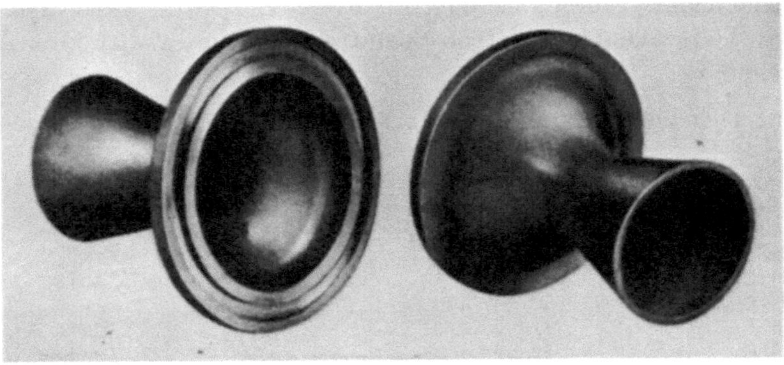

Fig. 137. Rocket nozzle inlet of an alloy of tantalum with 10% tungsten.

The small natural reserves of tantalum restrict its application. On the whole, tantalum alloys will evidently be used for small components used in electronic engineering (springs, capacitors, resistance films, etc.).

## Alloys Based on Titanium, Zirconium, and Hafnium

The group IVa metals, titanium, zirconium, and hafnium, are refractory metals but are dimorphic. The transition from the low-temperature hexagonal modification to the high-temperature cubic modification occurs in titanium at 882°C, in zirconium at 865°C, and in hafnium at 1760°C; at the same time the metal loses almost completely its ability to resist external forces. Consequently, titanium, zirconium, and hafnium cannot form the base for high-temperature-resistant alloys [42, 61]. However, on the basis of these metals, especially titanium and zirconium, a large number of different compositions of high-strength, heat-resistant (up to 500-700°C) constructional alloys have been developed which are capable of service in various chemically active media (sea water, steam at high parameters, gas turbine atmosphere) or at low temperatures, for example, in liquid oxygen. Alloys based on the $\alpha$-, $\beta$-, or ($\alpha + \beta$)-modifications are highly ductile, not prone to cold shortness, and may be used at quite low temperatures. In recent years, a start has been made on the development of alloys based on metallic compounds of titanium (titanium–nickel, titanium–aluminum); their technological properties are independent but these alloys are capable of service up to 1000°C.

The principal advantages of titanium alloys is the high ratio of tensile strength to yield point, good technological properties, and corrosion resistance. The tensile strength of a number of titanium alloys at the ordinary temperature exceeds 100 kg/mm$^2$. The principal shortcomings are relatively high cost, limited weldability, a tendency to absorb hydrogen from the air, and a tendency to spark formation.

Titanium has become a fairly widespread industrial nonferrous metal, being of interest in shipbuilding, in rocket and gas-turbine engineering, as well as in the chemical and metallurgical industries. According to [24], the bulk of the titanium produced at the present time is used for aircraft gas-turbine compressors, rocket bodies, and deep-diving submarines. In the USA in 1965 the output of titanium and semifinished products was 8160 tons [25]. The phase diagrams, nomenclature, and properties of titanium alloys are contained in [126-129].

Currently, considerable attention is being devoted abroad to the development of zirconium-based alloys for use in nuclear reactor construction and chemical engineering. Owing to the small cross section of capture of thermal neutrons (0.18 barn) and high corrosion resistance, zirconium and its alloys are used for the production of cooling tubes, sheaths of heat-liberating elements, and other constructions in nuclear power stations. An alloy of the zirconium–tin system, which has good welding properties, has found maximum application.

The demand of atomic engineering for metallic zirconium resulted in the organization of its industrial production and afforded the possibility of employing zirconium in other branches of engineering [42]. In 1962, more than 1200 tons of metallic zirconium was produced in the USA.

The high corrosion of zirconium in hot nitric acid of any concentration, in boiling sulfuric acid (up to 70%), and phosphoric acid (up to 60%) determines the possibility of extensive application of zirconium in chemical engineering for the construction of pumps, heat exchangers, and other components. The application of zirconium in the production of the new pyrophoric alloys is growing [130, 131]. The addition of zirconium to constructional steels refines austenitic grain, reduces the quenching temperature, increases the high-temperature strength of the steel, and permits the temperature of chemical heat treatment to be increased.

Zirconium is a most important alloying element in the production of superconducting alloys (this has been discussed in detail in the preceding section). The use of zirconium boride is being rapidly developed in military engineering in the USA. The principal field of application of zirconium is in the nuclear reactors of submarines. Currently, more than 180 fields of application of zirconium, its alloys, and compounds are known [132].

Because of its high cross section of capture of thermal neutrons (115 barns), hafnium shows promise for use in the elements of an atomic reactor control system [132]. Recent scientific research shows evidence of the high prospects for the use of hafnium as alloying addition to refractory metals, and also as addition to cathode alloys used in electronics. Hafnium considerably increases the high-temperature strength of tantalum and niobium. The demand for hafnium is constantly increasing in the USA amounting to 72 tons in 1964.

## Alloys Based on the Noble Metals

Iridium, ruthenium, palladium, and platinum, which, compared with tungsten, molybdenum, and other refractory metals, possess high corrosion resistance at elevated temperatures, are of considerable interest for the new technology. In addition to refractoriness and corrosion resistance, these noble metals also possess a number of interesting physical properties, permitting them to be used as precision alloys.

Until recently, little attention was paid to metals of the platinum group, although their natural reserves are comparatively large (osmium is an exception). The production of palladium at the present time appreciably exceeds the demand. The development of new alloys based on refractory metals of the platinum group, especially palladium, is one of the most important problems of metallurgy. Work is being conducted in this direction in the USSR, USA, Federal Republic of Germany, and other countries.

The noble metals of group VIII attract attention by their special properties, the stability of their electrical properties, oxidation resistance at high temperatures, ductility, and corrosion resistance. Particularly high technological ductility is a feature of palladium and platinum, on the basis of which a large number of alloys having special physical properties have been developed. A study of the phase diagrams of alloys of palladium with other elements, especially with transition metals, and the determination of the relationships governing their influence on the properties of palladium have made it possible to discover a number of important contact and potentiometer materials which have found industrial application.

The palladium–tungsten system is the most promising basis for the development of stable electrical alloys [133]. The high melting point and recrystallization temperature of alloys of palladium with tungsten, and the absence of transformations in the solid state up to 23 wt.% tungsten determine the stability of their properties. Alloying palladium with tungsten considerably increases the specific electrical resistance of the alloys; for a tungsten content of 20 wt.%, the value of the specific electrical resistance attains 110 $\mu\Omega \cdot$ cm with a temperature coefficient of $4\text{-}6 \cdot 10^{-5}$ [134]. A considerable increase is also found in the strength of palladium when the latter is alloyed with tungsten, a fairly high ductility value being retained. These properties have made it possible to develop on the basis of the palladium–tungsten system a number of alloys which are being used in the new technology as corrosion-resistant potentiometer materials, capable of giving service in a number of corrosive media.

Of the other alloys with the noble metals, the alloys of palladium with iridium are quite important. Owing to its high elasticity, the alloy Pd– 18%Ir is used in aircraft instrument construction as electric spring contacts [135]. Platinum–rhodium alloys are used for high-temperature thermocouples. Platinum/platinum–rhodium thermocouples are characterized by high stability of thermoelectric properties and good corrosion resistance, and can function up to 1500°C. For the measurement of higher temperatures, some foreign firms produce thermocouples of iridium–ruthenium alloys. For thermocouple alloys, Pt–20%Rh and Pt–40%Rh are also recommended. The composition of some thermocouple alloys and their thermo-emf's are given in Table 49.

Alloys of platinum with iridium (up to 30% Ir) are considered to be very reliable materials for power contacts [136]. As substitutes for them, alloys of platinum with ruthenium, platinum with osmium, and platinum with tungsten are used [136]. Alloys on a platinum basis are used as heaters for electric furnaces designed for temperatures of up to 1500°C; an alloy of 70% Pt–30% Rh is suitable for heating furnaces operating up to a temperature of 1800°C.

There are practically no alloys of industrial importance based on iridium, ruthenium, rhodium, or osmium, despite the fact that these alloys possess a number of unique properties (elastic, strength, thermoelectric, and other properties). The low technological ductility of these metals has so far impeded the production of alloys based on them. There is no doubt that the production of high-purity ruthenium, iridium, and osmium will enable alloys having a number of valuable properties to be made on the basis of these metals. Figure 138 shows the influence of alloying elements on the hardness of rhodium, iridium, ruthenium, and osmium [36]. Experience shows that temperatures of above 1500°C are required for forging and rolling these alloys.

Alloys containing osmium, iridium, and ruthenium are used as wear-resistant components of instruments instead of natural osmium–iridium (needles of ships' compasses, pivots of ro-

TABLE 49. Thermoelectromotive Forces (in mV) of Different Thermocouples as a Function of Temperature [136]

| Temperature, °C | (70% Pt + 30% Rh) — —(94% Pt + 6% Rh) | Rh— —(80% Pt+20% Rh) | Ir — —(40% Ir+60% Rh) | W — Ir |
|---|---|---|---|---|
| 100 | 0.03 | 0.07 | 0.55 | — |
| 200 | 0.18 | 0.17 | 0.10 | — |
| 300 | 0.44 | 0.27 | 1.65 | — |
| 400 | 0.82 | 0.41 | 2.20 | — |
| 500 | 1.32 | 0.58 | 2.75 | — |
| 600 | 1.85 | 0.81 | 3.30 | — |
| 700 | 2.47 | 1.09 | 3.85 | — |
| 800 | 4.07 | 1.83 | 4.25 | — |
| 1000 | 4.96 | 2.28 | 5.50 | 14.25 |
| 1100 | 5.93 | 2.82 | 6.05 | 16.56 |
| 1200 | 6.96 | 3.39 | 6.60 | 18.91 |
| 1300 | 8.07 | 3.99 | 7.10 | 21.30 |
| 1400 | 9.17 | 4.62 | 7.65 | 23.70 |
| 1500 | 10.27 | 5.30 | 8.18 | 26.15 |
| 1600 | 11.36 | 6.00 | 8.70 | 28.62 |
| 1700 | 12.43 | 6.74 | 9.25 | 31.12 |
| 1800 | 13.50 | 7.51 | 9.80 | 33.69 |
| 1900 | — | — | 10.30 | 36.28 |
| 2000 | — | — | 10.85 | 38.88 |

tating parts of measuring instruments, etc.). The addition of ruthenium effectively strengthens platinum and palladium, and also considerably increases the electrical resistance of these metals. Thus, for example, an addition of 5% Ru increases the hardness of annealed platinum from 40 to 130 Vickers units, and the tensile strength from 12.6 to 42.4 kg/mm$^2$ [137]. An alloy of Pt–10%Ru has a tensile strength of 60 kg/mm$^2$ and a hardness of 190 Vickers units. The influence of ruthenium on the hardness of palladium is examined in [138]. According to [137], 0.1% ruthenium increases by a factor of 100 the corrosion resistance of titanium in reducing media.

Ruthenium, together with palladium, is the most accessible metal of the platinum group. This circumstance, as well as the refractoriness, high-strength properties, and a number of other unique physical properties of ruthenium make it an extremely promising base for the development of alloys having special physical properties. One of such properties is the extremely high damping capacity of ruthenium. Ruthenium is resistant to the action of a number of corrosive media, for example phosphorus and nitrogen compounds (at temperatures of up to 1000°C), and resists the corrosive action of lead, tin, tellurium, gallium, and iridium when heated to temperatures 100° above the melting point of the respective metals [137]. It has recently been possible, by advanced purification from interstitial impurities, to obtain highly ductile ruthenium, which opens up considerable possibilities for its alloying and the production of easily workable highly alloyed ruthenium alloys having given physical properties. The ruthenium–rhenium–osmium system represents a unique example of a ternary system of hexagonal metals with unlimited solubility of the components in the liquid and solid states [139].

Platinum metals are being used in electronic engineering as antiemission coatings in electric vacuum instruments, supports, cathodes, cathode bases, etc. Coatings of platinum, palladium, and rhodium are widely used for the more important parts of various instruments. The development of alloys for high-temperature thermocouples is of considerable interest [140]. Platinum metals are also used as catalysts in chemical processes (palladium, platinum) [141] in glass fiber manufacture. In this case, the material of the debiteuse heater has to

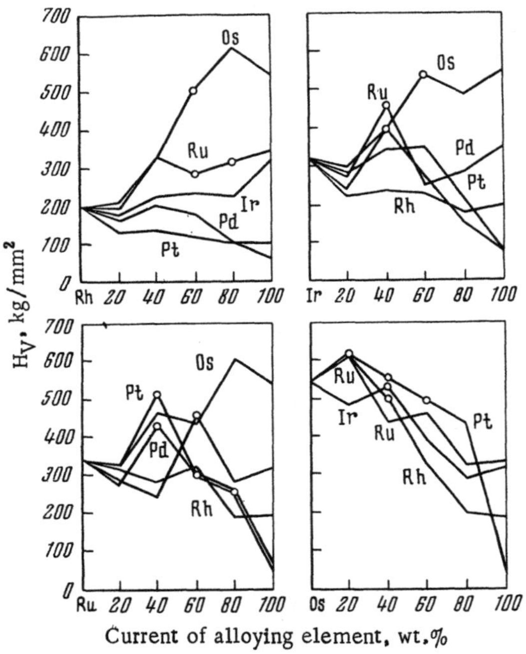


Fig. 138. Hardness of refractory noble metals in the as cast condition (alloys with osmium are two-phase alloys).

retain its strength at 1400°C and corrosion resistance in the molten glass mass; alloys of platinum with rhodium are used for this purpose.

## Metallic Compounds

In addition to alloys of refractory metals, there is considerable interest in the new technology for metallic compounds, including many refractory materials and materials with special physical properties. Several thousand compounds are known, the majority of which have either not been studied at all or only very slightly. This is a practically unexplored field, from which in the future, for satisfying the growing needs of the new technology, it will be possible to produce constantly new materials with better properties than the original metals.

The wide variety of the types of metallic bond and crystal structure in metallic compounds results in a wide range of electrical, emission, optical, and other physical properties. Some compounds are already beginning to find application in engineering [2, 2a].

Hafnium and tantalum carbides together with carbon are the most refractory substances in nature; the melting point of hafnium carbide is 3900°C, that the tantalum carbide is 3800°C. Carbides of other refractory metals also have high melting points: NbC has a melting point of 3760°C, VC 2830°C, $W_2C$ 2800°C, ZrC 3530°C, TiC 3150°C, etc.

By virtue of their high-temperature strength and also high-temperature oxidation resistance, carbides, borides, and to some extent nitrides are being used more and more abroad as high-temperature materials [142]; details of the properties of these compounds are given in [143].

Some refractory metals form with silicon silicides characterized by greater resistance to high-temperature oxidation than that of the constituent metals of these compounds [144-148]. For example, one such compound, molybdenum disilicide $MoSi_2$, has a melting point of 2030°C, its stress-rupture strength after 100 h at 980°C is 21 kg/mm$^2$. $MoSi_2$ retains its resistance to oxidation up to 1700°C; heaters made from it function for a long time at a temperature of 1700°C in air. Molybdenum disilicide is one of the promising coatings for niobium and molybdenum.

Metallic compounds for the type of $\sigma$- and $\chi$-phases which have not found applications because of their brittleness, have recently been the subject of intensive study as thermoemission materials; these include the $\sigma$-phases in molybdenum–rhenium ($T_{mp}$ = 2600°C) and tungsten–rhenium ($T_{mp}$ = 2800°C) systems, the $\chi$-phases in hafnium–rhenium ($T_{mp}$ = 3000°C), zirconium–rhenium ($T_{mp}$ = 2500°C), and niobium–rhenium ($T_{mp}$ = 2700°C) systems, and others [27]. There is information that compounds of the $\sigma$-phase type have high specific resistance; for example, the $\sigma$-phase in the tungsten–rhenium system has $\rho = 300\ \mu\Omega \cdot$ cm. Such a high specific electrical resistance is not possessed by any of the existing high-ohmic alloys.

On the metallic compounds, more than 900 are known to be superconductors [90]. Some of the superconducting compounds are given in Table 50.

TABLE 50. Critical Temperature of Some Superconducting Compounds Having Different Structures [90, 154]

| Compound | Critical temperature, °C | Compound | Critical temperature, °C |
|---|---|---|---|
| | $Cr_3Si$-type structure | | |
| $Nb_3Sn$ | 18.07 | $Nb_3Pt$ | 9.30 |
| $Nb_3Al$ | 17.50 | $Mo_3Os$ | 7.20 |
| $V_3Si$ | 17.00 | $V_3Sn$ | 6.00 |
| $V_3Ga$ | 16.50 | $V_3Ge$ | 6.00 |
| $Nb_3Ga$ | 14.50 | $V_3Pt$ | 2.83 |
| $Nb_3Au$ | 11.50 | $Mo_3Ge$ | 1.40 |
| | Laves-phase-type structure | | |
| $LaOs_2$ | 6.50 | $LuOs_2$ | 3.42 |
| $CaRh_2$ | 6.40 | $NfOs_2$ | 2.68 |
| $Ir_2Th$ | 6.50 | $YJr_2$ | 2.80 |
| $SrIr_2$ | 5.70 | $LaRu_2$ | 1.63 |
| | σ-phase-type structure | | |
| $Nb_2Al$ | 12.00 | $Nb_3Pd$ | 2.0G |
| $Nb_2Os$ | 9.80 | $Ta_3Rh$ | 2.00 |
| $NbRh$ | 4.10 | $NbOs$ | 1.40 |
| $Nb_5Pt_3$ | 3.73 | $Ge_2Mo$ | 1.20 |

The superconducting properties of compounds of rare-earth metals with noble metals have been reported [152]. By far the majority of superconducting compounds have a complex crystal lattice of low symmetry and are characterized by brittleness at room temperature and at the temperature of liquid helium. The high value of the critical magnetic field is a feature of the superconducting compounds; for example, the compounds $Nb_3Sn$ and $V_3Ga$ have critical magnetic fields of up to 250 and 350 kG, respectively [90]. Recently, the superconductivity of the monocarbides WC ($T_c \geq 10$°K) and MoC (from 2 to 14.3°K) has been established [153].

The mechanical properties of metallic compounds are peculiar, and have been found to be very temperature-sensitive [2, 155]. Under ordinary conditions, metallic compounds possess very high hardness, and brittleness, which retards their use in technology. Until recently, the view was held among metallurgists that metallic compounds, because of their brittleness, could not be produced in the form of any kind of articles and therefore had no industrial future whatsoever. It was found, however, that metallic compounds are brittle only at the ordinary temperature, but at temperatures of 70-90% of the melting point, all metallic compounds are quite plastic [2, 156].

Subsequently, another interesting phenomenon was discovered, the increase in strength of metallic compounds when the latter were heated to temperatures amounting to 50-80% of the melting point, whereas when ductile metals and alloys are heated, the strength always decreases according to an exponential law. The physicochemical causes of this effect have not yet been sufficiently elucidated, but it is indisputable that it is typical of compounds having a mixed interatomic bond, and depends on the increase in the proportion of metallic bond when the compounds are heated [191].

## Materials of SAP Type (Metal – Metal Oxide)

Intensive research is being conducted on the production of materials of the Me–MeO system, since precipitates of refractory compounds, including oxides, refine the grain and increase the strength and ductility of the alloys.

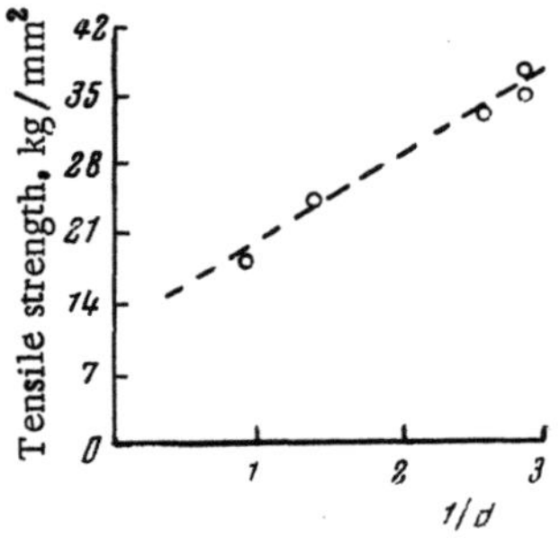


Fig. 139. Tensile strength of an aluminum alloy of SAP type as a function of distance d between particles.

There are two different types of Me–MeO systems: synthetic systems produced by powder metallurgy methods and natural types, in which the formation of precipitates of the second phase occurs in the subsequent treatment of the alloys, the rate of the process, the amount of precipitated phase, and the character of its distribution being determined by the chemical composition of the alloy and the conditions of the combined heat and mechanical treatment.

In addition to good technological ductility, materials based on Me–MeO systems have high tensile strength, stress-rupture strength, and creep strength. Fairly detailed descriptions have been published of researches into such systems based on metals of moderate refractoriness (nickel, copper) and based on refractory metals (titanium, chromium, molybdenum, tungsten, etc.) [157-161]. It should be noted that the disperse particles of oxides and other phases, noncoherently bonded with the matrix, suppress the grain growth and inhibit recrystallization processes; for example, $ThO_2$ and $ZrO_2$ with a particle size of 0.01 μ considerably refine the grain in tungsten and reduce the brittleness threshold on bending by 70-90°C [162]; the recrystallization temperature is raised by 200°C at the same time.

Materials which are highly ductile at room temperature have been produced on the basis of a chromium–magnesium oxide composition [163]. A feature of these materials is their high ductility in the recrystallized condition at room temperature, and also after heating in air. The materials were produced by hot pressing followed by rolling. Titanium was added to the mixture of chromium and MgO as gas absorbent. The ductile–brittle transition temperature of such a material is 12°C according to [163]. The material showed high resistance to oxidation and to erosion in a gas stream at 2650°C.

The precipitation hardening mechanism may also be used for molybdenum [164]. A considerable increase in strength is obtained in molybdenum by the introduction of $Al_2O_3$, $Cr_2O_3$, $ZrO_2$, or $TiO_2$. So far, however, optimum properties have not been obtained in the preparation of such materials by powder metallurgy methods. Better results are obtained by vacuum-arc melting of molybdenum with deoxidation of the latter by means of thermodynamically active metals, for example, titanium or zirconium. The vacuum-melted alloy Mo–0.45%Ti, according to Grant [164], has a stress-rupture strength after 100 h of about 28 kg/mm$^2$, whereas the stress-rupture strength of an alloy of molybdenum with zirconium oxide, produced by the powder metallurgy method, is only 17 kg/mm$^2$. Information is available on the production of a high-temperature strength alloy of the chromium–$Al_2O_3$ system [165]. The alloy showed good creep resistance at high temperatures, but low ductility on account of the large grain size.

Most investigations conducted on the development of the above-mentioned materials have been carried out with oxides; less attention has been devoted to carbides and nitrides. The problem of the type of the disperse phase in any case is connected with the features of its interaction with the metal base of the alloy. The maximum effect with regard to increasing the strength of the alloy as well as with regard to increasing its ductility has been achieved in the absence of any interaction between the disperse phase and the metal base, when the disperse particles are noncoherently bonded with the matrix. For most systems, the most suitable additions have been found to be oxides of thermodynamically active metals having a high heat of formation. The properties of the metal base and the individual operations of the preparation of the materials also exert a considerable influence on the properties of the resultant material. The mechanism of the influence of oxides and other inert inclusions has not been definitively studied. Individual views have been advanced by Mott [166], Orowan [167], Gensamer [168], Grant [164], Goetzel [169], and other investigators. The variation in tensile strength with interparticle distance has been studied by Gregory and Grant [170]. This relationship for an aluminum alloy of SAP type is shown in Fig. 139.

## Protection of Refractory Metals and Alloys from Oxidation

The main obstacle to the production of high heat-resistant alloys based on refractory metals is their considerable oxidizability even at a relatively not very considerable increase in temperature.

In connection with the high activity of refractory metals with regard to atmospheric constituents, one of the most important problems which has to be solved concerning their wider application is the protection of these metals from high-temperature gas corrosion. This problem is being attacked by different investigators by alloying the refractory metals and by means of heat-resistant coatings or by a combination of both methods. The influence of alloying additions on the oxidizability of niobium is dealt with in [172, 173]. In the opinion of the authors of [173], the efficacy of the influence of any element on the resistance of niobium to oxidation depends on the size of the ion of that element, and also on the character of the influence of the alloying element on the structure of the oxide film formed. Alloys of niobium with titanium, aluminum, and chromium were found to be the most promising. Alloying of niobium with up to 5 at.% tungsten and up to 20 at.% tantalum reduces the rate of oxidation but further increase of these components in the alloy leads to a considerable acceleration of oxidation. Molybdenum and zirconium have the same effect. The authors of [172] found a considerable increase in the resistance of niobium to oxidation by alloying with tungsten (17.5 at.%), titanium (17.5 at.%), and iron (5 at.%), although this increase was far from satisfying the requirements of the new technology. Figure 140 shows the resistance to oxidation of binary niobium alloys at 1200°C [172]. Smith's results on the reduction in the rate of oxidation by a factor of 100 on alloying niobium with titanium (20 at.%), tungsten (10%), and nickel (3-4%) [181] are very interesting.

An investigation of the oxidizability of niobium–tungsten alloys [4] showed that maximum resistance to oxidation was found in alloys with 15-30 wt.% W. Further increase in the tungsten content resulted in a sharp reduction in heat resistance; with 67-80% W the heat resistance of niobium–tungsten alloys rose again. Of the alloys of the molybdenum–niobium system, maximum oxidation resistance was possessed by alloys with 10-15 wt.% Mo [4].

It is quite clear that for a chemically active metal base, for example niobium, it is impossible just by alloying it with additions to obtain absolute protection from interaction with gases at high temperatures. Therefore, together with work on the protection of niobium from oxidation by alloying, research is being conducted on the development of protective coatings for niobium [172]. Coatings of niobium aluminide, produced by metallizing or alitizing, are considered to show promise. Niobium aluminide has good resistance to oxidation at 1200°C.

In [4, 174], zinc protective coatings were applied by dipping niobium and its alloys in molten zinc, followed by diffusion annealing in air. In this process, a metallic compound of niobium and zinc is formed on the surface. On oxidation, the metallic compounds are converted into mixed oxides, zinc niobates. Metallic compounds of the niobium–zinc system are brittle and prone to fracture under mechanical and thermal shocks. They bond well with the niobium and show high oxidation resistance when heated in air. Zinc-coated niobium also shows no signs of destruction when tested in a current of fuel combustion products (at 1090°C, flow velocity 152 m/sec)[174]. The phenomenon of "self-healing" of zinc coatings is observed;

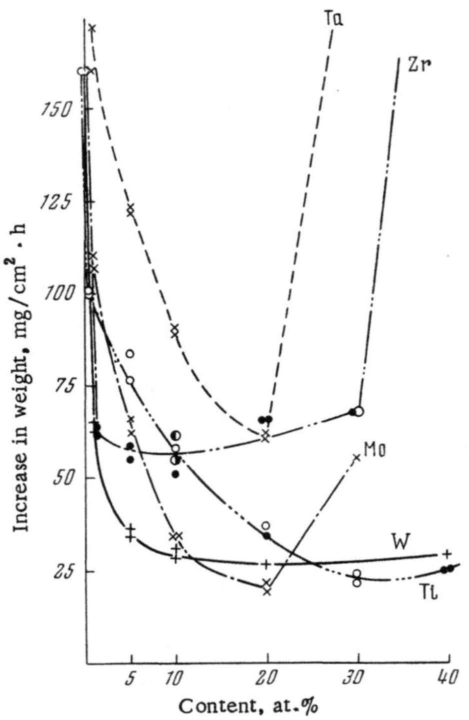


Fig. 140. Oxidation of binary niobium alloys at 1200°C as a function of alloying addition content.

this is due to the diffusion of zinc from the adjacent coating. A mixture of powdered aluminum oxide, thorium dioxide, and metallic niobium is used for protecting niobium from oxidation [175, 176]. The coating is applied in layers. The niobium content of the mixture increases with increase in distance from the surface of the metal base. Annealing results in the formation of a protective coating closely applied to the niobium. It is assumed that the formation of the compact layer is due to the oxidation of the niobium grains followed by reaction of the niobium pentoxide with $Al_2O_3$ and $ThO_2$ resulting in the formation of the niobates AlNbO and $ThNb_2O_7$. Attempts to produce two-layer coatings are of interest [115].

The methods which have been most developed for the protection of niobium alloys from oxidation at 1100-1200°C are coatings of molybdenum disilicide, special enamels, chromizing with a chromium–yttrium coating or a combination of these methods [115]. The development of bimetals, niobium alloy–chromium is very promising. The protective coatings on niobium and its alloys guarantee their service at high temperatures under tensile stress in oxidizing media and the combustion products of fuel for hundreds of hours, but they are not very reliable for service lasting thousands of hours.

Much attention has recently been given in American publications to protective coatings of zirconium diboride [177]. The Kawecki Chemical Co. is proposing to coat niobium and tantalum with zirconium diboride [178], which withstands the action of high temperatures from 1375 to 1925°C for a long time, and is also corrosion-resistant in some molten metals (aluminum, iron, copper).

The resistance to oxidation of another group Va metal, tantalum and its alloys, has been studied to a lesser degree than that of niobium. Additions of group Va metals (vanadium and niobium) result in some increased resistance to oxidation of tantalum [171]. It is also reported that increase in tungsten content reduces the rate of oxidation of tantalum at 1200 and 1700°C [171]; according to Rostoker's results, cited in this source, group IVa metals (titanium, zirconium, hafnium) have the most effective influence in increasing the resistance of tantalum to oxidation; a 30 at.% addition of Ti reduces the rate of oxidation of tantalum at 1095°C by nearly 90%; an addition of 10% Co or Ni or 30% Fe has the same effect. This influence on the oxidizability of tantalum is evidently due to the high thermodynamic activity of these elements and the stability of their oxides, in which diffusion processes are difficult. The resistance of tantalum to oxidation is sharply increased by its combined alloying with titanium and aluminum. According to [171], the alloy Ta–40%Ti–10%Al has the highest heat resistance of all the tantalum alloys so far developed. At 1400°C, the oxidation resistance of this alloy is 300 times higher than that of unalloyed tantalum at the same temperature. The alloy may be used at 1400°C for 6 h without any appreciable signs of gaseous corrosion.

There are few published data on the development of protective coatings for tantalum and its alloys. Evidently, in view of the similarity of the chemical properties, the protective coatings developed for niobium could also be used for tantalum.

With regard to molybdenum, the majority of investigators consider that its complete protection from high-temperature gas corrosion by alloying to be an unpromising problem, although some progress has recently been made [179, 180]. The principal efforts are here being directed to the development of heat-resistant coatings, which for molybdenum are divided into metallic, oxide, cermet, and mixed. Metallic protective coatings may be pure metals, solid solutions, or metallic compounds. Most of the coatings which have been proposed contain silicon. A coating of $MoSi_2$ has been found to be very resistant to oxidation [180]. A substantial drawback of this coating is its high sensitivity to mechanical damage. The protective properties of metallic coatings of nickel–chromium–boron and aluminum–chromium–silicon were studied in [172]. The coatings were applied by metallizing. The service life of a coating of aluminum–chromium–silicon alloy at 1100°C was 500 h. The use of a protective coating for molybdenum, consisting of a nickel–iron–cobalt–chromium alloy having a low coefficient of thermal expansion, close in value to that of molybdenum, has been reported [179]. Protective coatings of this alloy were applied directly to the molybdenum or to a substrate of titanium or chromium.

Nickel–aluminum alloys, applied to previously chromized molybdenum, have been tested [182]. According to the results, these alloys protect molybdenum from oxidation at 1100°C for 400 h.

The use of platinum coatings for the protection of molybdenum from oxidation has been reported; according to [183] such a coating 0.1 mm thick is able to ensure the protection of molybdenum from oxidation for several hundreds of hours at 1100-1200°C.

Considerable interest attaches to work on the application to molybdenum of protective oxide coatings, consisting of the oxides of one or more chemical elements. The initial mixtures, containing silicates or other compounds and showing a tendency to transformation to a vitreous condition, form enamels when applied to molybdenum [179]. A substantial shortcoming of all oxide coatings, making it difficult to use them despite their high protective properties, is their high sensitivity to mechanical and thermal shock. In view of their refractory properties and high resistance to oxidation, they may be effective in protecting parts made of molybdenum and also other refractory metals, operating for a short time at high temperatures.

A very interesting trend is the development of mixed coatings consisting of a combination of different fillers and binders, the melting point of the filler being higher than that of the binder. Refractory metals may be used as fillers, and vitreous materials, for example, chromium–glass coatings [179], may be used as binders; the filter may also be refractory oxides and the binder, metals or alloys (cermet coatings). A coating consisting of nickel and magnesia is known for molybdenum [184].

The high oxidizability of rhenium and its alloys gives rise to the necessity of protecting these materials under high-temperature conditions. Coatings on rhenium have been obtained by the methods of combined chemical and heat treatment: boriding, chromizing, siliconizing, and aluminizing [185, 186]. It has been shown that boriding increases the resistance of rhenium to oxidation at 800°C by a factor of five. Chromizing also results in a shape decrease in oxidizability of rhenium; siliconizing and aluminizing fail to show any substantial influence on resistance to scaling. It has been shown that alloying with silicon reduces the rate of oxidation of rhenium [187]. The resistance to scaling of alloys of tungsten with rhenium is higher than that of pure tungsten. Alloying with rhenium diminishes the oxidizability of tungsten by a factor of 6–7 [27].

We have briefly reviewed the methods of increasing the resistance to high-temperature oxidation of refractory metals by alloying and the production of oxidation-resistant coatings. Published experimental results as yet do not permit the formulation of any sufficiently definite fundamental relationships regarding the effect of alloying on the oxidizability of refractory metals. It is evident that alloying with elements forming durable oxides (more durable than the metal base) will help to increase the resistance to high-temperature oxidation of refractory metals. Views have also been expressed on the influence of valence and the size of the ion of the alloying element on the resistance to scaling of an alloy [173]. In the development of protective coatings, considerable significance attaches to the compatibility of certain physical properties of the alloy and the protective coating. The "life" of the coating is determined to a great extent by the difference in the coefficient of thermal expansion of the alloy and protective coating. Close values of the coefficients of thermal expansion ensure a strong bond of the coating to the base. In the case of a metal coating, diffusion of the coating from the metal to the alloy is possible, which may lead to the formation of brittle metallic compounds. The result is the destruction of the durable interlocking bond of the coating and metal. The character of the physicochemical interaction of the protective layer and alloy is therefore of very great importance. In the case of nonmetallic coatings, no active interaction occurs with the alloy, but nonmetallic coatings are of low ductility and have no tendency to self-heating.

The methods of applying coatings to refractory metals depends on their composition and also the shape, dimensions, and purpose of the coated articles. Usually cladding, deposition from the gaseous phase, immersion in a metal bath, diffusion saturation, melting-on, etc., are employed. There is information on the use of plasma torches for applying protective coatings to refractory metals [115, 188, 189]. Coatings applied to plasma torch are characterized by a good interlock between metal and base. The plasma torch may also be used for the application of melted-on coatings of refractory metals (tungsten, molybdenum, rhenium, tantalum, niobium) and refractory compounds (borides, carbides, oxides of some transition metals)[189].

The Boeing Co. has proposed a method of applying silicon coatings to refractory metals in a fluidized bed of silicon, whereby a uniform silicon film is deposited on the metal. The fluidized bed of silicon is formed in a special chamber in an atmosphere of argon and iodine vapor. The silicon tetraiodide formed reacts with the refractory metal placed in the chamber. The furnace temperature is 1000°C. The method described has been used for coating molybdenum, niobium, and other refractory metals.

The protection of refractory metals from high-temperature gas corrosion is one of the most pressing questions in the problem of refractory metals, the solution of which will considerably increase the scale of application of refractory metals and highly heat-resistant alloys based on them. The development of heat-resistant coatings for refractory metals is a very difficult problem, owing to the peculiarities of the physicochemical properties of refractory metals and their oxides (for example, high vapor tension of the oxides formed, low coefficient of linear expansion of the refractory metals, etc.). Considerable progress has already been made, however, in the development of heat-resistant coatings. A number of heat-resistant coatings on niobium and molybdenum alloys for a service life of not less than 100 h at a temperature of 1000°C and higher have been developed; means are available for protecting molybdenum from oxidation at a temperature of 1700°C in the absence of a dynamic load [171], etc. The development of a number of thin coatings for refractory metals and alloys, resisting oxidation at temperatures of 1100-1650°C is reported [190]; these include coatings based on silicon, aluminum, beryllium, platinum, Al–Si, Al–Sm, Si–Cr–Ti alloys, etc. However, the problem has not been finally solved. It is still necessary to solve the problem of the brittleness of the protective layer and also the strength of its bond with the metal base, etc.

## LITERATURE CITED

1. E. M. Savitskii, Vestnik Akad. Nauk SSSR, No. 3, p. 24 (1962).
2. E. M. Savitskii, Influence of Temperature on the Mechanical Properties of Metals and Alloys [in Russian], Izd. Akad. Nauk SSSR (1957).

2a. E. M. Savitskii, Vestnik Akad. Nauk SSSR, No. 11 (1966).
3. R. Kieffer, K. Seldachek, and H. Braun, Problemy Sovremennoi Metallurgii, No. 4, p. 164 (1959).
4. E. M. Savitskii, V. V. Baron, and K. N. Ivanova, Izv. Akad. Nauk SSSR, OTN, Metallurgiya i Toplivo, No. 2, p. 119 (1962).
5. E. M. Savitskii and G. E. Chuprikov, Izv. Akad. Nauk SSSR, OTN, Metallurgiya i Toplivo, No. 4, p. 137 (1962).
6. E. M. Savitskii, M. A. Tylkina, and G. E. Chuprikov, Zh. Neorgan. Khim., No. 9, p. 2272 (1962).
7. E. M. Savitskii and A. M. Zakharov, Metall. i Term. Obrabotka, No. 3, p. 8 (1965).
8. B. Lement and I. Perlmutter, Problemy Sovremennoi Metallurgii, No. 1, p. 95 (1961).
9. E. M. Savitskii, G. S. Burkhanov, and Ch. V. Kopetskii, Izv. Akad. Nauk SSSR, Metallurgiya i Gornoe Delo, No. 6, p. 12 (1963).
10. R. M. Parke and F. P. Bens, Symposium on Materials for Gas Turbines (ASTM), p. 80 (1941).
11. A. Salli, Chromium [in Russian], Metallurgizdat, Moscow (1958).
12. F. Henderson, S. G. Kvass, and H. L. Wein, Problemy Sovremennoi Metallurgii, No. 3, p. 62 (1955).
13. C. T. Sims and J. W. Clark, Trans. Met. Soc. AIME, 230(5):1168 (1964).
14. E. M. Savitskii et al., Alloys of Rare-Earth Metals [in Russian], Izd. Akad. Nauk SSSR (1962).
15. E. M. Savitskii and V. F. Terekhova, in: Research on Heat-Resistant Alloys [in Russian], Izd. Akad. Nauk SSSR, Vol. 2, p. 148 (1957).
16. D. M. Scruggs, Mater. Design Engng., 56(7):115 (1962).
17. E. M. Savitskii and V. V. Baron, in: Strength of Metals [in Russian], Izd. Akad. Nauk SSSR, p. 144 (1956).
18. In: Molybdenum [Russian translation] (1962).
19. M. V. Maltzev and V. I. Shulepov, in: Physical Nature of the Brittle Fracture of Metals [in Russian], Kiev, Izd. "Naukova Dumka," p. 125 (1965).
20. R. T. Begley, SAE Journal, 65(7):70 (1957).
21. N. N. Morgunova, Heat-Resistant Alloys Based on Molybdenum [in Russian], VINITI (1959).
22. M. Semchyshen and R. Q. Barr, in: Refractory Metals and Alloys [Russian translation], Moscow, Izd. "Metallurgiya," p. 149 (1965).
23. R. M. Parke and S. L. Ham, Trans. AIME, 171:416 (1947).
24. Express Information, Metall. i Term., VINITI, No. 16 (Ref. 1) (1965).
25. Express Information, Tsvet. Metall., VINITI, No. 21 (Ref. 13) (1966).
26. H. Braun, M. Semchyshen, and R. Q. Barr, Metallwerk Plansee A. G. Reutte/Tyrol, 351 (1965).
27. E. M. Savitskii, M. A. Tylkina, and K. B. Povarova, Rhenium Alloys [in Russian], Izd. "Nauka" (1965).
28. G. H. Rendal, S. M. Johnston, and W. I. Harrington, Problemy Sovremennoi Metallurgii, No. 4, p. 59 (1955).
29. E. G. Czarnecki, J. T. Stacy, and D. K. Zimmerman, Refractory Metals and Alloys, Vol. II, p. 395 (1963).
30. N. L. Deuble, Metal Progr., 68(2):77 (1955).
31. R. H. Hassner, Refractory Metals for Aircraft and Spacecraft, in: Some Problems of Refractory Metals and Alloys [Russian translation], IL, p. 11 (1963).
32. H. H. Hausner and W. Wolkowitz, Metalle für die Raumfahrt, Metallwerk Plansee A. G. Reutte/Tyrol, p. 65 (1965).
33. M. Semchysen and E. Kalns, Climax Molybdenum Company of Michigan, Ann Arbor, Michigan, p. 1 (1965).

34. Metallurgie, 73(435):8 (1966).
35. Express Information, Metall. i Term., VINITI, No. 32 (Ref. 1) (1964).
36. R. I. Jaffee, D. J. Maykuth, and R. U. Douglass, Rhenium and Refractory Metals of the Platinum Group [Russian translation], IL (1963).
37. D. Cobins, J. Less-Common Metals, 1:396 (1959).
38. D. J. Maykuth and H. R. Ogden, High Temperature Materials II, 18:81 (1963).
39. W. W. Austin, Problemy Sovremennoi Metallurgii, No. 5, p. 99 (1960).
40. D. J. Maykuth and H. R. Ogden, in: Some Problems of Refractory Metals and Alloys [Russian translation], IL, p. 57 (1963).
41. J. T. Taylor, J. Less-Common Metals, 7(4):278 (1964).
42. E. M. Savitskii, Rare Metals and Alloys, Moscow, Izd. Dom Tekhniki (1959).
43. Prospects of the development of refractory materials for service at high temperatures, in: Some Problems of Refractory Metals and Alloys [Russian translation], IL, p. 47 (1963).
44. V. E. Heil, P. C. Murphy, and L. F. Neely, Rhenium, New York, p. 163 (1962).
45. F. D. Calvert and R. A. Beall, J. Metals, 18(1):38 (1966).
46. U. S. Patent Specification No. 3160501, Class 75-176.
47. E. M. Savitskii, M. A. Tylkina, and K. B. Povarova, Author's Certificate No. 165310, Byull. Izobretenii i Tovarnykh Znakov, No. 18 (1964).
48. E. M. Savitskii, Izv. Akad. Nauk SSSR, Metally, No. 5, p. 3 (1965).
49. R. I. Jaffee, J. Metals, No. 5, p. 410 (1964).
50. Metal Progress, 85:83 (May 1964).
51. In: Thermionic Energy Conversion [in Russian], Atomizdat (1964).
52. M. Semchyshen, R. Q. Barr, McArdle, and D. Gordon, Development and Evaluation of Tungsten-Base Alloys, Climax Molybdenum Company of Michigan, Contract No. NNO W 62-04-0419-C (1963).
53. M. A. Tylkina and E. M. Savitskii, Rhenium, Trans. Second All-Union Conference on the Problem of Rhenium [in Russian], Izd. "Nauka" (1964).
54. E. M. Savitskii and M. A. Tylkina, Izv. Akad. Nauk SSSR, No. 3, p. 33 (1956).
55. E. M. Savitskii, M. A. Tylkina, and K. B. Povarova, Zh. Neorgan. Khim., 4(2):424 (1959).
56. E. M. Savitskii, M. A. Tylkina, and K. B. Povarova, Atomnaya Energiya, 7(5):470 (1959).
57. M. A. Tylkina, I. A. Tsyganova, E. M. Savitskii, E. I. Gladyshevskii, and M. P. Mulyaeva, Zh. Neorg. Khim., 7(8):1608 (1962).
58. E. M. Savitskii, M. A. Tylkina, and V. P. Polyakova, Zh. Neorgan. Khim., 7(2):439 (1962).
59. E. M. Savitskii, M. A. Tylkina, L. L. Zhdanova, L. A. Zubkova et al., in: Research on Heat-Resistant Alloys [in Russian], Vol. 8, Izd. Akad. Nauk SSSR, p. 194 (1962).
60. E. M. Savitskii, M. A. Tylkina, K. B. Povarova, and V. P. Vinokurov, Proizvodstvenno-tekhn. Byull., No. 5, p. 1 (1962).
61. E. M. Savitskii, Izv. Akad. Nauk SSSR, OTN, Metallurgiya i Toplivo, No. 5, p. 52 (1960).
62. E. M. Savitskii, T'ao Ts'u-tsung, Fiz. Metall. i Metalloved., 14(1):148 (1962).
63. D. J. Maykuth, F. C. Holden, and R. I. Jaffee, Rhenium, New York, p. 114 (1962).
64. L. Northcott, Problemy Sovremennoi Metallurgii, No. 5, p. 52 (1961).
65. M. A. Tylkina, K. B. Povarova, and E. M. Savitskii, Zh. Neorgan. Khim., No. 11, p. 2458 (1960).
66. P. N. Flagella and C. O. Tarr, AIME Technical Meeting, French Lick, Indiana (October 1965).
67. R. A. Ul'yanov and N. D. Tarasov, Izv. Vyssh. Uch. Zaved. Tsvetnaya Metallurgiya, No. 2, p. 140 (1964).
68. J. W. Pugh, L. H. Arma, and D. T. Hurd, Trans. AIME, 55(3):451 (1962).
69. C. N. Reid and A. Gilbert, J. Less-Common Metals, 10(2):77 (1966).
70. M. A. Tylkina and E. M. Savitskii, in: Rare Metals and Alloys [in Russian], Moscow, Metallurgizdat, p. 80 (1959).

71. M. A. Tylkina and E. M. Savitskii, in: Rhenium [in Russian], Izd. Akad. Nauk SSSR, p. 108 (1961).
72. S. K. Danishevskii, in: Rhenium [in Russian], Izd. Akad. Nauk SSSR, p. 162 (1961).
73. J. C. Lachmann and F. W. Kulther, ISA J., 7(3):67 (1960).
74. J. T. Muller, Instrum. Practice, 14(5):510 (1960).
75. B. P. Nikonov and N. S. Otmakhova, Voprosy radioelektronika, Seriya "Elektronika," No. 5, p. 91 (1962).
76. T. F. Kuleshova and V. P. Nikonov, in: Rhenium [in Russian], Izd. "Nauka," p. 186 (1964).
77. M. V. Kaganovich and R. M. Makarova, Voprosy radioelektronika, Seriya "Elektronika," No. 7, p. 73 (1961).
78. J. H. Port, Metallwerk Plansee A. G. Reutte/Tyrol, 613 (1965).
79. B.T. Matthias, V. B. Compton, H. Sahl, and E. Corenzwit, Phys. Rev., 115:1597 (1959).
80. J. H. Kunzler, E. Buchler, F. S. Hsu, B. T. Matthias, and C. Wahl, J. Appl. Phys., 38:325 (1960).
81. I. E. Dekabrun, in: Rhenium [in Russian], Izd. Akad. Nauk SSSR, p. 180 (1961).
82. V. V. Usov and M. D. Dovolotskaya, in: Rare Metals and Alloys [in Russian], Metallurgizdat, p. 123 (1960).
83. E. M. Savitskii, Z. A. Timofeeva, and M. A. Tylkina, in: New Materials in Electrical Instrument Manufacture [in Russian], Izd. Otd. Nauchno-tekhn. Informatsii po Priborostroeniyu, Sredstbvam Avtomatizatsii i Sistemam Upravleniya (1965).
84. Z. A. Timofeeva, E. M. Savitskii, and M. A. Tylkina, Obmen Opytom v Radioelektronnoi Promyshlennosti, No. 10, p. 5 (1964).
85. Yu. A. Polyakov, Principles of Metallurgy of Vanadium [in Russian], Metallurgizdat, Moscow (1959).
86. Yu. V. Efimov, E. M. Savitskii, and V. V. Baron, Author's Certificate No. 148521 of October 9th, 1961, Byull. Isobretenii i Tovarnykh Znakov, No. 13 (1962).
87. W. Rostoker, Metallurgy of Vanadium, Wiley, New York (1958).
88. In: Superconducting Materials [in Russian], Izd. "Mir" (1965).
89. E. M. Savitskii, V. V. Baron, Yu. V. Efimov, and E. I. Gladyshevskii, Izv. Akad. Nauk SSSR, Neorganicheskii Materialy, 1(2):208 and 1(3):334 (1965).
90. E. M. Savitskii, in: Physical Metallurgy and Metal Physics of Superconductors, Izd. "Nauka" (1965).
91. T. F. Fedorov and E. I. Gladyshevskii, Teplofizika Vysokikh Temperature, 1(3):449 (1963).
92. E. M. Savitskii and A. M. Zakharov, Zh. Neorgan. Khim., 9(10):2429 (1964).
93. V. V. Baron, K. N. Ivanova, and E. M. Savitskii, Izv. Akad. Nauk SSSR, OTN, Metallurgiya i Toplivo, No. 4, p. 143 (1960).
94. G. D. Gemmell, Problemy Sovremennoi Metallurgii, No. 4, p. 157 (1960).
95. E. M. Savitskii and A. M. Zakharov, Zh. Neorgan. Khim., 7(11):2575 (1962).
96. E. M. Savitskii and A. M. Zakharov, in: Research on Nonferrous Metal Alloys [in Russian], No. 4, p. 108, Izd. Akad. Nauk SSSR (1960).
97. R. T. Begley and J. H. Bechtold, J. Less-Common Metals, 3(1):1 (1961).
98. E. M. Savitskii and A. M. Zukharov, Izv. Vyssh. Uch. Zaved. Chernaya Metallurgiya, No. 1, p. 104 (1965).
99. E. M. Savitskii and A. M. Zakharov, Zavod. Lab., No. 12, p. 1501 (1964).
100. D. Peckner, in: Some Problems of Refractory Metals and Alloys [Russian translation], IL, p. 92 (1963).
101. Express Information, Metallovedenii i Termoobrabotka, VINITI, No. 19 (Ref. 69) (1966).
102. D. J. Maykuth and R. I. Jaffee, in: Problems of Modern Metallurgy [Russian translation], No. 6, p. 154 (1960).

103. R. W. Fountain and C. K. McKinsey, in: Columbium and Tantalum, Wiley, New York, p. 198 (1963).
104. W. F. Shelley and J. L. Wilson, in: Some Problems of Refractory Metals and Alloys [Russian translation], IL, p. 120 (1963).
105. E. M. Savitskii, V. V. Baron, and M. I. Bychkova, Transactions of the A. A. Baikov Inst. of Metallurgy [in Russian], No. 12, p. 179, Izd. Akad. Nauk SSSR (1962).
106. E. M. Savitskii, V. V. Baron, M. I. Bychkova, S. A. Bakuta, and E. I. Gladyshevskii, Izv. Akad. Nauk SSSR, Metally, No. 2, p. 159 (1965).
107. E. M. Savitskii and A. M. Zakharov, Zh. Neorgan. Khim., 9(10):2427 (1964).
108. E. M. Savitskii and A. M. Zakharov, Izv. Vyssh. Uch. Zaved. Tsvetnaya Metallurgiya, No. 5, p. 117 (1965).
109. M. Shussler, A. L. Mincher, T. L. Wilson, and H. E. Dufendach, in: Columbium and Tantalum, Wiley, New York, p. 32 (1963).
110. B. B. Gulyaev, O. N. Magnitskii, and A. A. Demidova, Refractory Metal Coatings [in Russian], Izd. "Mashinostroenie," Moscow (1964).
111. O. Winkler, Metalle für die Raumfahrt, Metallwerk Plansee, A. G. Reutte/Tyrol, p. 199 (1965).
112. Metal Progress, 76(6):78 (1960).
113. E. M. Savitskii and A. M. Zakharov, Izv. Akad. Nauk SSSR, Metally, No. 1, p. 151 (1965).
114. M. V. Mal'tsev, A. I. Baikov, and V. Ya. Solov'ev, Technology of the Production of Niobium and Its Alloys [in Russian], Izd. "Metallurgiya" (1966).
115. A. I. Borisenko, Protection of Niobium from High-Temperature Gas Corrosion [in Russian], Izd. Akad. Nauk SSSR (1961).
116. V. K. Grishin, M. G. Glazunov, A. G. Arakelov, A. V. Vol'dent, and G. S. Makedonskaya, Properties of Lithium [in Russian], Metallurgizdat (1963).
117. E. M. Savitskii, V. V. Baron, M. I. Bychkova, V. Ya. Pakhomov, and V. R. Karasik, Author's Cert. No. 168250, Byull. Izobretenii i Tovarnykh Markov, No. 4 (1965).
118. E. M. Savitskii, V. V. Baron, V. R. Karasik, S. Sh. Akhmedov, V. Ya. Pakhomov, and M. I. Bychkova, Pribory i Tekhnika Eksperimenta, No. 1, p. 182 (1963).
119. V. L. Martin, M. G. Benz, L. A. Bruch, and K. Kh. Rosner, Superconducting Materials [Russian translation], Izd. "Mir," p. 254 (1965).
120. V. L. Newhouse, Applied Superconductivity, Wiley, New York (1964).
121. R. Kieffer and H. Braun, Vanadin-Niob-Tantal., Berlin (1963).
122. E. M. Savitskii, V. P. Polyakova, M. A. Tylkina, and G. S. Burkhanov, Zh. Neorgan. Khim., 11(7):1645 (1964).
123. M. A. Tylkina, I. A. Tsyganova, and E. M. Savitskii, Zh. Neorgan. Khim., 5(8):1905 (1960).
124. F. F. Schmidt, F. C. Holden, H. R. Ogden, and R. I. Jaffee, Investigation of the Properties of Tantalum and Its Alloys, WADD Tech. Rep., 61–106 (1961).
125. E. M. Savitskii and M. A. Tylkina, Byull. Izobretenii, No. 6, p. 45 (1960).
126. In: Titanium and Its Alloys, Physical Metallurgy of Titanium [in Russian], Izd. Akad. Nauk SSSR, No. 3 (1960).
127. In: Titanium in Industry [in Russian], Oborongiz (1961).
128. V. N. Eremenko, Titanium and Its Alloys [in Russian], Izd. Akad. Nauk Ukrainian SSR (1960).
129. V. N. Eremenko, Multicomponent Titanium Alloys [in Russian], Izd. Akad. Nauk Ukrainian SSR (1962).
130. I. S. Stepanov, Rare Metals and Technical Progress [in Russian], Izd. "Metallurgiya" (1964).
131. G. L. Miller, Zirconium, Academic Press, New York (2nd ed.) (1957).
132. S. Ya. Kaganovich, Zirconium and Hafnium [in Russian], Izd. Akad. Nauk SSSR (1962).
133. E. M. Savitskii, M. A. Tylkina, V. P. Polyakova, I. A. Tsyganova, and Ch. B. Kopetskii, in: Elektricheskie Kontakty [in Russian], Izd. "Energiya," p. 395 (1964).

134. E. M. Savitskii, M. A. Tylkina, and V. P. Polyakova, New Materials in Electrical Instrument Manufacture [in Russian], ONTIpribor, Moscow, p. 31 (1965).
135. O. E. Zvyagintsev, Zh. Prikl. Khim., 32(9):1889 (1959).
136. A. A. Rudnitskii, Transactions of the A. A. Baikov Inst. of Metallurgy [in Russian], No. 4, p. 94, Izd. Akad. Nauk SSSR (1960).
137. Express Information, Metallovedenie Termoobrabotka, VINITI, No. 4 (Ref. 14–15) (1964).
138. G. Zwigmann, Z. Metallkunde, 54(5):286 (1963).
139. E. M. Savitskii, M. A. Tylkina, and V. P. Polyakova, Zh. Neorgan. Khim., 6(6):1471 (1961).
140. A. A. Rudnitskii, Transactions of the A. A. Baikov Inst. of Metallurgy [in Russian], No. 4, p. 8, Izd. Akad. Nauk SSSR (1960).
141. A. M. Rubinshtein, Izv. Akad. Nauk SSSR, Sektor Platiny, No. 19, p. 61 (1943).
142. J. H. Westbrook, Problemy Sovremennoi Metallurgii, No. 4, p. 111 (1960).
143. G. V. Samsonov and Ya. S. Umanskii, Hard Compounds of Refractory Metals [in Russian], Metallurgizdat (1957).
144. E. M. Savitskii, V. V. Baron, and M. I. Bychkova, Transactions of the A. A. Baikov Inst. of Metallurgy [in Russian], No. 12, p. 12, Izd. Akad. Nauk SSSR (1962).
145. E. M. Savitskii and V. V. Baron, Transactions of the A. A. Baikov Inst. of Metallurgy [in Russian], No. 5, p. 156, Izd. Akad. Nauk SSSR (1960).
146. V. V. Baron and E. M. Savitskii, Dokl. Akad. Nauk SSSR, 94(2):269 (1954).
147. R. Kieffer and Z. Cerwenka, Z. Metallkunde, 43:101 (1952).
148. R. Kieffer, F. Benesovsky, and E. Gallistl, Z. Metallkunde, 43(8):284 (1952).
149. Ch. Kirsten, Monatsberichte, 41(9):596 (1962).
150. E. M. Savitskii, in: Mechanical Properties of Metallic Compounds [in Russian], Metallurgizdat (1962).
151. Ch. Kirsten, P. Paufler, and G. E. R. Schulze, Monatsberichte, 6(2):140 (1964).
152. R. H. Gassner, Metal Progress, 78(5) (1960).
153. R. H. Willens and E. Buchler, Appl. Phys. Letters, 7(1):25 (1965).
154. E. M. Savitskii and V. V. Baron, Izv. Akad. Nauk SSSR, Metallurgiya i Gornoe Delo, No. 5, p. 3 (1963).
155. In: Metallic Properties of Metallic Compounds [in Russian], Metallurgizdat (1962).
156. E. M. Savitskii, Dokl. Akad. Nauk SSSR, 62(3):349 (1948).
157. N. J. Grant and O. Preston, Problemy Sovremennoi Metallurgii, No. 5, p. 111 (1957).
158. G. Gregory and N. J. Grant, J. Metals 6(2):247 (1954).
159. E. M. Savitskii and A. I. Vlasov, Tsvetnye Metally, No. 12, p. 77 (1961).
160. E. M. Savitskii, V. A. Livanov, P. A. Nuss, G. S. Burkhanov, M. I. Musatov, and A. D. Semonchuk, in: Titanium in Industry [in Russian], Oborongiz, p. 85 (1961).
161. R. H. Hiltz and N. J. Grant, J. Metals, 9(10):41 (1957).
162. J. L. Ratcliff, D. J. Maykuth, H. R. Ogen, and R. I. Jaffee, Trans. Metallurgy, Soc. AIME, 230(3):490 (1964).
163. G. C. Reed and W. L. Schalliol, J. Metals, 16(2):175 (1964).
164. K. M. Tsvilki and N. J. Grant, in: Some Problems of Refractory Metals and Alloys [Russian translation], IL, p. 150 (1963).
165. A. Gatty, in: Some Problems of Refractory Metals and Alloys [Russian translation], IL, p. 167 (1963).
166. N. F. Mott, Imperfections in Nearly Perfect Crystals, New York (1952).
167. E. Orowan, Discussion, Symposium on Internal Stresses in Metals and Alloys, Monograph and Report Series, No. 5, p. 451, Inst. of Metals, London (1958).
168. M. Gensamer et al., Trans. AIME, 30:983 (1942).
169. K. Goetzel, Problemy Sovremennoi Metallurgii, No. 5, p. 152 (1959).
170. E. Gregory and N.J.Grant, Trans. AIME, 200:247 (1954).

171. F. E. Bacon and P. M. Meanfeldt, in: Columbium and Tantalum, Wiley, New York, p. 347 (1963).
172. J. L. Miller and D. J. Cox, in: Protection of Refractory Metals from Oxidation [Russian translation], IL, p. 42 (1962).
173. E. M. Savitskii, Vestnik Akad. Nauk SSSR, No. 11 (1966).
174. G. Sandos, J. Metals, 12(4):34 (1960).
175. D. M. Dovey and K. S. Randle, Metal Treatment, 20(95):341 (1953).
176. E. V. Vasil'eva and D. A. Prokoshkin, Niobium Alloys [in Russian], Izd. "Nauka" (1964).
177. Rare Metals [in Russian], Izd. Giredmet, No. 29, p. 44 (1964).
178. Am. Metal Market, 71(199):13 (1964).
179. A. I. Borisenko, Protection of Molybdenum from High Temperature Gas Corrosion [in Russian], Izd. Akad. Nauk SSSR (1960).
180. I. N. Frantsevich, I. F. Voitovich, and I. A. Lavrenko, High-Temperature Oxidation of Metals and Alloys, Kiev, GOSONTI (1963).
181. R. Smith, in: Protection of Refractory Metals from Oxidation [Russian translation], IL, p. 89 (1962).
182. D. E. Couch, H. Shapiro, and A. Brenner, J. Electrochem. Soc., 105(8):485 (1958).
183. H. Buckle, Les alliages de molybdène et leur protection contre l'oxydation, La Recherch Heri natique, No. 61, p. 47 (1957).
184. "Cermet coatings for high temperature protection," Materials and Methods, 42(3):143 (1955).
185. A. N. Minkevich, Metall. i. Termoobrabotka, No. 8, p. 9 (1961).
186. A. N. Minkevich, M. A. Tylkina, L. N. Rastorguev, and G. P. Rodionova, in: Rhenium [in Russian], Izd. "Nauka," p. 221 (1964).
187. A. G. Knapton, Plansee Proceedings, 1958, London, Pergamon Press, London, p. 412 (1959).
188. P. Fossen, "New plasma tools handle ultrahard metals," Missiles and Rockets, 4(25):26 (1958).
189. "30,000 degrees with the plasma jet," J. Metals, 11(1):40 (1959).
190. E. G. Czarnecki, J. T. Stacy, and D. K. Zimmerman, Refractory Metals and Alloys, Vol. II, Chicago, p. 395 (1962).
191. E. M. Savitskii and O. B. Dashevskaya, Izv. Akad. Nauk SSSR, Metally, No. 3, p. 151 (1967).
192. E. M. Savitskii, V. V. Baron, and K. N. Ivanova, Atomnaya Energiya, 23(1):32 (1967).

CHAPTER VII

# MELTING AND TREATMENT OF REFRACTORY METALS AND ALLOYS

## Melting and Casting of Refractory Metals and Alloys

Melting, refining, and casting of chemically active refractory metals must be carried out under the following conditions. There must be no interaction with the crucible material and constituents of the air, a highly intense source of heat must be used, and the melting process must be carried out in a medium, precluding any interaction between it and the metal [1]. Melting is therefore carried out in a vacuum or in an atmosphere of purified inert gas, either argon or helium.

The following methods are used for the vacuum melting of refractory metals:

1) Electric-arc melting,

2) Electron-beam melting,

3) Electron-beam zone refining,

4) Levitation melting in a magnetic field,

5) Induction melting in "cold" crucibles,

6) Melting in solar furnaces.

The methods of vacuum melting enumerated ensure not only the protection of the refractory metals from contamination, but to a considerable degree their further purification from interstitial and metallic impurities. Electron-beam vacuum melting has been widely used in industry. In recent years, electron-beam melting [2] and also electron-beam zone refining have become widely used, due to the increasing demands of the new technology for the purest refractory materials. Sometimes the process known as the duplex process is used, which consists in electric-arc remelting of electron-beam melted ingots. Melting by means of an argon or argon–hydrogen plasma torch has appeared. Data on the use of the plasma torch for melting, cutting, and atomization of refractory metals may be carried out by means of a plasma torch, ensuring extremely high temperatures (up to 30,000°C) [3]. Melting in the levitated state in a magnetic field as yet holds out no promise in industrial production, mainly because of the low output (the weight of an ingot is some tens of grams). A start has been made in the development of methods of vacuum-induction heating of refractory metals in water-cooled crucibles [4]. Of the possible methods of melting in the near future, reference may also be made to the melting of refractory metals with the use of quantum generators – lasers. However, owing to the high cost of the apparatus and comparatively low power of this source of heat, it will evidently only be used in special cases. The advantages of melting in furnaces heated by solar radiation include the possibility of the concentrated focusing of cheap solar energy and the convenience of the introduction of solar rays in a vacuum space.

## 1. Electric-Arc Melting

Considerable progress has been made in the electric-arc melting of pure refractory metals. Molybdenum ingots 300 mm in diameter and weighing 1000 kg or more are being melted, while titanium ingots attain 5 to 10 tons. Niobium and tantalum may be arc-melting without particular difficulty. The diameter of the melted tungsten ingots attains 100 mm or more, and the weight more than 300 kg. The ingot weight of refractory metals is limited not so much by the capacity of the melting means as by the power of the pressing equipment.

Currently, electric-arc vacuum melting is the most extensively used and understood industrial method of producing cast refractory metals and alloys [199]. Methods of continuous casting in water-cooled copper molds are widely used, melting being carried out in a "bottomed" mold or by the continuous withdrawal of the ingot.

Furnaces for electric-arc vacuum melting are divided according to the electrode type into furnaces with consumable and nonconsumable electrodes. The construction of these furnaces and their working principle are described in detail in [5-9].

The method of melting using a nonconsumable electrode has been widely used in research laboratories, and has also been used in slag lining melting. The electric-arc furnace with nonconsumable electrode may be used for carrying out zone refining of refractory metals. Zone refining of molybdenum, vanadium, and iron, carried out in an electric-arc furnace, has given good results [8].

The consumable electrode melting method is used in the industrial production of large ingots of refractory metals. As is well known, in this method, the consumable electrode is the melted metal itself and the ingot is melted in a water-cooled copper mold. The size of the ingot is determined by the dimensions of the mold. During the feed of the electrode, the level of the bath of molten metal in the crucible alters. Constant arc length is ensured by an automatic device for raising the electrode and reacting to the variation in voltage drop with variation in length of the arc. A solenoid magnetic field is generally employed for stabilizing the arc discharge in the vacuum and preventing burn-through of the mold. This field (the radial component of the force vector) also produces rotation and mixing of the melt. In arc melting with a consumable electrode, there are always impurities present having a higher vapor tension than the metal base in the hot zone of the arc. The use of a high vacuum assists the more complete removal of gas from the melt.

The capacity of the vacuum pumping system attains tens of thousands of liters per second. According to [11], in the arc melting of tungsten, the pumping system used had a capacity of 40,000 liters/sec, the residual pressure before melting being $1 \cdot 10^{-6}$ mm Hg, and did not exceed $1 \cdot 10^{-5}$ mm Hg in the melting process. Refining may take place both on the electrode in the process of formation of drops during its fusion and in the pool of liquid metal.

During withdrawal of the ingot, the level of the liquid bath is always situated in the upper part of the mold, thus facilitating the process of degasification and increasing the efficiency of deoxidation. The authors of [9, 12] assume that a combination of the methods of repeated remelting and continuous withdrawal of the ingot ensures the production of the most homogeneous ingots of refractory metals and alloys.

In the electric-arc melting process, the crystallization of the ingot occurs simultaneously with the melting of the electrode. Therefore, at any instant, only a small part of the ingot is in the liquid state. Individual portions of metal are, as it were, fused onto the ingot. The shape and depth of the bath depend on the chemical composition of the melting alloy and the electrical conditions of the melt. We shall discuss the relationship between the quality of the ingot and the electrical parameters of the melt.

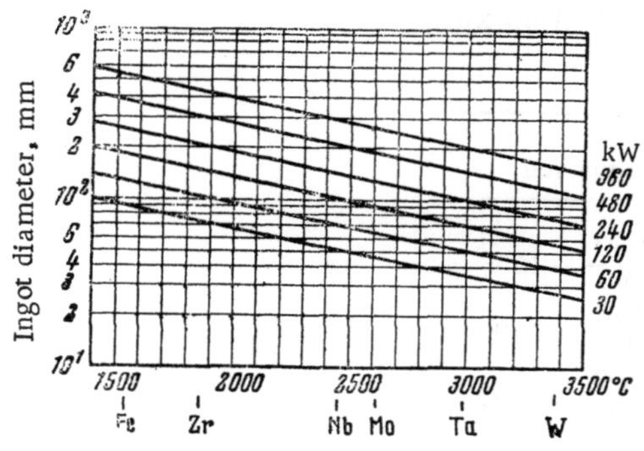


Fig. 141. Power required for melting ingots of refractory metals of different diameters.

In arc melting, a dc arc with either forward or reverse polarity, or an ac arc is used. The power distribution in the arc-melting process, i.e., a lower cathode temperature and a much higher scattering of energy from the surface of the anode, is related to the polarity [13].

The efficiency of the furnace is higher when less power is supplied to the anode. In this case, however, a shallow bath of molten metal is formed in the upper part of the ingot, which reduces the efficiency of impurity removal. The most complete removal of impurities occurs at minimum rate of melting of the metal. The quality of the resulting ingot is largely determined by the electrical parameters of the melting process, such as current density, arc stability, arc length, etc. Arc stability increases with increase in current density.

A study was made in [9] of the dependence of the porosity of molybdenum on the rate of melting, related to the value of the arc current. In melting nondeoxidized molybdenum using dc with negative polarity, the microporosity of the ingots at first decreased with increase in arc current, and then began to increase, passing through a minimum at 1600 A. In ac melting, the porosity decreased, while in dc melting with back or reverse polarity, it increased, this being attributed in [14] to the different size of the drops of molten metal produced in the different methods of supplying the current. According to [11], in electric-arc melting of tungsten, maximum density is found in ingots produced by dc melting with forward polarity. Sell [13] also considers that melting with forward polarity ensures a higher ingot quality (higher degree of purification, better surface quality).

The power of the current required for melting refractory metal ingots of different diameters is shown in Fig. 141 [22].

Ingots, melted in vacuum-arc furnaces with water-cooled molds, have a coarse columnar structure. At the same time, the more refractory the melted metal, the coarser is its structure after arc melting. Such a relationship is not coincidental; the higher the melting point, the greater is the temperature gradient in the crystallizing metal, and consequently the better are the conditions for crystal growth in the direction of the temperature gradient.

Ingot growth occurs upwardly. Water cooling of the walls of the copper mold produces a temperature gradient in a direction perpendicular to the growing surface. Considerable growth of columnar crystals is observed because of the steep temperature gradient. The crystals growing from the bottom of the mold are arranged parallel to the vertical sides; the crystals growing in the sides grow in a radial direction with an appreciable upward bend. The direction of growth of these crystals depends to a considerable extent on the rate of melting [15]. The directivity of the crystals is determined by the curvature of the crystallization front, i.e., the shape of the molten pool.

In large ingots of molybdenum and tungsten, the size of the crystallites may attain several millimeters in diameter and several tens of millimeters in length. For such dimensions of the crystallites, the ratio of the area of the intergranular surface to the volume of the grain is small. The elements forming interstitial solid solutions are precipitated at low temperatures on the grain boundaries and produce embrittlement of these metals. Group VIa metals are particularly sensitive to interstitial impurities, since the solubility of interstitial impurities in them is insignificant [16, 17]. For this reason, impurities having low solubility in

TABLE 51. Ratio of Vapor Pressures of Monoxide and Metal [18]

| Oxygen is removed | | Oxygen is not removed | |
|---|---|---|---|
| $\frac{MoO}{Mo} = 10^{0.5}$; | $\frac{ThO}{Th} = 10^{3}$; | $\frac{TiO}{Ti} = 1$; | $\frac{MnO}{Mn} = 10^{-5}$; |
| $\frac{NbO}{Nb} = 10$; | $\frac{HfO}{Hf} = 10^{4}$; | $\frac{VO}{V} = 10^{-2}$; | $\frac{FeO}{Fe} = 10^{-6}$; |
| $\frac{BO}{B} = 10^{2}$; | | $\frac{BeO}{Be} = 10^{-3}$; | $\frac{NiO}{Ni} = 10^{-7}$; |
| $\frac{WO}{W} = 10^{2}$; | $\frac{TaO}{Ta} = 10^{4}$; | $\frac{CrO}{Cr} = 10^{-4}$. | |
| $\frac{ZrO}{Zr} = 10^{2}$; | $\frac{YO}{Y} = 10^{5}$. | | |

the metal base are concentrated, on crystallization, in large quantities along the grain boundaries. These boundary precipitates produce intercrystalline brittleness. Therefore, brittle crystalline ingots of tungsten and molybdenum may sometimes be easily fractured into individual grains. The high-temperature gradient (when the ingot is solidified in a water-cooled copper mold) results in the formation of internal stresses and cracks. Cracking is caused by axial tensile stresses at the ingot center, due to the transverse temperature gradient followed by uneven shrinkage. The process of cracking is also due to a high interstitial impurity content, resulting in stacking defects during crystallization. Tungsten and molybdenum, which as already mentioned have maximum sensitivity to interstitial impurities, are particularly prone to cracking. In group Va refractory metals (vanadium, niobium, tantalum), cracking appears to a much lesser degree, this evidently being due to the higher solubility of interstitial impurities in them (in the solid phase). Well-deoxidized molybdenum and tungsten also show much less tendency to cracking. Another factor which affects cracking is the segregation of impurities and alloying elements. This is a very important factor in alloy melting. The melting of an ingot of uniform composition ensures not only stability and uniformity of the properties, but also the production of a sound metal.

Porosity is a frequently encountered defect of ingots produced by arc melting. It is quite obvious that ingot porosity is due to the liberation of gases during the melting process; particularly high porosity is to be found in nondeoxidized ingots.

The methods of improving the structure and properties of cast refractory metals include refining, modifying, repeated melting, semicontinuous casting, slag lining melting, centrifugal casting, the duplex process, etc.

Refining. In the simplest case, refinement of the metal from oxygen may occur in remelting the metal in an electric-arc or electron-beam furnace by evaporation of its monoxide in a vacuum. This process may occur when the vapor pressure of the oxide is higher than the vapor pressure of the metal base (Table 51).

The degree of purification which may be attained in electric-arc vacuum melting depends to a considerable extent on the melting conditions: holding time, temperature and pressure in the zone of the arc and liquid metal, purification occurring not only by removal of oxygen in the form of monoxides but also by removal of other impurities, including metallic impurities, having a low melting point and high vapor tension.

Deoxidation of refractory metals may be carried out by treatment of the melt with chemically active elements. Numerous investigations have shown that the deoxidation of refractory metals occurs not only by the formation and subsequent evaporation of fusible oxides, but also as the result of a corresponding reaction with the formation of gases [8, 9, 19-21]. For example, niobium may be deoxidized by zirconium owing to the formation of volatile zirconium monoxide ZrO. The vapor tension of ZrO is approximately ten times that of NbO and 100 times that of pure niobium. A deoxidation effect is also observed in the case of molybdenum on introducing zirconium, titanium, hafnium, or carbon in the melting process. In deoxidation by carbon, a chemical reaction occurs, with the formation of CO.

The quality of the carbon used is also important; satisfactory results are obtained by using spectrophotographically pure electrode graphite; the introduction of carbon in elementary form is inexpedient because of the low specific gravity of carbon. The removal of CO during melting results in a decrease in both carbon and oxygen contents. Although carbon is quite an effective deoxidizer for refractory metals, excessive carbon content has a detrimental effect on the ductility of these metals, particularly tungsten and molybdenum, because of carbide precipitation along the grain boundaries (see Chapter V). It is therefore necessary to proportion and control the carbon content carefully during melting. According to [22], improvement in the technology of the treatment of molybdenum powder enabled molybdenum to be melted and cast having a carbon content of less than 0.005%. By a more complete reduction of molybdenum powder forming the melting charge, the amount of carbon added to the charge may be diminished, whereby the residual carbon content of cast molybdenum may be decreased to 0.001-0.005%.

Owing to the considerable difficulties in the ultimate purification of the metal from excess carbon (practically the only method of purification in melting is the combustion of the carbon with oxygen) it is necessary to remove as much as possible the sources of contamination of the molten metal by carbon. The additional carburization of the metal during melting occurs mainly by migration of oil vapor from the oil-vapor pumps of the vacuum system to the melting chamber, and the products of the cracking of oil on metal surfaces. Freezing traps are used for preventing the entry of oil into the melting chamber. Recently, a start was made in the use of zeolite traps, which have been found to be quite an effective means of protecting the molten specimen from carburization. Obviously, a change-over to an oil-less system of pumping in electric-arc melting would be quite promising. The use of sorption-ionic, getter, condensation, electric discharge, and other pumps and traps will not only preclude carburization, but will also ensure a high pumping rate and low residual pressure in the melting process.

In melting refractory metals, nitrogen is removed in the form of $N_2$ molecules at a rate of approximately one-quarter the rate of oxygen removal. For a more thorough removal of nitrogen, correspondingly lower melting rates are required. Hydrogen is removed from the melting chamber in the form of $H_2$, $CH_4$, and water vapor.

Since in all the cases considered, the reaction products are removed continuously in the form of volatile monoxides or gases, there is a continuous shift of the reaction toward the formation of fresh volatile products, and equilibrium is never attained in melting. The completeness of the course of these reactions depends on the technological melting conditions, temperature, time, and residual pressure.

Deoxidation by Rare-Earth Metals. An investigation of the reaction between group Va and VIa refractory metals and rare-earth metals showed that there was a wide range of separation into liquid and solid states [23-28]. In the range of separation, the molten rare-earth metals, because of their high chemical activity, act as deoxidizers for the refractory metals. Compounds of the rare-earth metals with the nonmetallic impurities, in particular with oxygen, are characterized by a high value of the free energy of formation of these compounds, compared with the refractory metals (Table 52). Small additions of rare-

TABLE 52. Heats of Formation of Oxides, Nitrides, and Sulfides of Some Rare-Earth and Refractory Metals [17, 28, 29]

| Metal | Oxides | | Nitrides | | Sulfides | |
|---|---|---|---|---|---|---|
| | formula | $-\Delta H^0_{298}$, kcal/mole | formula | $-\Delta H^0_{298}$, kcal/mole | formula | $-\Delta H^0_{298}$, kcal/mole |
| Lanthanum . . | $La_2O_2$ | 457 | LaN | 72 | $La_2S_3$ | 351 |
| Cerium . . | CeO | 233 | — | — | CeS | 118 |
| Cerium . . | $Ce_2O_3$ | 445 | CeN | 78,2 | $Ce_2S_3$ | 300 |
| | | | | | $Ce_3S_4$ | 421 |
| Praseodymium. . | $Pr_2O_3$ | 445 | — | — | — | — |
| Neodymium . . | $Nd_2O_3$ | 442 | — | — | $Nd_2S_3$ | 282 |
| Yttrium . . | $Y_2O_3$ | 440 | YN | 71 | — | — |
| Vanadium . . | $V_2O_3$ | 290 | VN | 60 | VS | 45 |
| Chromium . . | $Cr_2O_3$ | 269 | $Cr_2N$ | 26.5 | $Cr_2S_3$ | 110 |
| Niobium . . | $NbO_2$ | 193 | NbN | 59 | — | — |
| Titanium . . | $TiO_2$ | 218 | TiN | 80 | TiS | 52 |
| Titanium . . | — | — | — | — | $TiS_2$ | 80 |
| Zirconium . . | $ZrO_2$ | 258 | ZrN | 82 | — | — |
| Molybdenum . . | $MoO_2$ | 130 | MoN | 17 | $MoS_2$ | 56 |
| Molybdenum . . | $MoO_3$ | 180 | — | — | $MoS_3$ | 61 |

earth metals to commercially pure niobium and tantalum remove nonmetallic impurities from these metals to an extent such that they can be deformed at room temperature [17, 25, 30, 31].

In [32], a study was made of the refining effect of cerium in electric-arc melting of vanadium using a monoconsumable electrode. The free energy of formation of cerium oxide is higher than that of vanadium [33]. This points to the thermodynamic probability of the reduction of vanadium oxide by cerium. In the arc-melting process, deoxidation of vanadium proceeds according to the reaction $2VO + Ce \rightleftharpoons 2V + CeO_2$. The cerium dioxide $CeO_2$ formed has a comparatively low vapor tension and its removal from the melt occurs mainly by floating in the slag. The removal of $CeO_2$ in the slag assists the shift of the reaction from left to right until equilibrium is established between the slag and molten metal. Further reduction in the oxygen content of the vanadium may be attained by removal of the slag and addition of cerium. Vanadium refined by cerium has high ductility and may be cold-rolled, without intermediate annealing, to a high degree of reduction (95%). The method developed for refining vanadium with cerium is already used in industry [34].

In [35, 36], dealing with the deoxidation of molybdenum by rare-earth metals, it was shown that slight additions of mischmetal, lanthanum, cerium, praseodymium, and neodymium (tenths of one percent) reduced the hardness and particularly the ductile–brittle transition temperature of commercially pure cast molybdenum. The residual rare-earth metal content in the metal did not exceed 0.05-0.15%. The ductile–brittle transition temperature of commercial molybdenum, melted with rare-earth metal additions, was close to or below room temperature, i.e., it was lowered by approximately 300-500°C. The occurrence of a second phase in the case of an excess of rare-earth metal led to a rise in the ductile–brittle transition temperature. The most effective deoxidizer was found to be lanthanum.

The introduction of rare-earth metals into molybdenum is made difficult by the considerable difference in the melting points of the initial metals. This circumstance, together with the high vapor tension of rare-earth metals, results in considerable evaporation of the latter. In our laboratory [36], a number of master alloys of lanthanum having a much higher melting point have been developed. With the introduction of rare-earth metals, dc melting in the majority of cases becomes impossible because of strong ionization and redistribution of the power on the electrode. In such cases, melting is usually carried out with alternating current [37].

Molybdenum, deoxidized in the melt with lanthanum hexaboride, after preliminary forging and rolling at 700°C, may be rolled at room temperature to strip and foil. Deoxidation by rare-earth metals is now being more extensively used in the industrial production of molybdenum. E. M. Savitskii and G. E. Chuprikova [38] obtained satisfactory results in the deoxidation of rhenium by rare-earth metals in electric-arc and electron-beam vacuum melting of the metal.

Work on the deoxidation of tungsten by rare-earth metals and other chemically active elements is of theoretical and practical interest, but so far very few experimental data have been acquired. Investigations which have been made show that the addition of thermodynamically active rare-earth metals (lanthanum, cerium) and also titanium, zirconium, and hafnium reduce the oxygen content of tungsten and also reduce the hardness and the ductile–brittle transition temperature. Owing to experimental difficulties, however, it has so far been impossible to obtain sufficiently reproducible results. There still remains much work to be done in this direction. The progress made in the removal of interstitial impurities from molybdenum and in the improvement in its structure and properties, after arc melting, during deoxidation by rare-earth metals give reason to hope that this problem will also be solved for tungsten.

Alloy Preparation. As already mentioned in Chapter VI, considerable alloying is applicable only to the ductile metal base (titanium, zirconium, niobium, tantalum): alloys of the systems niobium–tungsten, niobium–molybdenum, niobium–tungsten–molybdenum–zirconium, niobium–tantalum, niobium–zirconium–titanium, niobium–zirconium, and other binary and more complex systems. Of the tungsten- and molybdenum-based alloys, highly alloyed alloys of the tungsten–rhenium, molybdenum–rhenium, tungsten–molybdenum–rhenium, and tungsten–tantalum systems have acquired special significance. These alloys show maximum heat resistance and have complex physical properties, enabling them to be used in instruments and devices of the new technology. In this connection, a need arose for the development of alloying methods in electric-arc vacuum melting, ensuring uniform distribution of the alloying elements and minimum loss of expensive materials (rhenium, tantalum, etc.). In the practice of vacuum-arc melting of refractory metals, two alloying methods are used – the introduction of alloying elements in the consumable electrode, and the introduction of the alloying additions directly into the bath of liquid metal.

The introduction of alloying elements by means of a consumable electrode is most commonly employed in making alloys of refractory metals when the ingot weight is comparatively small. The preparation of electrodes for alloy melting is carried out by partial charging and extrusion. Partial charging is used for a more even distribution of the alloying elements in the metal base. This method is successfully employed in the preparation of titanium alloys. In melting niobium, molybdenum, and tungsten alloys, the electrode is usually made of forged or cermet bars or rods of the respective components of the alloy bound together.

For tungsten, a ratio of the cross section of the electrode to the area of the mold, equal to 0.2, is evidently the best.

When small ingots are being made, the alloying additions may also be introduced in the form of rods, located in the hollow electrode of the metal base, previously drilled along the vertical axis. Such a method of preparing the electrodes is used when making tungsten alloy with additions of niobium, molybdenum, and tantalum. Electric-arc melting of previously prepared "alloyed" electrodes ensures the production of alloys of uniform composition and structure, as well as the removal of interstitial impurities. For a more thorough removal of interstitial impurities, deoxidizers may also be introduced into the electrode along with the alloying elements. For the production of an alloy of more uniform composition throughout the ingot, a second melting may be carried out. After the first melting, the ingot is forged, machined to the required dimension, and remelted. The method of alloying the electrode, ensuring minimum segregation of impurities is the basic method in making tungsten–rhenium, molybdenum–rhenium, niobium–zirconium, tungsten–molybdenum, tantalum–tungsten, and other alloys [38].

As investigations have shown, a lower melting rate is required for the production of alloys than for the production of the unalloyed metals.

The method of adding the alloying elements (in granule or powder form) directly to the bath of liquid metal is more economical and is quite a convenient method of measuring out the proportions. It does not, however, ensure uniform distribution of the alloying elements. As pointed out in [9], such a method may be successfully used in the first melting; for its wider use, a number of constructional difficulties must be overcome, more particularly connected with the continuous introduction of the alloying additions into the vacuum space of the furnace. Powder should be fed continuously and uniformly. In the introduction of deoxidizer alone, or a small amount of alloying addition, however, it is difficult to ensure a uniform rate of addition. In such a case, the deoxidizer or the powder of the alloying component is mixed with powder of the metal base. An important factor is also the particle size of the powder. Very fine powder may be carried off by the stream of gas before it can enter the bath.

The method of introducing alloying additions directly into the bath of liquid metal has not been used extensively in vacuum-arc melting.

Casting Methods. For a more uniform distribution of the alloying elements in making high alloys, double remelting is used. In recent years, methods of slag lining melting have been developed, by means of which it is possible to produce ingots of uniform chemical composition and fine grain.

Basically, the "drip" or "melt through" method comprises gradual electric-arc melting of the top face of a previously prepared cylindrical blank, which also acts as melting crucible. The cast blank is previously produced by electric-arc melting using a consumable electrode. A depression is drilled in the bottom face of the bank. As the top face melts, the depth of the bath of liquid metal increases. The electric arc acts as a point source of heat of high power. The liquid–solid interface has a spherical shape. During melting of the layer of metal between the top face of the blank and the orifice, the metal runs through the orifice into a copper mold, intensely cooled externally by water. Figure 142 [8] shows the apparatus diagrammatically. As the result of the flow of metal, a hemispherical cavity is formed in the blank, and this cavity is filled up by means of a consumable electrode in readiness for subsequent use.

The melt-through method gives an ingot of fine-grain dense structure. This method is used for producing ingots of tungsten, molybdenum, and niobium and a number of niobium alloys.

Recently, for casting intricate shapes in refractory metals, the lining method of melting has been widely employed, in which a rotating crucible is used. In this method, a portion of liquid metal is poured into a copper, water-cooled crucible (or a noncooled graphite crucible). Discharge of the metal into the stationary or rotating mold is carried out during rotation of the crucible. Melting may be done with either a consumable or nonconsumable electrode. By means of this method of melting, it is possible to produce castings of intricate shape, and also centrifugal castings having a fine coaxial structure.

A fine-grain structure in a casting considerably improves its technological properties. Molybdenum, tungsten, and their alloys, cast by the method of slag lining melting with a rotating

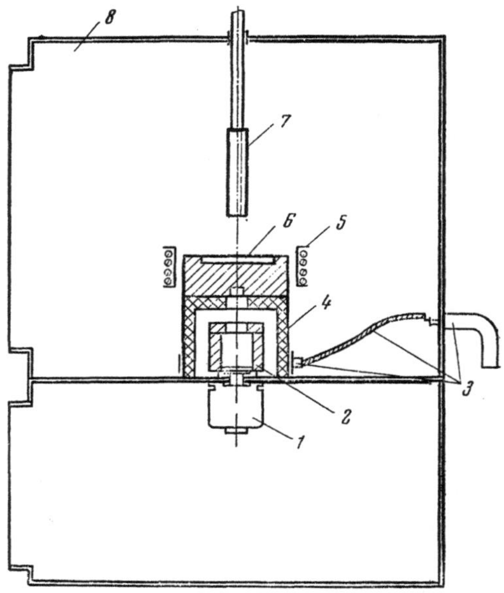


Fig. 142. Diagram of vacuum apparatus for the production of castings of refractory metals by the melt-through or drip method. 1) Electric motor; 2) copper mold; 3) current supply system; 4) graphite ring support; 5) dc solenoid; 6) melt-through blank; 7) tungsten electrode; 8) melting chamber.

crucible, have shown increased ductility in the cast state and higher and stable mechanical properties in deformed, semifinished products, compared with alloys of similar composition produced by the methods of ordinary electric-arc surface melting [39].

Centrifugal casting is of considerable interest with regard to the production of blanks for tubes of refractory metals and their alloys. According to published data [40], this method has been used for the production of pipes of molybdenum, niobium, and tungsten 370 mm in diameter, 415 mm long, and up to 220 kg in weight. A report is given in [41] on the production of tungsten pipes of various dimensions by the centrifugal casting method.

Annular castings having an external diameter of 254 mm and weighing 99.8 kg have been produced from pure tungsten. The technology of the production of castings from tungsten, molybdenum, tantalum, niobium, vanadium, hafnium, and alloys of these metals has been developed.

## 2. Electron-Beam Melting

Electron-beam vacuum melting has a number of indisputable advantages over other melting methods – arc, induction, etc. Among these advantages must be included the possibility of melting in a high vacuum, smooth regulation of the thermal conditions of melting, heating the bath of liquid metal several hundred degrees above the melting point, and keeping the metal for a long time under definite thermal conditions. These features of electron-beam melting enable the metal or alloy to be subjected to profound refining, this being of especial importance for refractory metals having a body-centered cubic lattice.

In electron-beam furnaces, the ingots may be produced from powder or scrap, thereby eliminating the laborious operations of electrode preparation, and also offering the possibility of the treatment of scrap and sometimes of the reduction of oxides or concentrates [2, 8, 21].

Electron-beam melting, however, cannot entirely replace other forms of melting, especially arc melting, which is still the principal method of producing cast refractory metals and particularly their alloys.

The need for melting in a high vacuum in electron-beam heating precludes the melting of metals having a high vapor tension, or alloys containing volatile or liquating components. One of the chief drawbacks of electron-beam melting and one which will certainly be overcome in the future development of the method is the production of a coarse-grained ingot structure. This has a negative effect on the mechanical properties of group VIa metals – tungsten and molybdenum – which are particularly sensitive to grain size. Group Va refractory metals – vanadium, niobium, and tantalum – are less sensitive to grain size because of the higher solubility of interstitial impurities in them in the solid state. The ductility of these metals increases sharply after electron-beam melting, due to the considerable removal of interstitial

TABLE 53. Vapor Pressure and Loss on Evaporation of Some Metals at the Melting Point [70]

| Metal | Melting point | Vapor pressure at melting point, mm Hg | Rate of evaporation, $g/cm^2 \cdot sec$ | Relative rate of evaporation, % |
|---|---|---|---|---|
| Tungsten . . . . . | 3380 | 20—30 | $2 \cdot 10^{-5}$ | 10 |
| Rhenium . . . . . | 3180 | 25 | $3.8 \cdot 10^{-4}$ | 200 |
| Tantalum . . . . . | 3000 | 4—6 | $8 \cdot 10^{-5}$ | 50 |
| Molybdenum . . . . . | 2610 | 10—30 | $2 \cdot 10^{-4}$ | 100 |
| Niobium . . . . . | 2468 | 0,2—0,6 | $2 \cdot 10^{-6}$ | 1 |

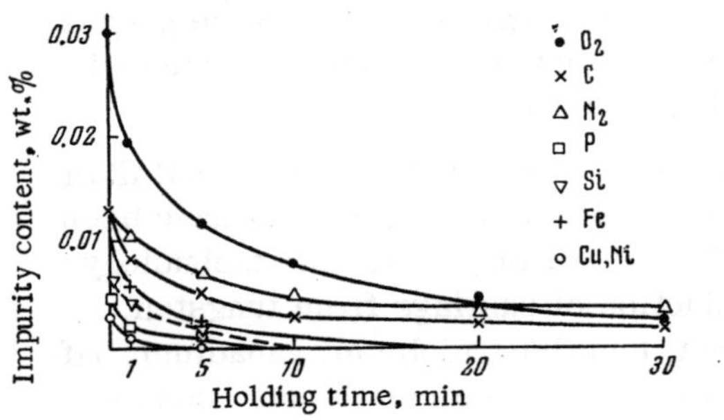


Fig. 143. Variation in the concentration of some impurities with the length of time that molybdenum held in the molten state in electron-beam melting.

and metallic impurities, despite the coarse-grained structure. As shown in [18, 53-55], refinement of group Va metals in the process of electron-beam melting occurs by the removal of oxygen in the form of lower oxides, CO, and metallic impurities and their oxides; nitrogen and hydrogen are removed in the form of gaseous molecules of $N_2$ and $H_2$. It should be pointed out that in the melting process of niobium and tantalum, the removal of oxygen in the form of lower oxides (NbO and TaO) takes place four times more rapidly than in the form of CO, and also four times faster than the removal of nitrogen. The rate of removal of CO and $N_2$ is limited by the diffusion of carbon, nitrogen, and oxygen in the liquid bath. A more effective reduction in the nitrogen content is obtained by reducing the rate of melting.

The process of electron-beam melting is also accompanied by the evaporation of metallic components having a higher vapor tension than the metal base. The final content of metallic components in the alloy may be calculated by taking into account values for the vapor pressure of the pure metals at different temperatures [56]. It is pointed out in [55] that the effective removal of any impurity in electron-beam melting occurs when the vapor pressure of the impurity is higher than the vapor pressure of the metal base by a factor of 10. Evaporation of the metal base occurs at the same time as that of the impurities. Data on vapor pressure at melting point and on evaporation losses of some refractory metals are given in Table 53.

It has been found that the rate of removal of a number of impurities from molten refractory metals under vacuum-melting conditions is a function of the concentration of those impurities. Figure 143 shows the variation on concentration of impurities as a function of the time of holding molybdenum in the molten state during electron-beam melting [54]. Oxygen and carbon are removed simultaneously during melting, but the oxygen content falls more rapidly. This circumstance is due to the fact that oxygen is removed from the bath not only in the form of CO, but also by the evaporation of volatile lower oxides of the refractory metals. Niobium, tantalum, vanadium, tungsten, and molybdenum, melted in the electron-beam furnace, contain a lower impurity content than electric-arc-melted metals (Table 54).

Work on the removal of interstitial impurities from tungsten has shown that the degree of purification from carbon is determined by the oxygen content of the metal [58]. The same

TABLE 54. Impurity Content of Initial Metal and Ingots Melted by Different Methods [57]

| Metal | Impurities | Impurity content, $10^{-4}$% | | | |
|---|---|---|---|---|---|
| | | in initial material | after electric-arc melting using consumable electrode | | after melting in electron-beam furnace |
| | | | in argon * | in vacuum | |
| Tungsten | Oxygen ....... | 540 | 40 | 10 | <5 |
| | Carbon ....... | 260 | 30 | 20 | <5 |
| | Nitrogen ...... | 60 | 6 | 3 | 1 |
| | Hydrogen...... | 10 | 1 | 1 | 1 |
| Tantalum | Oxygen ....... | 600 | 400 | 300 | 60 |
| | Carbon ....... | 300 | 300 | 200 | 50 |
| | Nitrogen ...... | 130 | 100 | 60 | 50 |
| | Hydrogen...... | 130 | 10 | <10 | 4 |
| Niobium | Oxygen ....... | 500 | 500 | 500 | 150 |
| | Carbon ....... | 200 | 190 | 200 | 70 |
| | Nitrogen ..... | 210 | 200 | 20 | 60 |
| | Hydrogen...... | 20 | 10 | <10 | <10 |

*Pressure above 60 mm Hg.

relationship was demonstrated by the authors of [59] with regard to niobium. If the oxygen content of niobium is less than 0.005%, their opinion is that the removal of carbon is arrested. For this reason, the oxygen content of the initial charge should be increased to ensure thorough purification from carbon. For reducing the carbon content of the metal, it is also necessary to remove the possibility of carbon contamination in the process of the chemical technological conversion of the initial raw material and melting of the metal.

Active additions are made in the process of electron-beam melting to ensure more thorough purification. The addition of zirconium (1%) in niobium melting leads to an additional purification from oxygen, due to its removal from the melt in the form of ZrO, the hardness of the niobium being reduced to 40-45 HB, evidence of the very effective removal of the impurities. The relationship between high vacuum and hardness of niobium in electron-beam melting is shown in Fig. 144 (according to data in [60]).

Electron-beam melting of vanadium, niobium, and tantalum is an industrial method of producing the cast metals; as regards group VIa metals (tungsten and molybdenum), electron-beam melting of these metals is practically still in the laboratory stage of research and experimental work. There is, as yet, no published information on the production of high-grade tungsten ingots by electron-beam melting. It has been found that as the result of remelting molybdenum in the electron-beam furnace, the content of oxygen, nitrogen, carbon, and other impurities is diminished [61, 62]. However, despite the thorough purification and reduction in hardness, no considerable reduction in the ductile–brittle transition temperature or improvement in technological ductility of the molybdenum was observed. This may have been due to the coarse-grain structure of cast molybdenum after electron-beam melting and the low solubility of interstitial impurities in it. Therefore, owing to the small total surface of the grains, as in electric-arc melting, a considerable quantity of impurities collects on the boundaries, weakening the bond between the grains. Similar results are obtained in molybdenum melting. According to data in [63], electron-beam melting of tungsten leads to a reduction in hardness of from 380 to 260 HB, indicating a very effective removal of impurities. The ductility of the tungsten, however, was not improved.

Despite the fact that the interstitial impurity content of tungsten is reduced by one order by electron-beam melting, such an ingot cannot be subjected to plastic deformation because of

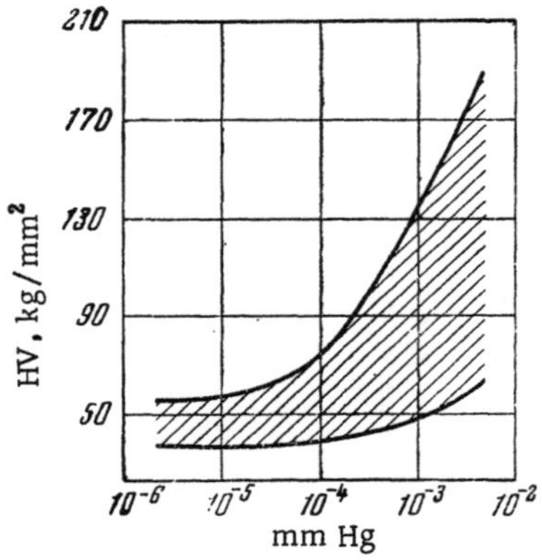


Fig. 144. Dependence between high vacuum and hardness of niobium in electron-beam melting.

its coarsely crystalline structure. These results, however, do not mean that electron-beam melting is without promise for tungsten and molybdenum. The results of the chemical analysis of some impurities and the considerable drop in hardness are evidence of the extensive removal of volatile metallic impurities and interstitial impurities from these metals. Individual crystals, cut from a tungsten ingot produced in the electron-beam furnace, show fairly high plasticity. They may be bent and subjected to various forms of deformation at room temperature. The cause of the coarsely crystalline structure of ingots of refractory metals melted in electron-beam furnaces primarily resides in the specific crystallization conditions of the ingots (high heat loss from the open metal bath, intense dissipation of heat to the water-cooled walls of the mold, absence of sufficiently effective stirring). The development of methods of refining the structure of refractory metals, particularly tungsten and molybdenum, is a most important problem, the solution of which will at once increase the scale of application of electron-beam melting.

Work on the refinement of the structure of refractory metals by the introduction of modifiers on the application of centrifugal casting methods shows considerable promise. Experiments carried out in the Refractory and Rare Metal Alloys Laboratory of the A. A. Baikov Institute of Metals on the treatment of tungsten and molybdenum, in the electron-beam melting process, by titanium, zirconium, rare-earth metals, boron, and other additions, and also by refractory compounds (TaC, HfC, etc.) are showing promising results. The modifying influence of boron on the structure of molybdenum is shown in [64-66]. It is not impossible that, to some extent, grain refinement is caused by the formation of disperse borides acting as additional crystallization centers. Molybdenum ingots containing boron forged well at 1200°C. In [67] a comparison was made between the influence of boron, thorium, yttrium, titanium, and carbon on the machinability of electron-beam-melted molybdenum ingots. The best results were obtained by the addition of boron. Reference has already been made to modification by boron–lanthanum master alloys [36]. Much work still remains to be done on the development of methods of modifying tungsten and molybdenum in the electron-beam melting process.

Treatment of the crystallizing melt by ultrasonic vibrations is deserving of note. As yet, however, there is no information on the effectiveness of this method. Quite promising, evidently, is also the duplex vacuum-melting process in which the refractory metals are purified by electron-beam vacuum melting, followed by remelting in an electric-arc furnace for grain refinement.

The electron-beam melting method has been found quite effective for the purification of titanium, zirconium, and hafnium. The results of the chemical analysis of titanium ingots, produced by electron-beam melting, showed a sharp drop in the hydrogen content as well as a reduction in the metal impurity content, compared with electric-arc-melted titanium. The oxygen, nitrogen, and carbon contents of titanium melted in the electric-arc and electron-beam furnaces were found to be approximately the same. A slight variation in the contents of these impurities is attributed to the high vapor tension of titanium [70]. Comparison of the tensile stress properties, according to these data, shows almost the same effect in both melting processes, except that electron-beam-melted titanium had an impact strength (27 $kgm/cm^2$ at −40°C) almost twice that of electric-arc-melted titanium. The increase in the impact strength of electron-beam-melted titanium is due to a reduction in hydrogen content [70].

According to [71], electron-beam-melted zirconium has a much lower impurity content and lower hardness than electric-arc-melted zirconium:

| | Electric-arc melting | Electron-beam melting |
|---|---|---|
| Impurity content, $10^{-4}$% | | |
| Hydrogen | 24 | 3 |
| Nitrogen | 71 | 22 |
| Oxygen | 735 | 690 |
| HV, $kg/mm^2$ | 180 | 140 |

In the electron-beam melting process, zirconium is purified from oxygen, nitrogen, and metallic impurities having a high vapor tension. The view has been expressed that the expensive process of iodide refinement of zirconium may be successfully replaced by electron-beam melting of magnesiothermic or electrolytic zirconium [72].

Very advanced purification of hafnium from oxygen and some metallic impurities in electron-beam melting is reported in [67, 73, 74].

Electron-beam melting ensures very effective removal of interstitial impurities from metals of the platinum group owing to the high melting point and low vapor pressure of these metals.

The work of our laboratory has demonstrated the effectiveness of electron-beam melting for the purification of hexagonal ruthenium from interstitial impurities; at the same time, the ruthenium acquires high plasticity, enabling it to be deformed at room temperature.

The evaporation of alloying elements having a high vapor tension compared with the metal base, and also the liquation of additions along the ingot, make it difficult to prepare alloys by the electron-beam melting method. However, the high purity with respect to interstitial impurities, and in consequently the adequate ductility of alloys produced by electron-beam melting of refractory metal alloys makes this quite a promising method. Alloys based on tungsten, molybdenum, rhenium, vanadium, niobium, tantalum, and titanium may be prepared by the electron-beam melting method.

Machinable alloys of tantalum with a high tungsten content (up to 15%) have been obtained by electron-beam melting [75]. This method may be used for melting alloys of niobium with tungsten and molybdenum possessing high ductility.

The properties of alloys based on niobium–tantalum with additions of tungsten and molybdenum, melted in the electron-beam furnace, have been investigated [75]. Most of these alloys have fairly high ductility. The production of a niobium alloy containing 10% W and 5% Zr is reported [76]. Sheets of this alloy may be bent cold at 180°C without cracking. High ductility is shown by superconducting alloys of the niobium–zirconium system, melted in the electron-beam furnace [77]. Titanium alloys, alloyed with niobium, tantalum, vanadium, and other refractory metals are made by electron-beam melting [70].

One of the main difficulties found in the electron-beam melting of refractory metals has been the deviation (up to 25%) of the content of alloying components compared with the nominal composition, especially in the case of a considerable difference between the specific gravities of the metal base and alloying addition.

In the scope of the present, there is no need to dwell on the details of the constructional features of electron-beam installations. These problems are discussed in detail in appropriate

Fig. 145. 500-kW electron-beam installation with vacuum pumping system.

articles and books, for example, in [44, 72, 78]. Figure 145 is the external view of a 500-kW electron-beam installation with vacuum pumping system [79]. Information is available on projected electron-beam installations of a power of up to 2000 kW [80].

Electron-beam installations are of two types: installations having an annular cathode and installations having axial guns. The advantages of the first type of installation are simplicity of construction, easy cathode replacement, and uniform energy distribution. However, the presence of an electrostatic field directly in the melting space results in the formation of a glow discharge at high residual pressures of up to $1 \cdot 10^{-4}$ mm Hg, due to the intensity of evolution of gas from the liquid bath and surfaces of the furnace construction. The need for protecting the electric supply unit from overloading necessitates interrupting the melting process. Such overloads are removed by means of special current limiters. An electron-beam furnace stabilizing system also includes an emission current limiter. Another drawback of the annular cathode furnace is the danger of the cathode being splashed with melting metal, resulting in a reduction in emission. Removal of the cathode from the bath reduces the efficiency.

The drawbacks inherent in annular cathode systems are eliminated by the use of installations with axial electron guns. In these installations, melting takes place in a region free from strong electric fields. The electron gun is located in a separate chamber, which communicates with the melting chamber by means of diaphragms for the exit of the focused electron beam. The sharper is the focusing of the electron beam, the lower is the flow of gas entering the gun chamber, and the higher is the efficiency.

It would be of considerable interest if a method could be devised for pouring the metal in electron-beam melting which would result in a considerable increase in the ingot quality. This process may be carried out by using a number of electron guns. Gruber [44] mentions several possible modifications of electron-beam melting furnaces. Figure 146a shows the construction of an electron-beam installation in which pouring is carried out by melting a plug closing the pouring orifice in the bottom part of the crucible. The electron gun for melting the orifice is switched on immediately before pouring. Obviously, this method may be improved by pouring the metal into a rotating mold as has been done in electric-arc melting.

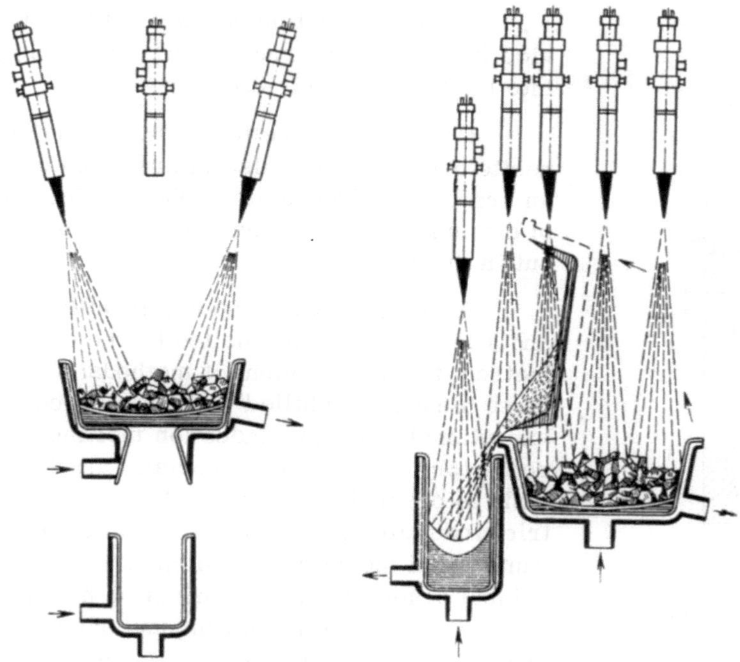


Fig. 146. Electron-beam casting furnaces. a) With bottom-tapped metal; b) with tilting crucible (with metal heated in the stream and in the mold).

Figure 146b shows a system with tilting crucible. During the pouring process, the pouring stream and also the surface of the metal in the mold, are heated by electron beams. Currently, a large proportion of ingots in electron-beam melting are poured into circular molds, but in principle molds of different cross-sectional shapes may be used. For example, ingots may be produced in the form of slabs for rolling, or any other form, depending on the subsequent treatment.

It is to be hoped that after the necessary experience in the construction of furnaces and a more profound study of the physicochemical processes involved in electron-beam melting, casting of refractory metals in electron-beam installations, especially with the application of the duplex process, will permit the production of ingots having a fine-grain structure and will, therefore, considerably increase their capacity for plastic deformation.

A further step in the production of specially pure refractory metals was the development of equipment and methods of electron-beam vacuum zone refining (see Chapter V).

## 3. Induction Melting

Induction heating is used successfully in melting refractory metals of the platinum group and chromium-based alloys. Melting of platinum refractory metals is carried out in a vacuum or an inert gas atmosphere (argon, helium). As refractory material, crucibles of $Al_2O_3$, BeO, $ThO_2$, or $ZnO_2$ are usually employed. "Corundis" crucibles (made of $Al_2O_3$) and zirconium oxide crucibles are most frequently used. When thorium oxide crucibles are used, precautionary measures must be taken for the protection of personnel. The noble metals and their alloys are inert to the refractory materials enumerated.

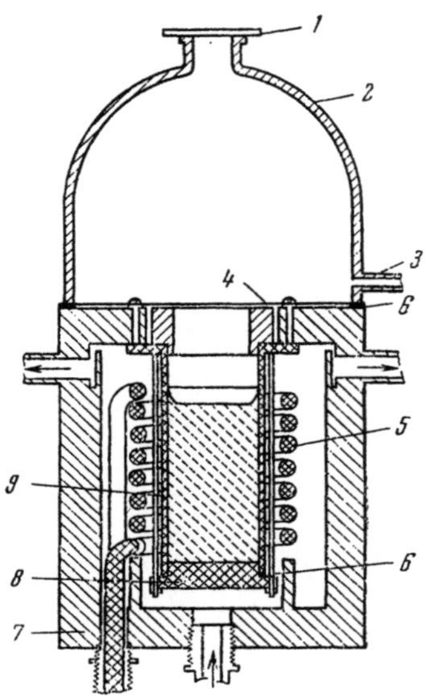


Fig. 147. Diagram of vacuum furnace for the induction melting in a water-cooled metal crucible. 1) Inspection window; 2) Pyrex vacuum hood; 3) vacuum pump connection; 4) screen; 5) inductor; 6) lining; 7) water jacket; 8) removal bottom; 9) metal crucible.

Usually induction furnaces are provided with a device for tilting the crucible and for pouring the molten metal into metal molds [42, 43].

Chromium is melted in "Corundis" crucibles or magnesite crucibles, and also in crucibles of beryllium oxide, in an argon or hydrogen atmosphere. Because of the resistance of chromium to oxidation, chromium melting is often carried out in the air.

Although most of the highly active refractory metals and alloys are usually melted by the electric-arc vacuum method, the development of methods of induction melting for these metals in "chilled" (i.e., water-cooled) crucibles is of exceptional interest. Induction melting of chemically active refractory metals, and especially alloys, would make it possible to preclude the principal drawbacks inherent in electric-arc melting. As is well known, one of the main advantages of induction heating is the intense agitation of the volume of molten metal. Induction heating affords the possibility of conducting melting under any pressure, and of producing ingots uniform in composition and structure from one melt, the necessity for preparing the electrodes being dispensed with, and the use of a water-cooled metal crucible precludes interaction between the alloy and the ceramic of the crucible. The combination of induction heating and a water-cooled metal crucible is ideal for vacuum melting. However, a number of difficulties must be surmounted.

An investigation of the possibility of conducting induction melting in a water-cooled copper crucible showed it to be technically possible to carry out this process [4]. An apparatus was constructed in which titanium ingots 50 mm in diameter and 130 mm in length were melted. The apparatus consisted of a split water-cooled copper crucible assembled from four insulated sections (Fig. 147), thereby excluding the possibility of the flow of circular currents. The current was supplied from a 100-kW machine generator with a frequency of 2000 cs. Melting did not take longer than 2 min. Vigorous mixing of the melt occurred during the melting process. The Battelle Institute (Ohio, USA) produced a similar furnace for ingots 350 mm in diameter. In an apparatus for the induction melting of refractory metals constructed by the "Heraeus" firm [44], an ingot withdrawal device is provided.

Dropwise melting with induction heating was proposed in [46]. The end of the electrode was melted in the inductor and the metal flowed into a water-cooled copper crucible. In such a melting method, however, the density of the resultant ingot is inadequate; in addition, the possibility of using scrap is precluded.

Obviously, the possibilities of the method considered are not restricted to titanium melting, and with further improvement, it may be successfully employed for melting more refractory metals, for example niobium and molybdenum. The use of induction melting in a "cold" crucible in the zone refining of metals is reported in [46]. The crucible was a water-cooled silver boat. Experiments on zone refining in cold crucibles was successfully carried out on silicon, iron, titanium, vanadium, zirconium, and molybdenum. The temperature of the molten material could be raised to 3000°C [46]. Work is being done on the induction melting of refractory metallic compounds in "cold crucibles."

It is possible that the main advantage of the induction method of melting, consisting in the production of a homogeneous ingot after one remelt, provided a reliably operating apparatus is produced, will compensate some disadvantages, due to increased power and water consumption.

Very cautious views are expressed in the literature regarding the prospects of the industrial application of the induction method of melting chemically active refractory metals (especially for the production of heavy ingots). The possibility of induction heating in noncrucible zone refining has also been pointed out [47, 48]. In industry, however, such a method of melting has been mainly applied to materials of moderate refractoriness (copper, nickel, manganese, and chromium), and also in the production of silicon, germanium, and rare-earth-metal single crystals. One of the difficulties of using induction heating in zone refining is the short-circuiting of the turns of the inductor through vapor of the melting metal condensing on them.

### 4. Levitation Melting in a Magnetic Field

The principle on which this method is based (Fig. 148) [49-51] is that the specimen to be melted is located between closely situated inductors; the forces of the electromagnetic field support the metal in the suspended state, and metal is heated and melted by the currents induced in it. In such a method, a crucible is not needed, and the liquid metal is held together and prevented from spreading by surface tension. However, the surface tension is capable of holding together only a very small quantity of metal of the order of 10-15 g. Because of its very small output, this method will not have any industrial significance in the foreseeable future and as yet is employed only in laboratory practice. Levitation melting may be carried out in a vacuum or in an inert gas. It is stated [52] that for the levitation melting of refractory metals, alloys, and metallic compounds, a 10-kW inductor was used; tungsten (up to 10 g) was melted and also alloys of iron with copper and platinum, and the compounds $YFe_2$, $YFe_3$, $YCo_5$, and $PrFe_5$ were produced.

## Pressure Working of Refractory Metals and Alloys

Before proceeding to describe existing methods of the working of refractory metals and alloys by pressure, it is necessary to discuss modern concepts of the mechanisms of plastic deformation of crystalline bodies. As is well known, the conditions of the pressure working of different materials are determined by their plasticity, i.e., the property of materials under known conditions to change within wide limits their shape and dimensions without fracture. Despite its apparent simplicity, the plastic deformation of solids is a complex physicochemical process, resulting in profound changes in crystal structure and the physicochemical properties of the deformed solid [81, 85].

The nature of the plastic deformation of materials, the fundamental laws governing this phenomenon, have been studied in various aspects by mathematicians, engineers, physicists, and physical chemists. Currently, the nature and mechanism of plasticity has been elucidated to a considerable extent, but a number of unsolved problems will remain.

Outside the Soviet Union, in recent years the dislocation theory of plastic deformation has been particularly fruitfully developed (Taylor, Orowan, Read, Frank, Shockley, Cottrell, Mott, Seitz, Seger, Sherby, Rostoker, Nabarro, etc.). In the USSR, a school of physicists and physicochemists have been engaged on research into the nature of plasticity (Kurnakov, Ioffe, Kuznetsov, Gubkin, Bochvar, Davidenkov). Soviet scientists have performed a great service in establishing the theory of the plastic working of metals.

The mechanism of plastic deformation and fracture is ultimately determined by the chemical composition of the substance, arrangement of the structural constituents, crystal

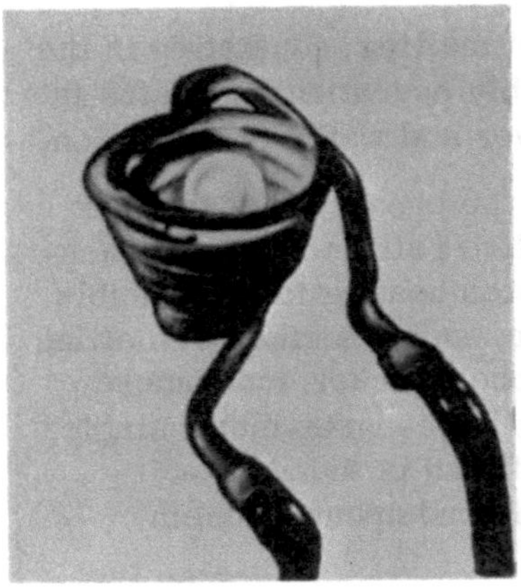

Fig. 148. Levitation melting in a magnetic field.

structure of the individual phases and also the external conditions of deformation (system of stressed state, temperature and rate of deformation, surrounding medium). Depending on the nature of the crystals and character of their stressed state, plastic deformation may occur either by slip or by mechanical twinning or by both simultaneously. Metals having a hexagonal or other noncubic crystal lattice show the greatest tendency to twinning in deformation. Reduction in temperature and increase in the rate of deformation promote twinning. The view has been expressed that in polycrystalline body-centered cubic metals, a close relationship is to be found between the characteristics of the twinning process and grain size [82]. This was found even earlier as applied to the magnesium type of hexagonal metals [83]. The larger the grain, the lower is the yield stress for twinning and the higher is the temperature of the transition from deformation by slip to deformation by twinning. Twinning may be preceded by slip, accompany it, or succeed it. The rate of formation of twins is very high and is often accompanied by propagation of sound waves and elastic stresses (crackling). A considerable amount of work has been devoted to the study of twinning mechanism by Soviet and other researchers. Twinning is discussed at length in the book by M. V. Klassen-Neklyudova [84] and also in [88].

Plastic deformation by slip is produced by the action of forces situated in the slip plane, i.e., tangential stresses; stresses normal to the slip plane may either assist or impede plastic deformation. Slip begins only when a shear stress of a definite magnitude is reached in the plane along which slip is occurring. The critical shear stresses are small, and for most commercially pure metals are of the order of tenths of a kilogram per square millimeter [85]. Slip, having commenced in some plane, is very soon retarded and is then transferred to other planes, an increase in strength occurs and for reproducing fresh acts of deformation, the value of the shear stress must be greater than the original value. The mechanism of slip is not yet completely clear. On the basis that small shear stresses in a metal single crystal are sufficient to produce slip, it may be assumed that at any given moment only a small portion of the crystal is involved in the slip process. For slip to occur throughout the entire crystal simultaneously, stresses are necessary which are hundreds and thousands of times greater. Thus, plastic deformation by slip does not amount to slip in the ordinary meaning of this word, but represents a "stepped" movement of one part of the crystal relative to another part by gradual transposition of atoms in layers having maximum packing density. On the basis of the concepts of the dislocation theory, plastic deformation occurs by multiplication and movement of dislocations. Plastic deformation can occur only in certain crystallographic planes (those most closely packed with atoms) and directions. The presence of a large number of slip planes is the main prerequisite for good plasticity at ordinary and low temperatures.

The typically metal crystal lattices are the face-centered cubic (fcc), hexagonal closely packed (hcp), and body-centered cubic (bcc).

In metals having a fcc lattice, slip occurs along the octahedral $\{111\}$ planes, in the [110] direction. Since there are three such planes and four directions, the total number of possible systems is 12 [86]. The following metals crystallize in such a lattice: copper, silver, gold, aluminum, nickel, platinum, palladium, iridium, $\gamma$-iron, high-temperature modifications of thallium, cobalt, rhodium, plutonium, and the polymorphic rare-earth metals – $\beta$-lanthanum, $\gamma$-caesium [87]. Metals with a fcc lattice have high plasticity at all temperatures owing to the large number of slip systems.

Metals having a hexagonal closely packed lattice include metals of group two of the periodic system (beryllium, magnesium, zinc, cadmium), rhenium, technicium, osmium, ruthenium, the rare-earth metals, dysprosium, holmium, erbium, thulium, lutecium, as well as the low-temperature modifications of the following polymorphic metals: alkaline earth metals ($\beta$-calcium, $\beta$-strontium), rare-earth metals ($\alpha$-scandium, $\alpha$-yttrium, $\alpha$-lanthanum, $\beta$-cerium, $\alpha$-praseodymium, $\alpha$-neodymium, $\alpha$-gadolinium, $\alpha$-ytterbium), and group IVa metals ($\alpha$-titanium, $\alpha$-zirconium, $\alpha$-hafnium, $\alpha$-thallium, $\alpha$-cobalt) [87].

In hexagonal metals, depending on the nature and temperature of the metal, slip is possible in three planes: base, prism, and pyramid planes. At room temperature, slip along the (0001) base plane is typical (metals of group II of the periodic system). Since, in this plane, there are only three closely packed directions (sides of the prism), only three possible methods of slip may be effective. Owing to the difficulty of deformation by shear in hexagonal metals, the ability to twin along closely packed planes is strongly developed in hexagonal metals, the characteristic crackling sound being audible at the same time. This was discussed at greater length in Chapter V.

The prism plane has only one closely packed direction, but there are three such planes, and consequently there are as many slip methods. The same may be said of the pyramid plane (six planes, one direction, six slip systems). Therefore, the plasticity of metals having a hexagonal, closely packed lattice depends on the number of slip systems involved in the deformation; if few systems are involved, the metal is deformed with difficulty, and if sufficient systems are involved, the metal will not be inferior in plasticity to fcc metals. The number of slip systems involved in deformation increases with temperature (magnesium, zinc) and also with increase in purity as regards interstitial impurities, which in turn is associated with the change in the ratio c/a of the axes in the hexagonal lattice [85].

Therefore, hexagonal metals in a particularly pure state are highly plastic ($\alpha$-thallium, $\alpha$-titanium, $\alpha$-zirconium, $\alpha$-hafnium, $\alpha$-yttrium, $\alpha$-scandium, etc.).

The following metals have a body-centered cubic lattice: alkali metals (lithium, sodium, potassium, rubidium, cesium, as well as barium, europium); group Va and VIa refractory metals (vanadium, niobium, tantalum, chromium, molybdenum, tungsten); high-temperature modifications of polymorphic metals ($\gamma$-potassium, $\gamma$-strontium, $\beta$-titanium, $\beta$-zirconium, $\beta$-uranium, $\beta$-neptunium, $\delta$-manganese, $\varepsilon$-plutonium, $\beta$-ytterbium, $\beta$-thorium, and $\alpha$-iron).

The crystal structure of metals according to the summary by E. Savitskii [85, 133] are given in Table 55.

Forty-eight slip systems are assumed for the bcc lattice. These include 6 (110) planes of the rhombic dodecahedron with two slip directions in each plane, 12 (112) planes, and 24 (123) planes with one slip direction in each. It has been found that slip occurs in $\alpha$-iron along all three planes enumerated [86]. In view of the existence of 48 slip systems, bcc metals should potentially be very plastic. This is indeed the case if they are pure or are worked at high temperatures. The presence of interstitial impurities in metals of commercial purity, however, results in low plasticity and makes body-centered cubic metals cold-short. The more impure the metal, the higher is the transition temperature. As already pointed out in Chapter IV, body-centered cubic metals on cooling do not alter the type of intermetallic bond in the crystal structure. From the physical standpoint, therefore, there is no reason for the existence in their case of temperature zones of brittle state. Cold brittleness is a property of bcc metals of commercial purity only, and is due to the difficulties in producing such metals in the pure state, the presence of impurities, and also spoiling of the metal and primarily of the grain boundaries caused by physicochemical effects of the surrounding gaseous medium during tests and treatment. In the purification of metals, therefore, there are two

TABLE 55. Crystal Structures of Metals [85, 133]

| Face-centered cubic | Hexagonal close-packed | Body-centered cubic | Rhombohedral |
|---|---|---|---|
| Copper | Beryllium | Lithium | Mercury |
| Silver | Magnesium | Sodium | Bismuth |
| Gold | Zinc | Potassium | Antimony |
| Aluminum | Rhenium | Cesium | |
| Platinum | Technetium | Barium | |
| Palladium | Osmium | Europium | |
| Iridium | Ruthenium | Vanadium | |
| γ-Iron | Dysprosium | Niobium | |
| β-Thallium | Holmium | Tantalum | |
| β-Cobalt | Erbium | Chromium | |
| β-Rhodium | Thulium | Molybdenum | |
| γ-Plutonium | Lutecium | Tungsten | |
| β-Lanthanum | β-Calcium | γ-Calcium | |
| γ-Cerium | β-Strontium | γ-Strontium | |
| γ-Manganese | α-Scandium | β-Titanium | |
| | α-Yttrium | β-Zirconium | |
| | α-Lanthanum | β-Hafnium | |
| | β-Cerium | γ-Uranium | |
| | α-Praseodymium | γ-Neptunium | |
| | α-Neodymium | δ-Manganese | |
| | α-Gadolinium | ε-Plutonium | |
| | α-Terbium | γ-Lanthanum | |
| | α-Ytterbium | δ-Cerium | |
| | α-Titanium | β-Praseodymium | |
| | α-Zirconium | β-Neodymium | |
| | α-Hafnium | β-Samarium | |
| | α-Thallium | β-Ytterbium | |
| | α-Cobalt | β-Thorium | |
| | | α-Iron | |

fundamental problems. The first is to make the metal pure and the second is to preserve its purity. Impurities introduced in the process of alloying, treatment, or service can easily nullify any effect resulting from the use of an initial metal of high purity [87].

Table 55 also presents data concerning metals having a rhombohedral lattice. It is here possible to include semiconductors having diamond lattice (germanium, silicon, α-tin, boron, carbon in diamond form). Strong covalent interatomic bonds and the difficulty of movement of dislocations situated in (111) planes result in low-temperature brittleness of these substances. Materials having a diamond lattice can only be deformed plastically when heated to $0.8\,T_{mp}$ or above, and at a low rate of deformation.

The mechanisms of slip and twinning are examined for a number of metals in [88]. As already pointed out, slip and twinning usually occur along the closest-packed planes. With increase in temperature other less closely packed atoms of a plane may be involved in slip. Thus, for example, in single crystals of aluminum, copper, gold, silver, and lead up to 450°C, slip occurs along (111) planes of the octahedron, and from 450°C slip also proceeds along the (100) planes of the cube. Deformation of tungsten single crystals at 77 and 20°K occurs by slip along the (011) planes and by twinning along the (112) planes, and at room temperature along the (110) and (112) planes.

In single crystals of hexagonal metals – magnesium, zinc, cadmium – slip occurs along the (0001) basal plane and, commencing from 225°C, there is also slip along the $(10\bar{1}1)$ and $(10\bar{1}2)$ planes in addition. The characteristics of the deformation mechanism in the refractory hexagonal metals rhenium and titanium was considered in Chapter V.

The following may be included in the fundamental factors distinguishing the plastic deformation of a polycrystal from a single crystal: much smaller grain size, influence of some grains on the deformation of other (neighboring) grains, which produces inhomogeneity of the stressed state in the body of the metal, and also determines the special part played by grain boundaries. The orientation of the grains of a polycrystal has a considerable bearing on the course of the deformation process.

The process of the plastic change in shape on the deformation of a polycrystalline material with full strengthening is characterized by the following phenomena: change in shape and orientation of the grains, texture formation, accumulation of bound potential energy (elastic stresses), and intracrystalline and intercrystalline damage. At the same time, an irreversible change in shape of a polycrystal may occur by slip produced in crystals along certain crystallographic planes and directions, twinning of crystals, directional diffusion of atoms at high temperature, or displacement of some crystals relative to others [81]. Sometimes all these deformation mechanisms are operative simultaneously. Slip, however, is the principle deformation mechanism of polycrystals. Slip in the grains of polycrystals obeys the same laws as does slip in single crystals. It was recently reported [89] that when single-crystal tungsten is rolled, the ingot elongates without widening, whereas widening is observed when a polycrystalline ingot is rolled, due to the action of several slip systems. The displacement of some grains relative to others results in the occurrence of intercrystalline cracks and fracture of the body before such displacements attain any considerable magnitude. In deformation with recrystallization, the intercrystalline disturbances caused by the displacement of crystallites relative to each other are less dangerous as regards the strength of the polycrystalline aggregate (healing of microcracks).

In plastic deformation under the effect of a field of force and temperature, there is a continuous change in the structure and physicochemical properties of the material.

The approach, described in the foregoing, to the study of the processes of plastic deformation may be termed the physicochemical approach as distinct from the purely physical or mechanicomathematical approach. Reference must also be made to the line of research on the mechanism of the plastic deformation of materials by the action of surface-active substances developed by the school of Academician P. A. Rebinder [90].

The process of plastic deformation is accompanied by the phenomenon of strengthening (hardening), in which an increase in the degree of deformation increases the resistance to deformation, in particular there is an increase in the yield point. In single crystals, the resistance may be increased by a factor of several tens. The strengthening intensity of the material is estimated according to the expression [81]:

$$U = \frac{\tau_{fc} - \tau_{ic}}{a},$$

where U is the intensity of strengthening (coefficient of strengthening, $\tau_{ic}$ is the initial critical shear stress, $\tau_{fc}$ is the final critical shear stress (on fracture), and $a$ is the crystallographic shear produced during elongation of the single crystals. Strengthening is often estimated by the increase in hardness.

The physical pattern of work hardening consists in the distortion of the lattice, reduction in size of the crystals, and the fixing of their orientation in relation to the direction of the deforming force (texture).

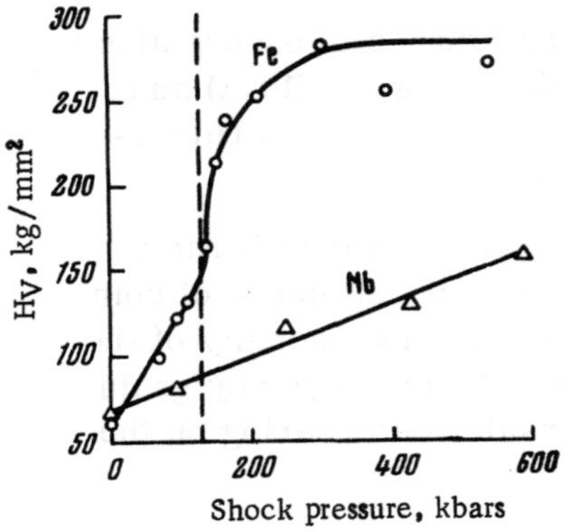


Fig. 149. Variation in hardness of iron and niobium as a function of shock pressure.

Strengthening may be accompanied by stress relief owing to recovery and recrystallization processes. The effect of strengthening at a given deformation temperature is determined by the ratio of the rates of strengthening and stress relief. A number of hypotheses have been advanced to explain the nature of strengthening. Reference may be made to the pinning and slip interference hypothesis [91], the interlocking and lattice distortion hypothesis [92-95], the hypothesis concerning a change in the type of chemical bond in the deformation process [96], and, finally, dislocation concepts of the mechanism of strain hardening (99-103), so fruitfully developed in recent years. According to dislocation concepts [99-103], strain hardening or work hardening is the result of interaction between dislocations or the interaction of the latter with interstitial impurities.

In connection with the development of high-speed methods of the shaping of metals and alloys, increasing attention is being paid to the study of the action of shock or explosive stressing on the structure and mechanical properties of the treated metallic materials, and to research on the effects associated with high-speed shaping. A detailed review and analysis of investigations on the strengthening (hardening) effect of shock stressing has been given by G. E. Ditter [106]. Figure 149 shows the variation in hardness of iron and niobium for different values of the shock pressure [106]. For both metals, twin density and hardness increase with increase in the shock pressure, but while the relationship for niobium is linear, for iron at a pressure of 130 kilobars, the hardness increases at a considerable rate. The maximum increase in the hardness of iron occurs in the range 130-220 kilobars, after which the hardness remains approximately constant. Some authors attribute the sharp increase in hardness of iron in the above-mentioned pressure range to the existence of a polymorphic transformation. According to Smith [107], the microstructure of iron, produced at pressures above 130 kilobars, resembles martensite formed when iron is quenched rapidly from the $\gamma$-region. Smith expresses the opinion that, in shock stressing, $\alpha$-iron is transformed into $\gamma$-iron, then on stress relief to the critical pressure of 160 kilobars the reverse transition $\gamma$-iron $\rightarrow$ $\alpha$-iron occurs and a twinning process takes place at the same time. Table 56 gives some values of the strength of bcc metals obtained in shock stressing.

As will be seen from the table monomorphic tantalum and niobium undergo approximately the same degree of hardening under shock stressing and mechanical twins only are found in their microstructure. Molybdenum under shock-stressing conditions behaves like tantalum and niobium [108].

The influence of high-speed deformation on the structure and mechanical properties of molybdenum are also studied in [109]. It was shown that deformation by explosion promotes the production of a large number of point defects in the metal. The high concentration of point defects (in the view of the authors of this article) is the cause of a considerable intensification of strain-aging effect in metals and alloys subjected to deformation by an explosion wave.

There is information available on the production of an additional slip system which is not found in ordinary methods of deformation. According to [110], the intensification of the strain-aging effect and twinning process in explosive deformation facilitate the development of brittle fracture. In addition, the considerable heating effect, the occurrence of additional slip planes and also the triaxial compression produced in explosive stressing help to increase the deformability of the material. Bearing in mind that twinning is a characteristic mechanism

TABLE 56. Mechanical Properties Obtained in bcc Metals in Explosive Machining [106]

| Metal | Treatment | Hv for a load of 1 kg | Yield point, $kg/mm^2$ | Tensile stress, $kg/mm^2$ | Elongation, % |
|---|---|---|---|---|---|
| Iron | Annealed. . . . . . . . | 64 | 13.3 | 21.8 | 68 |
| | Pressure 95 kbars . . . | 121 | 21.8 | 25.3 | 17 |
| | 220 kbars . . . | 250 | 76.5 | 86.6 | 16 |
| Tantalum | Annealed . . . . . . | 160 | 43.6 | 50.6 | 10 |
| | Pressure 275 kbars . . . | 260 | 81.0 | 88.0 | 7 |
| Niobium | Annealed . . . . . | 68 | 18.3 | 31.0 | 25 |
| | Pressure 250 kbars . . . | 115 | 33.1 | 35.8 | 5 |

of deformation for all bcc metals, it may be assumed that an increase in temperature will assist in reducing the tendency of a metal to twinning in high-speed deformation for most metals having a bcc lattice. The examples of molybdenum [109] and steel [111, 112] have shown that preliminary work hardening also reduces the tendency of a metal to twinning under shock stressing.

The view that twinning is one of the causes of brittle fracture is supported by the fact that conditions favorable to twinning resemble the conditions which promote brittle fracture [113]. Many investigators consider the presence of twins in material that has been subjected to brittle fracture to be indirect evidence of the connection between brittle fracture and twinning. In a number of cases, cracks are initiated at the intersection of twinning layers [114]. However, many investigators consider that twinning does not affect the cold brittleness of bcc metals. Without dwelling on a discussion of this work, we would remark that the last-mentioned view is convincingly supported by the high ductility of tungsten–rhenium and molybdenum–rhenium alloys which are deformed both by slip and by twinning; in this case, twinning considerably facilitates the process of plastic deformation.

An analysis of published data shows that in a number of cases high-speed deformation may be an indispensable method of producing semifinished products from refractory metals and alloys having an optimum combination of strength and ductility.

An important practical question, which must be solved in the problem of refractory metals, is the production of semifinished products and parts by the methods of plastic deformation. Most semifinished products of tungsten and molybdenum are produced in all countries by cermet methods and by hot deformation with heating in a hydrogen atmosphere, which does not ensure the necessary purity and ductility of the metals. The production of semifinished products of cast tungsten, molybdenum, and chromium still presents considerable difficulties. In the case of refractory metals of the platinum group, it has not yet been possible to produce osmium and ruthenium in the form of suitable semifinished products; there is a report [116] of the hot forging of sintered ruthenium at 1400-1800°C, which was then rolled into strip at 1800°C, but any attempt to deform osmium failed. The deformation of rhodium and iridium gives rise to considerable difficulties. However, in a state of high purity, after electron-beam zone refining, ruthenium, rhodium, and iridium withstand a considerable degree of cold deformation.

To date, the problem of the production of semifinished products by plastic deformation of vacuum-melted niobium, vanadium, and tantalum has been successfully solved. Electron-beam vacuum melting ensures the production of these metals in a state of high purity and ductility; they are deformed with comparative ease even without heating. The treatment of a number of high alloys based on these metals already presents considerable difficulties owing to the sharp reduction in ductility when these metals are alloyed with other refractory metals. The behavior of niobium, vanadium, tantalum, and their alloys on deformation is largely determined by their interstitial impurity content (hydrogen, oxygen, nitrogen, carbon).

The cold brittleness of tungsten, molybdenum, and chromium of commercial purity necessitates their deformation at elevated temperatures, during which, however, they are contaminated by interstitial impurities, catastrophically affecting the mechanical properties and deformability of these metals. The surface of molybdenum and tungsten sheets, after hot rolling, is subjected to grinding and etching to remove the contaminated layer.

Although the production of semifinished products of cast refractory metals is rapidly developing, this process is still exceptionally laborious and results in a high loss of metal (up to 25%). The severe demands of modern technology with regard to mechanical properties and dimensional tolerances considerably complicates the problem of their production. Intensive research is in progress with the object of increasing the ductility of refractory metals (deoxidation, modification, single-crystal production) and reducing the temperature at which they may be worked. The influence of vibration on a deformed metal is being studied [195]; it is found that this treatment gives a uniform, fine-grain structure. Work is being done on the hot deformation of refractory metals in a vacuum or an atmosphere of inert gas, on the development of new technological systems of working refractory metals by pressure, etc. There are reports of industrial workshops for the pressure working of refractory metals by pressure in an argon atmosphere, where the attendants work in pressurized suits in underground workshops with an artificial climate [117]. At certain scientific conferences, the questions under discussion include the future possibility of melting and working chemically active metals on the moon's surface.

## 1. Choice of Optimum Conditions for Working by Pressure

S. I. Gubkin and E. M. Savitskii [85, 118, 119] formulated the principles for the choice of the conditions of the hot deformation of metallic materials, which also apply to refractory metals and alloys. The conditions of working by pressure are determined by the influence of temperature, the rate and degree of deformation, and the system of stressed condition on the structure and mechanical properties of the worked metal. The principal information on the influence of these factors may be gathered from phase diagrams, ductility, and recrystallization [85]. The phase diagrams indicate the permissible working temperature range. The upper limit of the working temperature range is determined by the temperature situated in the limits of 0.9-0.95 of the solidus temperature. It is essential to avoid increasing the working temperature to a region where the liquid phase may appear, since this results in rapid intergranular fracture. The lower working temperature limit is usually bounded by the solubility line in the solid state (Fig. 150). The aim is to conduct working by pressure in a one-phase region, since the occurrence of a fresh phase results in nonuniform deformation and additional stresses. The temperature curves of different mechanical properties (elongation, impact strength, reduction in area, angle or bend, tensile strength and yield point, hardness, specific pressure, etc.) make it possible to define more precisely the temperature range of the working of metals and alloys by pressure [85]. Unfortunately, such data are not yet available in sufficient amounts for refractory metals and particularly alloys. A knowledge of all these diagrams makes it possible to select for the deformation a temperature region in which, for given metals and alloys, minimum resistance to deformation is combined with good plasticity.

An estimate of the resistance to deformation of molybdenum and some of its alloys may be made from the temperature dependence of hardness, yield point, and strength [120]. Considerable softening of tungsten and molybdenum is found at 1600-2000°C. However, the high

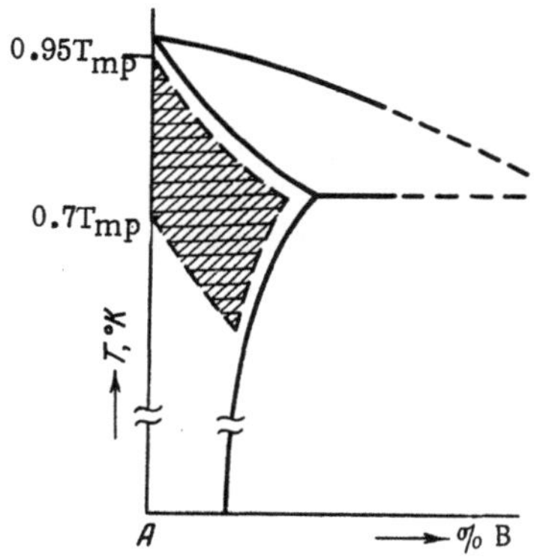


Fig. 150. Temperature range of the deformability of an alloy.

activity of molybdenum, tungsten, and other refractory metals with respect to gases, and the lack of suitable heating equipment compel a reduction in their working temperature, although this means a considerable increase in the specific pressures in the deformation process.

For establishing the optimum deformation and annealing conditions, a knowledge of the recrystallization diagrams is very important, since recrystallization is accompanied by a sharp change in the mechanical properties. Above the recrystallization temperature (hot deformation mechanism) metals and alloys are comparatively easily hot worked by pressure (rolling, forging, extrusion), and it is possible to produce from them semifinished products with considerable reduction for relatively small efforts.

A knowledge of the recrystallization diagrams also makes it possible to work out the optimum annealing conditions of semifinished products for the purpose of obtaining the necessary grain sizes and requisite properties. The recrystallization temperature is one of the characteristic forces of the interatomic bond in metals and alloys. Refractory metals and alloys are characterized by the highest values of the temperature of the commencement of recrystallization. The principal recrystallization diagrams of refractory metals were discussed in Chapter II. The recrystallization diagrams of refractory metals of groups IVa, Va, VIa, and rhenium were constructed by E. M. Savitskii and co-workers in the period 1955-1962 [121-131].

The plasticity rule of high-temperature modifications established in [85, 119, 132] indicates directly the optimum temperature ranges of the technology of hot working by pressure of polymorphic metals and alloys. It is obvious that, other conditions being the same, working is carried out more advantageously at the temperatures at which the metal is plastic, i.e., the range of existence of the most plastic high-temperature modification [132]. An increase in plasticity of the metal in a polymorphic transformation is particularly effective in cases where the low-temperature form has a complex crystal structure. Three typical examples may be cited. $\alpha$-Manganese, stable below 705°C, has a complex crystal structure with 58 atoms in the unit cell, and evidently has a mixed interatomic bond; it is therefore hard and brittle; on the other hand, $\gamma$-manganese existing at high temperatures has a bcc lattice and is quite plastic. The same occurs with uranium. The low-temperature $\alpha$-modification of uranium with a complex orthorhombic lattice is semibrittle, while the high-temperature cubic $\gamma$-modification in its capacity to change its shape is reminiscent of a dough-like substance [134]. Finally, ductile white tin, familiar to everybody, is its high-temperature form, while the gray modification of tin formed below 18°C as the result of "tin pest" does not belong to metals but to semiconductors and has the form of a solid, brittle powder. This is a striking example of the influence of the type of interatomic bond on the physical properties of metals [133].

The plasticity rule of high-temperature modifications has helped considerably in establishing a rational technology for the hot working by pressure of semifinished products of the polymorphic metals and alloys necessary for the new technology. The most refractory polymorphic metal, hafnium (melting point 2222 ± 30°C, temperature of polymorphic transformation 1760 ± 35°C), is readily worked by pressure in the temperature range of the existence of the bcc lattice. This applies wholly also to two other polymorphic refractory metals, titanium and zirconium.

## 2. Extrusion

The characteristics of the cast structure and brittle—ductile behavior of refractory metals and alloys introduce important characteristics in the technology of working them by pressure. The preliminary working of tungsten and molybdenum ingots is usually carried out by hot extrusion, by pressing them through a die, i.e., according to a system of nonuniform hydrostatic compression. A system of hydrostatic compression, by increasing the difficulty of intercrystalline shear and crack development, lends maximum assistance to the plastic deformation of the substance. Preliminary working is carried out not only with the object of imparting a definite shape to the metal but also of refining the cast structure of tungsten, molybdenum, and other refractory metals and their alloys. As previously mentioned, the area of intergranular surface per unit volume is an important factor in influencing the plasticity of refractory metals having a bcc lattice.

In working refractory metals by pressure, it is very important to select the temperature range of the treatment. The upper limit of the temperature range of working refractory metals is limited by the potentialities of the equipment for heating the ingots and (this is very important) also by the high-chemical reactivity of refractory metals with regard to interstitial impurities at high temperatures. Working is carried out at temperatures which do not ensure completion of recrystallization during the treatment. In this sense, the term "hot" working as applied to refractory metals is arbitrary. Working is carried out by a "hot-like" deformation mechanism, or by a mechanism of "mild" hot working when, under the deformation conditions, only a recovery (relief of elastic stresses) but not recrystallization takes place [85]. The lower limit of the temperature range is set by the brittleness threshold of the refractory metals and the power of the equipment.

For cast tungsten of commercial purity, the heating temperature during extrusion is 1500-2000°C; extrusion is completed at 800-1000°C. For molybdenum, the maximum extrusion temperature is 1300°C. Transition to higher temperatures calls for a protective atmosphere. Treatment at temperatures below the limits mentioned reduces contamination by interstitial impurities, but assists cracking and lamination during subsequent treatment. Extrusion at low temperatures results in a nonuniform cross-sectional stress distribution; extrusion temperatures are therefore selected at which the rate of the relaxation process is comparable with the rate of deformation. The ingots are usually extruded into rods or billets in slab form for further hot rolling to sheet (attempts to produce ingots in slab form in a three-electrode furnace without extrusion are described in [135]). The starting billets for the production of tubes from cast tungsten and molybdenum may be tubular centrifugal coatings.

Extrusion is a most important preparatory operation in the pressure working of tungsten and molybdenum and serves to reduce liquation, refines the coarse cast structure, and imparts a suitable shape to the ingot for its subsequent retreatment. As already pointed out, arc-melted or electron-beam vacuum-melted tungsten and molybdenum have a coarse-grained structure, and the grain boundaries of these metals are the sites of crack inception in strain under tensile stresses. The destruction and refinement of the cast structure are most easily achieved by hot extrusion through a die, i.e., under conditions of nonuniform hydrostatic pressure. The greater is the reduction per pass, the more thorough is the removal of the cast structure, the greater is the grain refinement and the more uniform are the mechanical properties over the cross section of the rod. With increase in the degree of reduction, however, the specific pressure of the metal on the die increases. The maximum degree of reduction is therefore limited by the power of the equipment and the strength of the extrusion die. Extrusion of tungsten in a 750-ton press with a reduction of 87% and a rate of deformation of 0.38 m/sec are reported in [136]. The ingots were heated in an induction furnace to 1760°-2200°C in an inert atmosphere and were rapidly transferred to a container previously heated to 300°C.

Before extrusion, the ingots are first stripped. For reducing the wear on the die, extrusion is carried out with lubrication. The use of steel shells as protective and lubricating material has been found to be very effective. Coating the inner surface of the die with a layer of $Al_2O_3$ has been used for wear reduction [120, 136]; the temperature resistance of the die was increased to 1880°C. Dies with ceramic coatings ($Al_2O_3$) gave up to six extrusions [137]; the use of such dies diminished the resistance to extrusion and increased the output of sound material; in addition, ceramic coating of the dies resulted in the production of rods having a higher surface quality than high-temperature extrusion in steel dies. It is reported that some success was also obtained by the use of water-cooled dies [135].

The extrusion of tungsten alloys gives rise to stil greater difficulties than that of tungsten. Extrusion of high alloys demands heating to high temperatures (above 2000°C), with a die capable of withstanding the greater specific pressures occurring during extrusion. It is reported [136] that an alloy of tungsten with 11% Ta could not be extruded even at 2000°C and with the maximum pressure 127 $kg/mm^2$ attainable in the press. An alloy of tungsten with 30% Mo is one of the most difficult alloys to deform [137].

The hot extrusion of molybdenum and some of its alloys does not give rise to any particular difficulties. In the extrusion of an alloy of molybdenum with 1.25% Ti, 0.3% Zr, and 0.15% C, the maximum percentage of sound material was obtained for rods with an elongation of 4:1 at a temperature of 1427°C [137]. This article also reports the successful extrusion of an alloy of molybdenum with 25% W and 0.1% Zr at temperatures from 1093° to 1760°C on an elongation of 8:1.

Niobium alloys may be extruded or forged at temperatures of 1000-800°C [140]. Pressure working of pure niobium does not encounter any difficulties. Niobium, as well as vanadium and tantalum, are easily purified in the electron-beam remelting process and become highly ductile metals which may be worked cold, being capable of withstanding considerable deformation without fracture. Ingots of vanadium of commercial purity, up to 150 mm in diameter, produced by electric-arc vacuum melting may be forged into rods of a diameter of up to 10 mm and bars up to 10 mm in thickness. The working temperature is 980-1200°C. Semifinished products of smaller cross sections may be obtained at working temperatures of 370-595°C [141]. This is due to the rapid oxidizability of vanadium above 600°C.

The treatment of niobium ingots in the first stage consists in the destruction of the cast structure by extrusion or forging. Ingots having a very low content of interstitial impurities may be worked cold. The total nitrogen, oxygen, carbon, and hydrogen content should not exceed hundredths of one percent (the exact limit content has not been established). After destruction of the cast structure, the hot-deformed niobium is subjected to roughing for removing the surface layers of high interstitial impurity content, and is vacuum annealed to produce a recrystallized structure. The oxidized layer may be removed by pickling with a mixture of hydrofluoric and nitric acids (25% HF + 75% $HNO_3$). Vacuum annealing of the ingots is carried out at 1000-1400°C. Heat treatment in this temperature range ensures recrystallization and the production of ductile material capable of cold deformation. Further working of the niobium (rolling or drawing) is carried out in the cold. Cast niobium of commercial purity is worked in the temperature range 1000-650°C; short-term heating of the ingots before working may be carried out in the air. There are reports of the heating of niobium alloys in helium and argon atmospheres, in molten glass, molten salts, in a steel shell [142]. For niobium alloys also, hot extrusion is the most promising method of working for the destruction of the cast structure. This refers more particularly to highly alloyed niobium alloys. Molten glass may serve as a good lubricant in extrusion. Some niobium alloys (niobium–titanium, niobium–tantalum systems) have high ductility and in a highly pure state with regard to interstitial impurities are worked approximately under the same technological conditions as pure niobium.

However, a number of highly alloyed niobium alloys containing tungsten, molybdenum, and other refractory metals require the application of special working conditions for the production of semifinished products from them. For the working of these alloys by pressure, hot extrusion, forging, explosive forming, extrusion in a shell, rolling with grooved rolls, etc., are used. For the destruction of the cast structure of high-strength niobium alloys, they are worked in a high temperature range (1300-1100°C). Some highly alloyed niobium alloys, particularly those containing tungsten, require heating to 1650°C and above before extrusion. To avoid contamination during hot working, an inert atmosphere, protective shells, coatings, etc., are used [140].

An important factor in the extrusion of refractory metals and alloys is the rate of deformation. At high extrusion rates, the metal has no time to cool appreciably and lose its plasticity. At high deformation rates, there is also additional heating of the deformed body through the heat effect of deformation. The effect of deformation rate on plasticity is examined in [85, 118]. In deformation at extremely high rates, the metal is able to withstand greater deformation than at ordinary deformation rates. According to [143], at high temperatures, the plasticity of tungsten increases with the temperature.

High-speed presses (23 tons/min) are employed in the USA for extruding refractory metals and alloys. For primary reduction of niobium alloys, presses are used in which the plunger is driven by a hydraulic system, a flywheel, or compressed gas; in the last-mentioned case, the rate of extrusion may reach 25,400 mm/min [142].

One of the problems encountered in high-speed extrusion is the wear on the dies. The degree of wear depends on the design and material of the die, and also on the lubricant. Molten glass is used as lubricant for the extrusion of molybdenum billets. As already pointed out, tungsten and tungsten alloys are often extruded in steel shells of low-carbon steel which acts as a good lubricant.

Considerable interest attaches to experiments [136] on the extrusion of tungsten ingots by explosion (rate of extrusion 50.8 m/sec). The wear on the die was not high despite the fact that extrusion was carried out without lubricant; the reduction was about 98%.

## 3. Forging

Extrusion does not exclude other forms of the preliminary working of refractory metals, for example forging. Chromium ingots, after annealing for the relief of internal stresses (at a temperature of 1200°C) are subjected to forging, the ingots being contained in a steel shell to reduce heat loss. Forging is carried out at 800-850°C and moderate reductions for refining the cast structure. Subsequently, forging of chromium is carried out at considerable reductions and a temperature of 500°C. Chromium rolling is carried out at approximately the same temperatures and reductions. In forging, as also in rolling chromium ingots, intermediate recrystallization annealing is carried out for precluding the production of a coarse-grain structure. It is pointed out in [139] that successful forging of remelted chromium is determined to a considerable extent by the deformation rate. The best results are obtained by upsetting the ingot to form a round or square rod under the press.

The forging of ingots of molybdenum and its alloys has a limited application, owing to the coarsely crystalline structure, sharply reducing the plasticity of the cast metal, and the high oxidizability of molybdenum at elevated temperatures. Forging is carried out in steam-air, pneumatic, and also hydraulic forging presses at 1400°C (for pure molybdenum); forging is completed at 900-1000°C [201]. For molybdenum alloys, the heating temperature for forging may attain 1600-1800°C. More often, preliminarily deformed molybdenum is subjected to forging. In the free forging of ingots of molybdenum and its alloys into discs, the method of upsetting in steel or nickel shells is used, with separate heating of the ingot to 1400-1800°C and the shell to 500-1000°C [201]. In wire production, forging of molybdenum rods in rotary forging machines is widely used.

Tungsten and its alloys, as a rule, are subjected to forging in the preliminarily deformed state. As for molybdenum, forging of tungsten to rods in rotary forging machines at temperatures of 1200-1600°C is extensively used.

As already pointed out, niobium alloys may be forged at 1000-800°C, while electron-beam melted niobium may be forged at room temperature.

Tantalum ingots are forged at room temperature without preliminary annealing. Alloys of tantalum and tungsten require heating to 1100-1500°C for forging.

## 4. Rolling

Following the destruction of the cast structure, subsequent working of deformed ingots of refractory metals is carried out by rolling, wire-drawing, pressing, rotary forging, or further extrusion.

After extrusion, the initial operations of rolling tungsten and molybdenum are conducted at 1200-1300°C [135]. Rolling is carried out on two-high mills. With the reduction in cross section of the metal the heating temperature before rolling is reduced. Tungsten and molybdenum sheets of 1-2 mm thickness are rolled at 700-800°C. The maximum reductions are made in the initial rolling stages (20-40% per pass) for working the full cross section of the billet. The initial rolling stages are essentially the continuation of the primary working stage. Their object is to refine the grain still further. In addition, in these stages, the width of the billet is regulated by cross rolling.

Tungsten and molybdenum sheets of a thickness of less than 1 mm are produced by pack rolling in a duo mill or by rolling individual sheets in a four-high mill. Rolling of molybdenum sheets in a four-high mill having working rolls 305 mm in diameter (barrel width 762 mm) and backing rolls 660 mm in diameter (barrel width also 762 mm) is reported. The rolling mill is equipped with an electrically driven screwdown [137].

The gradual reduction in rolling temperature promotes the production of a fibrous structure, which reduces the ductile–brittleness transition temperature and increases the ductility of tungsten and molybdenum. The favorable effect of work hardening is evidently due to destruction of the brittle films on the grain boundaries. The presence of a deformed structure with elongated grains is particularly important in group VIa metals, since in these metals of commercial purity the interstitial impurity content exceeds the solubility limit. In group Va and VIa metals, the presence of deformation texture enables the permitted quantities of interstitial impurities to be higher.

Molybdenum sheets of a thickness of less than 0.5 mm may be worked cold. Surface contamination of the sheets occurs in the process of hot rolling and intermediate annealing. Interstitial impurities are dissolved in the surface layers and diffuse into the depth of the metal. The depth of penetration depends on the impurity content of the surrounding atmosphere, and also the holding temperature and time of the molybdenum sheets (for sheets of a thickness of 0.5-1.3 mm, the depth of the contaminated layer lies within the limits 0.025-0.075 mm). Surface contamination is detrimental to the mechanical properties of the sheets. The final cold working of molybdenum sheets is therefore carried out after descaling and annealing in a vacuum or inert atmosphere. Sheets of small thickness are even back-rolled. In some cases, rolling of molybdenum (and also tungsten) strip is done in a cluster mill with front and rear tension [116].

Fig. 151. Rolling refractory metals in an argon atmosphere.

Considerable progress has been made in recent years in the production of tungsten and molybdenum sheets. In 1964, eleven American firms produced sheets from 23 different alloys based on refractory metals. The molybdenum sheets were produced in a width of up to 1000 mm and a length of up to 3-5 m; according to [144], tungsten sheets are made in a width of 500 mm and a length of 1200 mm. It is reported that the technology of the production of tungsten sheet having the dimensions 914 × 2438 mm from electric-arc vacuum-melted ingots is under development.

Electric-arc vacuum-melted molybdenum may be rolled in roll passes into rods without previous extrusion. Rolling of molybdenum ingots 41.3 mm in diameter to square rods 25.4 × 25.4 mm is reported [145]. Rolling temperature is 1350°C, reduction per pass 30%. For recrystallization and the relief of internal stresses, the rods are subjected to annealing at 1350°C for 1 h. The mechanical properties of tungsten and molybdenum in the rolled state are largely determined by the working conditions.

Chromium rolls well in the 900-450°C range, the metal being rolled in a steel shell. The use of a combined coating of low-carbon and stainless steel is recommended [146], the stainless steel being removed mechanically from the combined coating and the low-carbon steel by pickling. As in the case of tungsten and molybdenum, rolling of chromium is carried out with high reductions. The microstructure of chromium after rolling is fine-grained and has characteristic, elongated grains. Annealing at 900-950°C leads to recrystallization.

Hot rolling of niobium alloys is sometimes carried out in a titanium shell [140].

The high resistance of refractory metals to deformation demands the use of powerful rolling equipment. The resistance to plastic deformation may be considerably reduced by heating to high temperatures. As already pointed out, however, the upper heating temperature is limited by the high chemical activity of refractory metals and their alloys.

Some firms in the USA are developing new furnace, rolling, and forging equipment for working refractory metals in a shop having an inert atmosphere [147]. Basically, the remote

control of this equipment is envisaged. There are other designs for enabling service personnel in space suits to work directly in an argon-filled chamber (Fig. 151) [148]. Entry of personnel into the chamber and assembly of the necessary equipment are effected through special air locks. The development of constructions of argon chambers for pressure working, heat treatment, and welding is a costly but promising direction in the production of refractory metals and alloys. Working in an inert medium provides the possibility of considerably raising the deformation temperature of refractory metals and alloys, thereby reducing the specific pressure on the tool, increasing the quality of the semifinished products, and increasing the output. The operations connected with the intermediate and final heat treatment of refractory metals and alloys in the process of their plastic deformation are simplified considerably. There is interesting information on the working of molybdenum and some alloys of molybdenum with niobium at a temperature of 2000°C [149]. The forge hammer was enclosed in a shell. The experiments showed that at this temperature cast molybdenum and alloys based on it become extremely yielding and plastic.

An experiment conducted by Jaffee (Battelle Institute, USA) is reported [150]. The laboratory rolling mill and the induction furnace for heating the billet were placed in a box filled with argon. The apparatus was controlled from the outside through special sleeves.

Alloying of tungsten and molybdenum with rhenium, as already mentioned, leads to an increase in their plasticity and workability. An alloy of molybdenum with 40 and 50 wt.% Re is easily deformed cold with 90% reduction without intermediate annealing. Alloys of the tungsten–rhenium system are readily deformed at temperatures of 1000-1200°C. The vacuum-melted alloys VR-27 VP and MR-47 VP can be rolled to strip and foil of micron thicknesses. For refining the coarsely crystalline structure of the material as cast, the cast metal is first forged hot followed by hot rolling and then cold rolling. Cold rolling of these alloys may be carried out without intermediate annealing [38, 151].

Sheet rolling of niobium and tantalum after extrusion, and production of tubes of these metals by drawing or rolling tube blanks produced by preliminary extrusion does not give rise to any considerable difficulty. This operation is carried out cold using specially selected lubricants. In the cold-working process, intermediate annealing (in a vacuum) is carried out for removing internal stresses and the production of fine grain. Alloys of the niobium–tantalum and niobium–titanium systems allow of 90-95% cold reduction followed by stress-relief annealing. After annealing, the sheet may be rolled to foil in a four-high mill. Sheet rolling of highly alloyed niobium alloys (for example, alloys of the niobium–tungsten–molybdenum–zirconium system ) presents considerable difficulty. The exceptional strength and relatively low plasticity of these alloys at room temperature necessitate the rolling of these alloys at high temperature in protective shells. The method of rolling in a shell is very promising for some alloys of refractory metals. There should be compatibility in the strength and plastic properties of the material of the shell and the rolled alloy, otherwise considerable tensile stresses are set up, leading to cracking and fracture of the alloy. For high-strength niobium alloys, stainless steel, titanium, or molybdenum may be used as shell material in rolling. The reduction per pass amounts to 20%. An example of the sequence of operations in working the alloy Nb–15%W–5%Mo–1%Zr is given in [142]:

1) Extrusion of a sheet billet 101.6 × 31.7 mm

2) Annealing for 1 h at 1550°C in argon

3) Cross rolling at 1370°C

4) Annealing for 1 h at 1300°C in argon

5) Rolling at 1200°C

6) Annealing for 1 h at 1370°C in argon

7) Rolling at 675°C to sheet 2.54 mm thick (at 675°C rolling was done without a shell)

8) Annealing for 1 h at 1200°C

9) Individual rolling of sheets at 260-450°C

10) Final annealing for 1 h at 1200°C in a vacuum.

It is very important to choose correctly the conditions of recrystallization annealing for producing the required grain size, and to control the temperature. Overheating and the production of a coarse grain lead to embrittlement of the alloy. Problems of production technology (including rolling) of niobium and its alloys are examined in the recent monograph by M. V. Mal'tsev, A. I. Baikov, and V. Ya. Solov'ev [140].

In rolling most refractory metals and alloys, a characteristic feature is the gradual reduction in rolling temperature with decrease in thickness of the sheet. This is necessary for the following reasons: With decrease in thickness of the sheet, it becomes more difficult to maintain the high temperature during the transfer of the sheet from furnace to rolling mill and in the rolling process itself; the reduction in temperature and the gradual transition to cold working results in the production of a fibrous structure and, perhaps what is most important, protects the metal from the harmful effect of interstitial impurities. The final operation of rolling sheets of refractory metals and alloys, which is often carried out cold or with slight heating, is the most responsible, owing to the high requirements which the sheet must satisfy with regard to surface quality and dimensions. For final rolling, it is necessary to use four-high or cluster mills of high rigidity with forward and return tension. Information is also available on the development of new equipment for regulating the unevenness of rolling by means of a special preliminary stressing device [135].

Research on development of methods for the vacuum rolling of refractory metals is of definite interest. Up till now, however, vacuum rolling has not been extensively used in industry. Problems concerning the rolling of refractory metals and bimetals in a vacuum and in an atmosphere of inert gases are discussed in [198]. It should be remembered that the conditions of friction between the worked metal and the tool in a vacuum differ considerably from the conditions of working in air [152]. An interesting but as yet insufficiently studied effect of the degasification of a metal during vacuum rolling was discovered by V. M. Amonenko et al. [196]. Positive results were obtained in the vacuum rolling of rhenium. Hot rolling of rhenium in air is impossible because of its hot shortness, and cold rolling produces very considerable work hardening of the rhenium, requiring frequent intermediate annealing. Hot vacuum rolling of rhenium eliminates frequent annealing for the relief of work hardening, and protects the metal from the formation of fusible oxides. At the present time, however, the commonest method of producing semifinished products from rhenium is cold rolling with frequent intermediate recrystallization annealing after a slight degree of deformation.

## 5. Wire Drawing

The bulk of wire made from refractory metals and alloys comes from tungsten and molybdenum produced by cermet methods. The technology of the production of wire from cermet tungsten and molybdenum is described in detail in [153]. We shall not, therefore, discuss this question at any length in the scope of this monograph. The production of wire from vacuum-melted refractory metals and alloys has only just commenced the development stage. It is here necessary to mention the production of wire from tungsten and molybdenum single crystals grown by electron-beam vacuum-zone refining. Due to the high purity of these materials, wire

drawing is carried out cold or with slight heating in the first reduction stages [154-156]. The tungsten and molybdenum single crystals are first forged in rotary forging machines into rods 2-3 mm in diameter, used as stock for wire drawing. The forging temperature of tungsten single crystals does not exceed 600°C; molybdenum single crystals are forged at a temperature not exceeding 300°C. The process of producing wire from molybdenum single crystals is carried out in the USSR on an industrial scale.

The Moscow refractory metals plant produces wire from the vacuum-melted alloys VR-27 VP and MR-47 VP. The high ductility of these alloys enables them to be cold-drawn. In cold drawing of wire from a cast billet of the alloy MP-47 VP, intermediate annealing was carried out after a reduction of the order of 60-70% [157]. The starting billets for the production of wire from tungsten–rhenium and molybdenum–rhenium alloys are produced by hot forging of ingots with a degree of reduction of 60%. The ingots are heated to 1400-1500°C in a hydrogen atmosphere.

The production of wire from superconducting alloys of the niobium–zirconium system has also been mastered. Ingots of niobium–zirconium alloys are first hot-forged or extruded at 1000-1500°C with a reduction of 60-70% (in a vacuum, inert gas, or protective shell of other metals). Before being worked further, the billet is subjected to an intermediate heat treatment, annealing for 30-60 min and quenching from 1200°C [158]. Then follows cold rolling in grooved rolls or forging in rotary forging machines to a diameter of 2-4 mm, followed by low-temperature annealing for 15-60 min and quenching at a temperature of 800°C. The resulting rods are cold-drawn with a slight reduction per pass (0.04-0.01 mm). Wire made by this method can be used for producing magnetic fields of a strength of up to 80,000 Oe at 4.2°K [158].

Aquadag is the lubricant usually employed in drawing wire from refractory metals and alloys. Recently, molybdenum disulfide or zirconium diboride have also been used for this purpose in cold drawing.

## 6. Stamping

According to S. I. Gubkin [81], stampability of a material is determined by its ductility and strengthening (hardening). The capacity of a metal for sheet stamping increases with increase in its uniform elongation and the ratio $\sigma_y/p_n$ where $\sigma_y$ is the yield point and $p_n$ is the true stress at the moment of the appearance of necking in plastic tension. In the case of deep drawing, a complex stressed condition is produced; the material is subjected simultaneously to the action of tensile, compressive, and bending stresses. As yet, there is no generally accepted criterion for the stamping capacity of a material. The Erichsen test defines the capacity of a material for linear elongation or stretching rather than its stampability. In individual cases, the stamping capacity of a material is preferably determined by a special technological test.

Deep drawing of refractory metal sheets is limited both by the dimensions and to some extent by the power of the equipment. The possibility of the deep drawing of tungsten and molybdenum in ordinary equipment is reported [160]. No detailed information has been published on the technological conditions of deep drawing. As a rule, stamping of most refractory metals and alloys is carried out hot at a temperature above the ductile–brittle transition point. Data are available on the deep drawing of caps 63 mm in depth and 102 mm in diameter from tungsten sheet at 1300-1400°C (wall thickness 1.5 mm). Currently, tungsten caps with a maximum diameter of 230 mm are being made.

Niobium and tantalum may be subjected to stamping, drawing, or forming into articles of complicated shape at room temperature, thus completely eliminating the possibility of oxygen

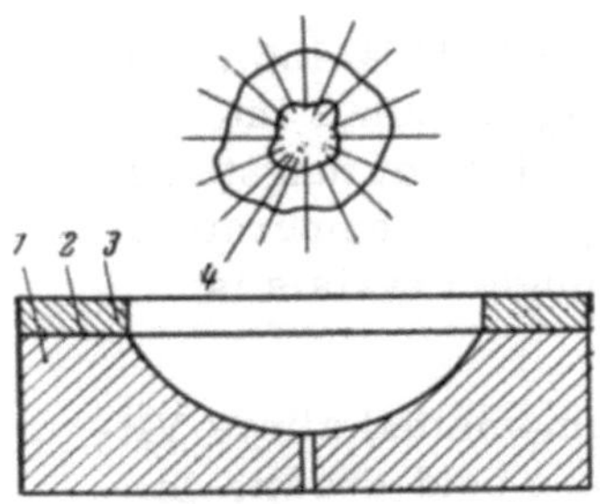


Fig. 152. Diagram of apparatus for shaping by explosion. 1) Die; 2) blank; 3) clamping plate; 4) explosive charge.

Fig. 153. Spun tungsten component.

embrittlement. Attention has been drawn to the need for the use of special lubricants in the production of articles from niobium and tantalum by stamping [142]. Stamping of high-strength niobium and tantalum alloys requires heating of the blank; the heating temperature should be approximately 500°C below the temperature used for destroying the primary structure of the ingot [142].

Another method of working, explosion forming [161], is of considerable interest for refractory metals and alloys. Research has shown that this technological process may be used successfully for the production of large-size components. The production of large-size components of some high-strength alloys by the method of ordinary pressing requires extremely high pressures which in practice cannot be attained in the most powerful presses. Explosion forming of a material imparts the necessary shape by the action of the pressure produced in the explosion of a chemical charge. The shock wave may be transmitted to the blank to be shaped through air, liquid, or some other transmitting medium. The energy of explosives is now being used for forming large-size articles from high-strength alloys. Figure 152 shows diagrammatically the apparatus used for high-explosive forming [161]. The principal advantage of this method is the simplicity of the equipment and the exceptionally high rate of deformation, whereby it may be possible to carry out very economical and productive machining of high-strength alloys, which cannot be worked by the ordinary methods. The powerful presses and bulky equipment for producing the necessary pressures in hydraulic forming are replaced in this case by a small quantity of explosive. Explosion forming of high-strength titanium alloys [161] and some niobium alloys [162] has been reported. Research conducted in the USSR and the USA has shown the effectiveness of the application of explosive power in the shaping of metals and alloys of high strength and low ductility, compared with the usual methods of machining. The disadvantages of explosion forming include the inapplicability under flow line industrial production conditions. The production of articles from refractory metal sheets by spinning (Fig. 153) has recently been adopted on an extensive scale in technology [147].

## 7. Tube Production

For tungsten and molybdenum tube, the starting materials are ingots (produced by centrifugal casting) or shells (turned on a lathe). Seamless and also welded tubes are made from niobium alloys by extrusion, the extrusion temperature being determined by the chemical composition of the alloys. For pure niobium and low alloys based on it, the extrusion temperature does not usually exceed 600-800°C.

The production of thin-walled, small-diameter tubes from niobium alloys and also tungsten and molybdenum is carried out by rolling on a mandrel and drawing. Internal stresses are relieved by intermediate vacuum annealing.

Tungsten and molybdenum welded tubes are not yet made industrially, owing to the embrittlement of these metals caused by welding; the production of welded tubes from tungsten–rhenium and molybdenum–rhenium alloys does not give rise to any particular difficulties. The use of roll welding in the production of welded molybdenum–rhenium tube is reported [38, 164]. Welding of niobium tube is carried out in an inert atmosphere. It is reported [163] that a special semiautomatic apparatus has been designed for arc welding of niobium tube, in which tubes having an external diameter of 610 mm and a wall thickness of 3.2 mm can be produced.

Hot working is used for the production of vanadium seamless tubes 50.8 mm in diameter and with a wall thickness of 6.36 mm. The production of vanadium tube with a coating of stainless steel is reported [141]. In the USSR and USA, seamless tube of tungsten–rhenium and molybdenum–rhenium alloy is produced. According to [164], in the production of seamless tungsten tubes, hollow blanks made by electric-spark machining of forged rod have been used. The tube blank produced in this way is forged on a molybdenum mandrel to an internal diameter of 3.17 mm and is then drawn to the required dimensions. According to the authors of [164], more economic methods are being developed in the USA for the production of seamless tubes from tungsten–rhenium alloy by extrusion.

### 8. Production of Manufactured Articles by Plasma Deposition

Articles of desired shape may also be produced by plasma deposition of metallic powders on a core of predetermined geometrical shape, which is subsequently removed. Articles so produced may if necessary be subjected to further mechanical and heat treatment. In the deposition process, both permanent and removable cores are used. Removal cores are made of graphite, copper, brass, and various fusible alloys. The removal of these cores is by chemical etching, fusion, or other methods. Permanent cores are made of brass or stainless steel. Plasma torch deposition may be used for producing articles of complicated configuration and relatively large dimensions. The possibility of using this method for the production of rocket nozzles and other articles of rocket engineering is reported [163]. General Motors has developed a process for the production of tungsten rings 305 mm in diameter and 44.5 mm thick by plasma deposition [165].

## Heat Treatment of Refractory Metals and Alloys

As is well known, the heat treatment of metals and alloys consists in imparting a change in structure and consequently in properties of alloys as the result of a predetermined thermal action. The nature of the structural charges taking place depends on the initial state of the alloy (chemical and phase composition) and the thermal conditions of the action (temperature and rate of heating or cooling, holding time at the given temperature, surrounding medium, etc.). The heat treatment of single-phase and multiphase alloys differs substantially.

In the case of single-phase alloys, heat treatment as a rule takes place after mechanical treatment (with the exception of ingot homogenization). As the result of heating previously deformed single-phase alloys, recovery, polygonization, and recrystallization occur. Polygonization is a special form of grain refinement to form subgrains [167]; it was discovered by S. T. Konobeevskii [168]. As the result of recovery, the internal stresses are relieved, while recrystallization leads to the removal of fibrous structure and the formation of the equiaxial structure existing before deformation. The anisotropy of the material is reduced at the same time. The physical basis of all these processes is the diffusion and self-diffusion of the atoms of the crystal lattice.

The main prerequisite for the choice of heat-treatment conditions of multiphase alloys is the constitutional or phase diagram. In the presence of phase transformations in alloys in the solid state, some phase is dissolved or a new phase appears on each transition through the phase transformation point. Phase transformation may be due either to ordinary diffusion or to crystallographically oriented displacement of atoms (for example, the martensitic transformation in steel). The annealing of multiphase alloys (full annealing) is used for changing the coarsely crystalline structure of cast alloys to a finely crystalline structure, removal of banded structure, stress relief before pressure treatment of alloys. The annealing temperature usually adopted is 30-50°C above the phase transformation point [169, 170]. The final structural state and the properties of the alloys are largely determined by the rate of cooling after the alloy has been held at the appropriate temperature. With a high cooling rate, the high-temperature state may be fully fixed or an intermediate state may be produced. Such a form of heat treatment (quench) results in a sharp change in physicomechanical properties. Classic examples of quenching to form supercooled solid solutions are the hardening of duralumінum and also of some sorts of steel (austenitic hardening). Intermediate structures are formed as the result of suppression of eutectoid transformation in the quenching of alloys based on polymorphic metals (martensite of carbon steel, $\omega$-phase in the quenching of some titanium alloys, etc.). Heating quenched alloys to temperatures below phase recrystallization, tempering, or aging leads to breakdown of the quenched or hardened state and to the formation of a thermodynamically more stable structure of the alloy. This process takes place in several stages and is accompanied by a considerable change in the properties of the alloy. Considerable hardening is found in the aging of alloys representing supersaturated solid solutions after quenching. As the final result, the degree of hardening of such alloys after aging is determined by the dispersion and by the character of the distribution and shape of the particles of the precipitated phase.

Heat treatment may be one of the effective means of controlling the structure and properties of alloys of refractory metals. However, the high quenching temperatures of refractory metals, the necessity for long-continued holding at these temperatures in a protective medium (inert gas, vacuum) create considerable difficulties and require the construction of special equipment for carrying out the heat treatment. This may partly explain the comparatively small-scale adoption in industry of the heat treatment of refractory metal alloys. Currently, in the Soviet Union and elsewhere, the construction of vacuum high-temperature annealing furnaces has been developed. Annealing is practically the only form of heat treatment of refractory metals and alloys widely employed on an industrial scale. After deformation, refractory metals and alloys are annealed for the relief of internal stresses. Annealing may also be an intermediate operation for the relief of internal stresses in the process of the plastic deformation of refractory metals and alloys, and a final operation for imparting the necessary physicomechanical properties to the finished article. In the last case, semifinished products or components are subjected to recovery annealing or recrystallization annealing. For a number of alloys based on group Va metals, aging is being used for sharply increasing the strength properties at room temperature or elevated temperatures.

Scientific and practical interest is presented in the heat treatment of refractory metals in a high vacuum ($10^{-8}$-$10^{-10}$ mm Hg), which is one of the effective means of increasing ductility especially on account of evaporation of oxides. The investigation of the structure and properties of single-crystal and polycrystalline specimens of refractory metals and alloys, subjected to different heat-treatment conditions in a high vacuum, enables a reliable assessment to be made of the influence of various factors (impurities, grain boundaries, dislocation structure, etc.) on the ductility of the metals.

## 1. Tungsten and Molybdenum Alloys

Commercially pure tungsten and molybdenum are brittle in the recrystallized state. Recrystallization brittleness is not a natural property of these metals but, as already pointed out, is due to the redistribution of impurities during recrystallization annealing. Recrystallization of tungsten and molybdenum of high purity with regard to interstitial impurities ought not to result in embrittlement but in increased ductility, as is found in all other metals. The authors have been able to confirm this assumption experimentally in the recrystallization annealing of foil and wire of molybdenum single crystals [156].

The recrystallization annealing of commercially pure tungsten and molybdenum comprises heating to the recrystallization temperature, holding at that temperature, and slow cooling; annealing is carried out in a hydrogen atmosphere. Stresses in commercially pure tungsten and molybdenum are relieved at 800-1100°C. Tungsten extruded with a reduction of 60-75% is annealed at a temperature of 1200-1400°C. The annealing time is determined by the given temperature and the dimensions of the billet.

Analysis of the phase diagrams shows that there is a considerable temperature dependence on the limit solubility of many metals and interstitial impurities in tungsten and molybdenum in the solid state. This circumstance gives rise to the possibility of varying the mechanical properties of metals and alloys by quenching and aging. The influence of quenching conditions on the structure and properties of tungsten and molybdenum was studied in [171]. Heat treatment was conducted in a vacuum. It was shown that by suitable choice of heating temperature and cooling rate, nonmetallic compounds in the structure of tungsten and molybdenum could be transferred to the solid solution. Transfer of the impurities from the grain boundaries to the solid solution by quenching substantially increases the plastic properties and reduces the ductile–brittle transition temperature of tungsten and molybdenum.

Our experiments showed that high-temperature annealing (heating to 2000°C) in a vacuum of $10^{-6}$-$10^{-7}$ mm Hg sharply embrittled commercially pure molybdenum wire 0.2 mm in diameter, due to the precipitation of impurities on the grain boundaries. The structure of the annealed wire consisted of large grains (some grains occupied the entire cross section of the wire) and nonmetallic carbide inclusions, arranged along the boundaries. After annealing, the wire failed to withstand even one bend. Subsequent heating and rapid cooling (1000°C/sec) considerably increased the ductile properties of the wire, owing to the conversion of the boundary impurities into solid solution. The wire withstood 15 bends at 90°C.

An analogous effect occurs in the quenching and annealing of tungsten wire. The influence of the annealing temperature and cooling rate of tungsten on the ductile–brittle transition temperature is examined in [172]; an increase in the annealing temperature increases the ductile–brittle transition temperature of tungsten.

The structure and properties of the molybdenum after heat treatment (this may refer equally to tungsten, rhenium, and other metals) are determined to considerable degree by the rate of cooling from the region of single-phase solid solution and the rate of heating to the impurity solution temperature. The latter circumstance is due to the fact that the heating process is not only accompanied by the solution of fine precipitates but by the coagulation of coarser precipitates. As pointed out by Jaffee [173], in the USA heat treatment is not practiced in the industrial production of molybdenum articles. The disperse phase precipitated in the solidification of the alloy is not converted into a solid solution on annealing nor is it precipitated, although the effectiveness of such a process has been shown under laboratory conditions. In a number of publications, aging due to the precipitation of finely disperse carbide phase is accepted as a very effective means of increasing the high-temperature oxidation resistance of alloys based on refractory metals with a bcc lattice [174]; this is particularly important for alloys of the niobium–tungsten–molybdenum–zirconium–carbon system.

Problems of the heat treatment of molybdenum and tungsten alloys have been dealt with very little in the literature. Thus, in [38] it is shown that the molybdenum–rhenium (1:1) alloy belongs to a number of aging alloys. Sheet material was treated in a vacuum. The quenching temperature of the alloy was 2300°C and tempering was carried out at 800-1400°C. In the quenched state, the alloy was a single phase solid solution having a bcc lattice; in long-continued tempering (up to 200 h), precipitation of a disperse $\sigma$-phase was observed.

The high strength at elevated temperatures of a molybdenum alloy containing niobium is reported; the maximum strength at 1316-1650°C is attained by treatment at 2100°C to form a solid solution, followed by aging. The opinion is expressed that strengthening is due to the precipitation of the carbide NbC inside the grains [175].

It has been shown that the mechanical properties of the alloy TZM (molybdenum–titanium–zirconium–carbon) are determined largely by the quantity and character of the distribution of carbide phases [176]. The character of the distribution of carbide phase depends on the heat-treatment conditions of the alloy. On quenching from 2200°C, all the carbide phase is dissolved; on quenching from intermediate temperatures, there is a partial solution of the carbides, and undissolved inclusions are spheroidized. Due to the fact that extrusion and rolling of the alloy TZM occurs in the aging temperature range, a special tempering operation was carried out. Plastic deformation accelerated the aging process. Quenching from 1850°C and subsequent aging of the alloy in its deformation process considerably increased the strength properties of the resulting semifinished products without appreciably reducing the ductility, this being due to the formation, as the result of the heat treatment, of a disperse carbide phase uniformly distributed in the alloy matrix.

An investigation of the influence of heat-treatment conditions on the structure and phase transformations in molybdenum alloys containing carbon, titanium, and zirconium was also carried out in [177]. The mechanism of the hardening process of the alloy Mo–1%Ti–0.1%Zr–0.14%C consisted in the complete or partial solution of the carbides TiC, ZrC, and $Mo_2C$ at high temperatures and intense precipitation of TiC under annealing conditions below 1650°C. According to the authors of [177], ZrC does not play any substantial part in the age-hardening of the alloy. The alloys Mo–0.04%C, Mo–0.5%Ti–0.035%C, and Mo–1%Ti–0.1%Zr–0.008%C are not prone to aging; this is explained by the low carbon content and incomplete solution of the carbide phase.

The age-hardening alloys evidently include the carbon-containing alloys of tungsten with niobium, zirconium, and titanium. The pronounced hardening (strengthening) effect of these elements may be due to aging of the alloys as the result of the precipitation of disperse carbides of the corresponding added metals. The optimum heat-treatment conditions for these alloys, however, have not been worked out.

## 2. Niobium and Tantalum Alloys

Due to the high activity of niobium and tantalum with respect to hydrogen, oxygen, and nitrogen, heat treatment of alloys based on these metals must be carried out in a vacuum or in an atmosphere of previously purified inert gas (argon, helium).

High-temperature vacuum annealing of niobium and tantalum results not only in relief of the residual stresses after deformation, but to partial removal of gaseous impurities. According to [178], the oxygen content after annealing the alloys VN-1 and VN-2 at 1800°C for 10 h fell from 0.05 to 0.005%. Hydrogen was removed at 1000-1400°C. Niobium containing a large number of impurities (nitrogen, hydrogen, oxygen, carbon) may be subjected to quenching and aging, owing to the considerable temperature dependence of the limit solubility of interstitial elements in niobium. Evidently, the form and character of the distribution of nonmetallic inclusions in niobium, as in other refractory metals (tantalum, tungsten, molybdenum), are largely determined by the heat-treatment conditions. The influence of tempering temperature of niobium on the character of niobium oxides was investigated in [179]. At a temperature of 800-1000°C, precipitates were found only on the grain boundaries (in the form of NbO). On tempering in the range 700-550°C, together with those arranged along the boundaries, precipitates were observed in the form of globules and oriented plates inside the grains; on tempering below 550°C only intragranular precipitates in the form of oriented plates were found.

In contrast to tungsten and molybdenum, vacuum-melted niobium and tantalum do not become brittle after recrystallization. Before further pressure working, therefore, semi-

TABLE 57. Annealing Conditions of Niobium Alloys [180]

| Alloy and semifinished product | Temperature, °C | Duration, h | Completeness of recrystallization, % |
|---|---|---|---|
| Niobium rod | 1065 | 3 | 100 |
| D-31 rod | 1345 | 1 | 100 |
| Cb-65 rod | 1205 | 1 | 100 |
| Fs-82, sheet ....... | 1425 | 1 | 100 |
| Nb–1%Zr, rod ...... | 1425 | 1 | 100 |
| F-48, sheet ........ | 1370 | 1 | 50 |
| Cb-74, rod ........ | 1425-1500 | 1 | 100 |

finished products of niobium and niobium alloys are subjected to recrystallization annealing. In a number of cases, recrystallization annealing may also be the final operation in the production of semifinished products of niobium alloys. The annealing conditions of some niobium alloys, according to [180] are given in Table 57. The optimum recrystallization annealing temperatures are 1200-1300°C for the alloy VN-1, 1300-1400°C for the alloy VN-2, and 1400-1500°C for the alloy RN-6.

As already mentioned, in the deformation process, niobium and tantalum alloys are subjected to intermediate annealing for internal stress relief.

The influence of heat treatment on the properties of niobium alloys was studied in [181]. The alloy Nb–28%W–7%Ti was vacuum-annealed for 30 min at different temperatures. Annealing at 1100°C merely resulted in internal stress relief; after annealing at 1500 and 1700°C, the alloys had a recrystallization structure. The influence of heat treatment on the form of the tensile stress curves on this alloy is shown in Fig. 154. In the case of recrystallization of the specimen, the curves show a yield point inflection. Recrystallization of the alloys is accompanied by an increase in ductility accompanied by a slight decrease in strength. In [181], the influence of heat treatment on the properties of niobium–tungsten–titanium alloys with additions of molybdenum, vanadium, and zirconium was also investigated. The hardening effect of small additions of zirconium was shown to be due to the formation of disperse inclusions of zirconium carbides, nitrides, and oxides. The hardening of alloys observed in high-temperature annealing is due to the solution of complex carbides of alloying elements and their reprecipitation on the grain and subgrain boundaries during slow cooling of the alloy. The addition of zirconium retards this process.

Appreciable hardening was also observed in alloys of niobium with zirconium and hafnium with increased oxygen content. This effect is retained at elevated temperatures. It is assumed that hardening is due to disperse particles of oxides formed by the internal oxidation of zirconium and hafnium additions. Increase in the oxygen content results in an appreciable hardening of the alloy Nb–1%Ti–1%Zr–5%Hf at room temperature and elevated temperatures. The alloy was found to contain the oxides $ZrO_2$ and $HfO_2$ [182]. It was shown that an alloy of niobium with 1% Zr was hardened by disperse ZrC and $ZrO_2$ particles. The formation of supersaturated solid solutions and the "wedging action" of fresh precipitates making plastic deformation difficult, resulted in hardening, accompanied by a reduction in ductility of the alloy. At the same time, there was an increase in both tensile strength and stress-rupture strength at room temperature and elevated temperature.

The size of the particles, their distribution, and the composition of the carbide phase in niobium alloys depend on the overall composition of the alloy [183] and also on the heat-treatment conditions.

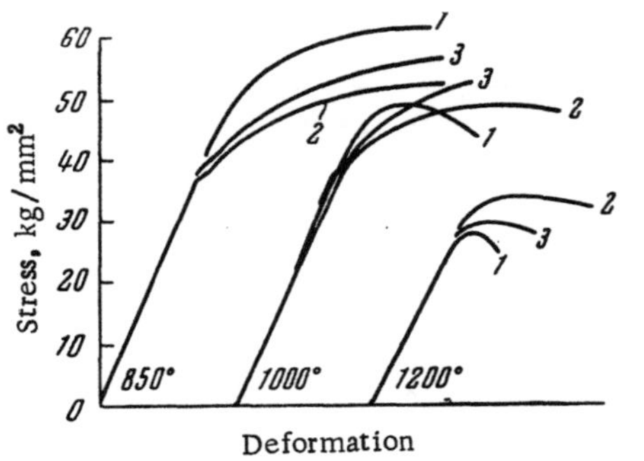


Fig. 154. Form of tensile strength curves of niobium with 28% W and 7% Ti at test temperatures of 850, 1000, and 1200°C as a function of annealing temperature. Annealing temperature, °C: 1) 1100; 2) 1500; 3) 1700.

It has been shown by the methods of microstructural, electrochemical, and x-ray analysis that alloys based on niobium (niobium–tungsten–molybdenum–zirconium–carbon systems), containing 1-1.5% zirconium, age at 900-1100°C with the formation of oxide, nitride, and carbide phases [200]. The possibility of an additional increase in the characteristic stress-rupture strength due to an increase in carbon content to 0.09-0.15% has also been shown [200]. This carbon forms a hardening phase of the type of (NbZrWMo)C. It is shown that the content of nitrogen and oxygen ought to be a minimum, since they bind the zirconium as oxides and nitrides, inhibiting the formation of the complex carbide. It is pointed out that the application of optimum quenching and aging conditions may increase the life of alloys of the niobium–tungsten–molybdenum–zirconium–carbon system by a factor of 15-20 [200].

Aging of tantalum alloys is found when tantalum is alloyed with small quantities of titanium and zirconium together with carbon.

In some niobium alloys, due to the choice of optimum quenching and aging conditions (precipitation of carbide phase), an increase in stress-rupture strength of many times may be obtained.

Thus, the results of experiments which have been made show that complex niobium alloys, containing additions of carbon of the order of 0.1 wt.%, are hardened on aging, i.e., on the decomposition of a supersaturated solid solution. The hardening mechanism consists in the precipitation from solid solution of the high zirconium monocarbide (Nb, Zr). A similar hardening of refractory metals is also possible as the result of the formation of nitride phases, but relatively little research has so far been done in this direction.

Precipitation hardening of niobium alloys with hafnium and nitrogen was studied in [184]. The susceptibility of these alloys to heat treatment is due to hardening on the decomposition of solid solutions, supersaturated with respect to nitrogen. A niobium alloy with 5% Hf containing 0.1 and 0.2% N was subjected to heat treatment. The preliminarily quenched specimens were subjected to aging in the range of 600-1800°C. Aging at 800 and 1000°C led to considerable increase in hardness; in aging above 1000°C, the hardness decreased, reaching a minimum at 1400°C; with further increase in the aging temperature an increase in hardness was again found, this being ascribed [184] to resolution of the nitride phase in the alloy base. Electron diffraction analysis showed the principal excess phase to be hafnium nitride HfN. The practical application of the above-mentioned alloy (Nb–5%Hf–0.2%N) is prevented by its low ductility due to the precipitation of nitrides on the grain boundaries. The supposition is made that a good combination of strength and ductility could be achieved with a lower nitrogen content.

The influence of the character of the distribution, shape, and size of the particles precipitated on aging on the strength properties of various materials has been studied by many investigators. Gensamer et al. [185] derived a semilogarithmic relationship between the yield point and the average distance between the particles in ferrite; according to these authors [185], the shape of the particles has no significant influence on yield point. It is pointed out in [186] that at different stages of aging, in addition to the factors already mentioned as influencing age-hardening, others should also be taken into consideration; the interaction of the stress field, surrounding coherent particles, with dislocations is a possible source of age-

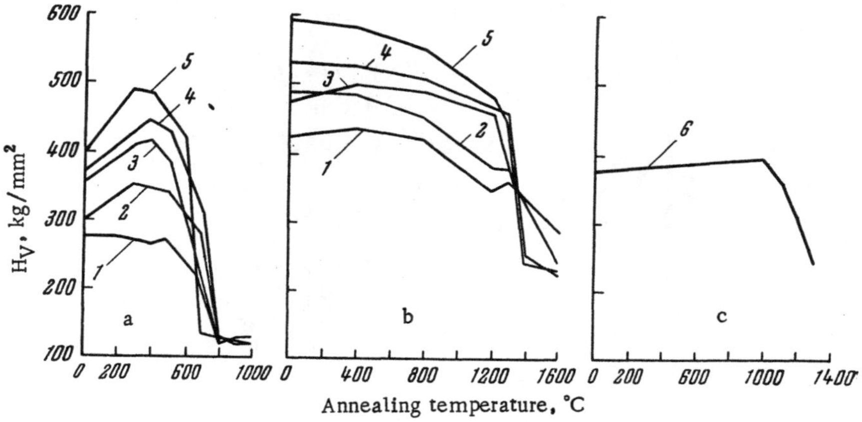


Fig. 155. Softening of rhodium (a), iridium (b), and ruthenium (c) as a function of annealing temperature. Degree of deformation, %: 1) 10; 2) 20; 3) 30; 4) 40; 5) 50; 6) 85.

hardening; an additional contribution is the energy necessary for the production of a fresh interface between zones. As emphasized in [186], these mechanisms play a leading part in precipitation hardening when the particles are very small.

### 3. Alloys of the Noble Metals

There is no detailed published information on the heat-treatment conditions of alloys based on group VIII metals. Since the majority of these alloys are solid solutions, the only form of treatment is annealing for the relief of residual stresses due to deformation and also recrystallization annealing.

Annealing of platinum alloys may be carried out in air owing to their high corrosion resistance. Palladium alloys as a rule are annealed in a vacuum or an inert gas atmosphere. The annealing temperature of the various alloys based on platinum and palladium is determined by their recrystallization temperature. These alloys are normally subjected to recrystallization annealing for complete internal stress relief and imparting the necessary ductility to the alloys.

Figure 155 shows the influence of work hardening on the softening of refractory metals of the platinum group; in the case of rhodium, a strain-aging process is observed during annealing, evidently due to inadequate purity of the rhodium [187]. These questions are examined in [197].

## Welding and Soldering of Refractory Metals

### 1. Welding

Alloys based on tantalum and some alloys based on niobium (ductile alloys) possess good weldability. Welding of alloys based on molybdenum and tungsten, and some also based on niobium, gives rise to considerable difficulties. The principal factors affecting the weldability of refractory metals and alloys are their high melting point and chemical activity at elevated temperatures. In welding these metals, therefore, high rates of heating and a pure, inert atmosphere are required.

In the case of niobium, tantalum, and their alloys, various forms of arc welding in argon or helium are successfully employed. In [188], reference is made to the high quality of the weld

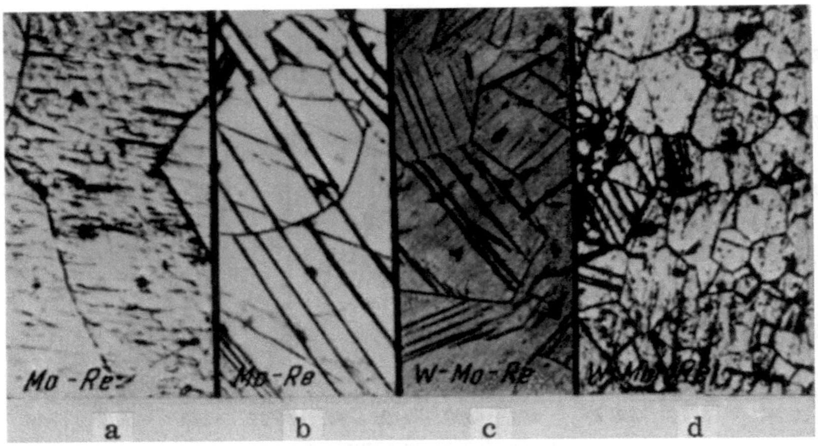


Fig. 156. Microstructure of welded sheets of alloys of tungsten and molybdenum with rhenium. a) Molybdenum–rhenium alloy, fusion zone; b) the same, position of bend; c) tungsten–molybdenum–rhenium alloy, position of bend; d) the same, recrystallization zone.

produced by electron-beam welding of refractory metals. The opinion is expressed [189] that, owing to a number of advantages, electron-beam welding will later become the principal method of welding refractory metals. The temperature conditions of welding can be easily varied by varying the filament-heating current and voltage. In electron-beam welding, the zone of thermal influence is narrowed. In vacuum electron-beam welding, the weld is purified by evaporation of volatile impurities. Welds of niobium, tantalum, and their alloys are sufficiently ductile owing to the relatively low sensitivity of these materials to interstitial impurities.

Welds of tungsten, molybdenum, and their alloys are brittle. It is evident that satisfactory welds can be produced only if the interstitial impurity content of these metals does not exceed the solubility limits. This assumption is confirmed by the authors' experiments on the electron-beam welding of foil produced from high-purity molybdenum single crystals. The ductility of the weld is determined not only by the purity of the metal but by the grain size, since the interstitial impurities in the form of compounds accumulate on the grain boundaries, weakening them. The production of ductile welds in welding tungsten and molybdenum is a complex technical problem, which is solved by far-reaching purification of these metals from interstitial impurities, and by the introduction, during welding, of chemically active additions acting as deoxidizers and structure modifiers.

A report on the arc welding of parts made of tungsten sheet is of interest [165].

For increasing the ductility of welded joints of tungsten and molybdenum, alloying of these metals has been found to be very promising. According to [190], the use of molybdenum–rhenium alloys as additive in the argon arc welding of thin molybdenum sheet considerably increases ductility of the welded joints (angle of bend for r = 3d attained 180°); fracture does not occur at the position of the weld but in the adjacent zone of the base metal. When molybdenum is welded under these conditions, embrittlement of both the base metal and the weld occurs. A welded specimen of molybdenum alloy with 50% rhenium withstood cold rolling from 1 to 0.3 mm in three passes [191]. Good weldability was shown by sheets 0.3-0.5 mm thick of tungsten–rhenium alloys and tungsten–molybdenum–rhenium ternary alloys [192]. But welding was carried out by means of an electron beam in a vacuum of $2 \cdot 10^{-5}$ mm Hg. The microstructure of welded sheets of vacuum-melted alloys of tungsten and molybdenum with rhenium is shown in Fig. 156.

TABLE 58. Solders for Molybdenum [194]

| Solder | Liquidus temp., °C | Solder | Liquidus temp., °C |
|---|---|---|---|
| Ag | 960 | Ni — Cr — Mo — Fe — W | 1304 |
| Cu | 1082 | Ni — Cu | 1349 |
| Ni | 1454 | Ni — Cr — Fe | 1427 |
| Pd | 1571 | Pd — Ni | 1204 |
| Pt | 1774 | Pd — Fe | 1316 |
| Ag'— Cu — Zn — Cd | 619—702 | Pd — Ag | 1316 |
| Ag — Cu — Zn | 719—788 | Co — Cr — Si — Ni | 1399 |
| Ag — Cu · | 780 | Co — Cr — W — Ni | 1427 |
| Ag — Mn | 981 | Mo — Ru | 1899 |
| Ni — Cr — B | 1066 | Cu — Mn | 871 |
| Ni — Cr — Fe — Si — C | 1066 | Mo — B | 1899 |
| Ni — Cr — Mo — Mn — Si | 1149 | Nb — Ni | 1191 |
| Ni — Ti | 1288 | | |

For producing a high-grade weld, correct choice of the heat-treatment conditions is very important. Thus, for example, in the case of niobium and tantalum alloys, heat treatment should ensure the production of a fine-grain structure in the weld (recrystallization annealing).

In addition to the above-mentioned welding methods, spot welding is also used for refractory metals and alloys. In [189] it is reported that spot welding is used for the production of honeycomb panels with tantalum cores and tungsten facing sheets. Very recently, welding by laser beams and plasma torch welding have been used.

## 2. Soldering

Owing to the fact that welding produces a brittle joint, it is often necessary to resort to soldering articles of refractory metals and alloys, that is to say, assembling parts using materials having a lower melting point than that of the parts to be assembled. Soldering of articles of refractory metals by means of special plastic solders has been widely used in assembling electrovacuum instruments and devices [193].

Soldering of refractory metals is carried out in an inert gas or in a vacuum. Soldering of tungsten and molybdenum articles may be carried out in a vacuum, inert gas, or hydrogen. Table 58 lists metals and alloys used as solders in soldering molybdenum articles.

In selecting a solder, it is necessary to take into account its compatibility with the material of the articles and antioxidation coatings, as well as the service conditions of the construction (temperature, medium, length of service, etc.).

## LITERATURE CITED

1. E. M. Savitskii, G. S. Burkhanov, and Ch. V. Kopetskii, Izv. Akad. Nauk SSSR, Metallurgiya i Gornoe Delo, No. 6, p. 12 (1963).
2. In: Electronic Melting of Metals [in Russian], Izd. "Mir" (1964).
3. R. I. McCullogh, J. Metals, 14(12):907 (1962).
4. G. Shipwright, A. Leserman, and D. Evers, Problemy Sovremennoi Metallurgii, No. 4, p. 99 (1961).
5. A. R. Moss, J. Less-Common Metals, 1:60 (1959).
6. M. Cook and E. Sxainson, J. Inst. Metals, 87:161 (1958).
7. S. L. Ham and S. V. Sillig, J. Metals, 9:976 (1957).

8. B. B. Gulyaev, O. N. Magnitskii, and A. A. Demidova, Refractory Metal Castings [in Russian], Izd. "Mashinostroenie" (1964).
9. E. R. Moss and D. T. Richards, Problemy Sovremennoi Metallurgii, No. 5, p. 79 (1961).
10. J. E. Geach and F. O. Johns, Problemy Sovremennoi Metallurgii, No. 3, p. 156 (1959).
11. F. A. Foyle, Lexis Research Center, National Aeronautics and Space Administration, Cleveland, Ohio (1964).
12. R. Kieffer and E. Pipitz, The Metal Molybdenum, ASM, Cleveland, Ohio (1958).
13. H. G. Sell and A. E. Lamarche, The Partition of Power in High Current Low Pressure Metal Vapor D. C. Arcs, A Theoretical Interpretation Based on Arc Melting Experiments on Tungsten, New York (1963).
14. A. R. Moss, Proc. Inst. Elect. Engrs. 102, P. A., No. 1, p. 45 (1955).
15. G. Robert, U. Barr, J. Daniel, and K. Klaphas, Climax Molybdenum Company of Michigan (1965).
16. G. T. Hahn, A. Gilbert, and R. I. Jaffee, Proc. Refractory Metals Symposium, Chicago (April, 1962).
17. E. M. Savitskii, Izv. Akad. Nauk SSSR, OTN, Metallurgiya i Toplivo, No. 5, p. 52 (1960).
18. H. Smith, C. Hunt, and C. Hanks, Problemy Sovremennoi Metallurgiya, No. 5, p. 140 (1959).
19. L. E. Olds and G. W. P. Rengstorff, J. Metals, 8:150 (1956).
20. S. J. Nesen, Problemy Sovremennoi Metallurgii, No. 3, p. 43 (1961).
21. Introduction into the Technology of Electron-Beam Processes [in Russian], Izd. "Metallurgiya" (1965).
22. H. Braun, M. Semchyshen, and R.Q. Barr, in: Columbium and Tantalum, Wiley, New York, p. 347 (1963).
23. E. M. Savitskii and V. V. Baron, Transactions of the A. A. Baikov Inst. of Metallurgy, No. 5, p. 156, Izd. Akad. Nauk SSSR (1960).
24. E. M. Savitskii, V. V. Baron, and Yu. V. Efimov, Zh. Neorg. Khim., 7(3):701 (1962).
25. E. M. Savitskii, V. F. Terekhova, and I. V. Burov, Zh. Neorg. Khim., 4(6):1462 (1959).
26. In: Rare-Earth Metals [in Russian], Izd. "Metallurgiya" (1965).
27. K. A. Gschneidner, Rare Earth Alloys, Van Nostrand, Princeton, New Jersey (1961).
28. E. M. Savitskii, V. F. Terekhova, I. V. Burov, I. A. Markova, and O. P. Naumkin, Alloys of Rare-Earth Metals [in Russian], Izd. Akad. Nauk SSSR (1962).
29. A. E. Vol, Structure and Properties of Binary Metal Systems [in Russian], Fizmatgiz (1959).
30. E. M. Savitskii and V. V. Baron, in: Alloys of Rare Metals [in Russian], Metallurgizdat (1960).
31. J. W. Semmel, The Effect of Rare Earth Metals on the Ductility of Arc-Melted Group Va Metals, Technol. Columbium (1958).
32. E. M. Savitskii, V. V. Baron, and Yu. V. Efimov, Izv. Akad. Nauk SSSR, OTN, Metallurgiya i Toplivo, No. 3, p. 107 (1962).
33. O. Kubaschewski and E. Evans, Metallurgical Thermochemistry, Pergamon, New York (3rd ed.) (1958).
34. Yu. V. Efimov, E. M. Savitskii, and V. V. Baron, Author's Certificate No. 148521 of 13th April, 1962, Byull. Izobretenii i Tovarnykh Znakov, No. 13 (1962).
35. E. M. Savitskii, V. V. Baron, and Tao-Tsu-ts'ung, Izv. Akad. Nauk SSSR, OTN, Metallurgiya i Toplivo, No. 1, p. 156 (1962).
36. E. M. Savitskii, V. V. Baron, and V. A. Frolov, in: Questions of the Theory and Application of Rare-Earth Metals [in Russian], Izd. "Nauka," p. 138 (1964).
37. S. F. Burmanov, V. L. Girshov, and L. N. Postnov, in: Vacuum Metallurgy [in Russian], No. 2, p. 9, Izd. Doma Nauchno-Tekhn. Propagandy, Leningrad (1964).

38. E. M. Savitskii, M. A. Tylkina, and K. B. Povarova, Rhenium Alloys [in Russian], Izd. "Nauka" (1965).
39. V. L. Girshov and L. N. Postnov, in: Vacuum Metallurgy [in Russian], No. 2, p. 33, Izd. Doma Nauchno-Tekhn. Propagandy, Leningrad (1964).
40. Refractory Metals Made Dense through Centrifugal Casting, Iron Age, 191(17):124 (1963).
41. R. G. Hardy, I. L. Gunter, and T. A. Hamm, Met. Progress, 72 (August 1962).
42. W. Hume-Rothery, J. Christian, and V. Pearson, Metallurgical Equilibrium Diagram and Experimental Methods for Their Determination, Reinhold, New York (1952).
43. S. G. Trekhsvyatskii and A. M. Cherepanov, Highly Refractory Materials and Articles Made of Oxides [in Russian], Metallurgizdat (1957).
44. H. Gruber, Z. Metallkunde, 52(5):291 (1961).
45. T. T. Magel, P. A. Kulin, and A. R. Kaufmann, J. Metals, 4:1286 (1952).
46. H. F. Sterling and R. W. Warren, Metallurgiya, 47(404):301 (1963).
47. H. W. Schadler, in: The Art and Science of Growing Crystals, New York, p. 343 (1963).
48. V. G. Epifanov and A. G. Lesnik, Voprosy Fiziki Metallov i Metallovedeniya, No. 20, p. 185 (1964).
49. D. M. Wroughton, E. C. Okress, P. H. Brace, G. Comenetz, and S. C. R. Kelly, J. Electrochem. Soc., 99:205 (1952).
50. R. T. Begley, G. Comenetz, P. A. Flinn, and S. W. Salatka, Rev. Sci. Instr., 30:38 (1959).
51. A. A. Fogel', Izv. Akad. Nauk SSSR, OTN, Metallurgiya i Toplivo, No. 2, p. 24 (1959).
52. Am. Metal Market, 71(131):18 (1964).
53. S. Siegl, R. Martin, and O. Bertea, in: Electronic Melting of Metals [in Russian], Izd. "Mir," p. 200 (1964).
54. J. Bero and H. Forestier, in: Electronic Melting of Metals [in Russian], Izd. "Mir," p. 131 (1964).
55. A. Calverley, M. Davis, and R. Lever, J. Sci. Instr., 34(4):142 (1957).
56. D. R. Stull and G. C. Sinke, Thermodynamic Properties of the Elements, ASM, Washington (1956).
57. T. E. Butler and R. P. Morgen, J. Metals, 14(3):200 (1962).
58. E. M. Savitskii and G. L. Tsarev, Izv. Akad. Nauk BSSR, No. 2, p. 84 (1965).
59. H. Smith, C. Hunt, and C. Hanks, in: Electronic Melting of Metals [in Russian], Izd. "Mir," p. 11 (1964).
60. E. Rekser, in: Electronic Melting of Metals [in Russian], Izd. "Mir," p. 293 (1964).
61. L. Verot and A. Forestier, Purification du molybdène par sussion sous bombardement electronique, Compt. Rend. Acad. Sci., 255(4) (1961).
62. H. R. Smith, I. R. Hum, A. Donlevy, and C. A. Hunt, J. Less-Common Metals, 2(2):69 (1960).
63. G. Ogierman and W. Scheibe, Metall, 15(1):3 (1961).
64. E. M. Savitskii and V. V. Baron, in: Strength of Metals [in Russian], Izd. Akad. Nauk SSSR, p. 144 (1956).
65. M. G. Manzone and J. Z. Briggs, Less-Common Alloys of Molybdenum, New York (1962).
66. A. D. Zumbrunnen and S. M. Fitzpatrick, J. Less-Common Metals, 7(5):356 (1964).
67. D. Peckner, Materials in Design Engineering, 53(3):101 (1961).
68. "Non-forgeable refractory metals," Metalworking Production, 104(138) (1960).
69. "Mechanical Properties of Hafnium, Melted in the Electronic Furnace," in: Electronic Melting of Metals [in Russian], Izd. "Mir," p. 331 (1964).
70. C. Dittmar and S. Abkovitz, in: Electronic Melting of Metals [in Russian], Izd. "Mir," p. 309 (1964).
71. L. Bangert and K. Hennemann, Metall, No. 7, p. 704 (1960).
72. G. F. Zaboronok, T. I. Zelentsov, A. S. Ronzhin, and B. G. Sokolov, Electronic Melting of Metals [in Russian], Izd. "Metallurgiya" (1965).

73. "Production and purification of high-purity hafnium," J. Metals, 12(1) (1960).
74. C. E. Armantrout and H. Kato, "The mechanical properties of arc-melting Kroll process hafnium," Reactive Metals, 2 (1958).
75. H. R. Smith, I. Hum, A. Donlevy, and C. Hunt, J. Less-Common Metals, 2(2–4):69 (1960).
76. "Large electron-beam furnace refines refractory metals," Metals Progress, 80(2) (1961).
77. E. M. Savitskii, V. V. Baron, and L. F. Myzenkova, in: Physical Metallurgy, Physical Chemistry and Metal Physics of Superconductors [in Russian], Izd. "Nauka" (1966).
78. H. Smith, in: Electronic Melting of Metals [in Russian], Izd. "Mir," p. 45 (1964).
79. R. Kieffer and H. Braun, Vanadin. Niob. Tantal., Springer Verlag, Berlin (1963).
80. J. Metals, 14(11):800 (1962).
81. S. I. Gubkin, Plastic Deformation of Metals [in Russian], Metallurgizdat (1960).
82. A. S. Drachinskii, V. F. Moiseev, and V. I. Trefilov, Fizika Metallov i Metallovedenie, 19(4):602 (1965).
83. E. M. Savitskii, Plastic Properties of Magnesium and Its Alloys [in Russian], Izd. Akad. Nauk SSSR (1941).
84. M. V. Klassen-Neklyudova, Mechanical Twinning of Crystals, Consultants Bureau, New York (1964).
85. E. M. Savitskii, Influence of Temperature on the Mechanical Properties of Metals and Alloys [in Russian], Izd. Akad. Nauk SSSR (1957).
86. O. D. Sherby, Am. Soc. Metals, 826:3 (May 1962).
87. E. M. Savitskii, Izv. Akad. Nauk SSSR, Metally, No. 5, p. 84 (1965).
88. R. Linther and E. Schmid, Werkstoffe des Reaktorbaues, Springer Verlag, Berlin (1962).
89. L. Raimondt and F. Chervet, Single-crystal tungsten sheets AIAA, 4(5) (1966).
90. V. I. Likhtman, E. D. Shchukin, and P. A. Rebinder, Physicochemical Mechanics of Metals, Izd. Akad. Nauk SSSR (1962).
91. J. Hollomon and K. Zinner, Uspekhi Fizicheskikh Nauk, 31(1):38 (1947).
92. J. Washborn, in: Mechanisms of the Hardening of Solids [in Russian], Izd. "Metallurgiya," p. 57 (1965).
93. P. Ludwik and R. Scheu, Stahl u. Eisen, No. 43, p. 999 (1923).
94. M. Polanyi, Z. Kristallogr., 61:49 (1925).
95. U. Dehlinger, Z. Metallkunde, 22:222 (1930).
96. G. V. Akimov, Dokl. Akad. Nauk SSSR, 47(8):567 (1945).
97. F. Seitz and T. A. Read, Appl. Phys., No. 2, p. 36 (1941).
98. F. Bragg, Nature, No. 3, p. 511 (1952).
99. A. Seeger, Phil. Mag., 46:1194 (1955).
100. W. M. Lomer, Phil. Mag., 42:1327 (1951).
101. A. H. Cottrell, Phil. Mag., 43:645 (1952).
102. U. R. Hibbard, in: Mechanisms of the Hardening of Solids [in Russian], Izd. "Metallurgiya," p. 9 (1965).
103. W. G. Johnston and S. S. Gilman, J. Appl. Phys., 31:632 (1960).
104. J. Washborn, in: Mechanisms of the Hardening of Solids [in Russian], Izd. "Metallurgiya," p. 57 (1965).
105. Express Information, Metallovedenie i Termoobrabotka, VINITI, No. 28 (Ref. 89) (1965).
106. G. E. Ditter, in: Mechanism of the Hardening of Solids [in Russian], Izd. "Metallurgiya," p. 245 (1965).
107. C. S. Smith, Trans. AIME, 214:574 (1958).
108. J. J. Kelly and P. S. Decarli, Bull. Am. Phys. Soc., 2:369 (1957).
109. F. E. van Wely and C. A. Verbraak, in: Mechanisms of the Hardening of Solids [in Russian], Izd. "Metallurgiya," pp. 382–394 (1965).
110. F. E. van Wely and C. A. Verbraak, Metalle für die Raumfahrt, Metallwerk Plansee A. G. Reutte/Tyrol, p. 382 (1965).

111. N. N. Davidenkov and G. N. Chuchman, Zh. Tekhn. Fiz., 3:2294 (1958).
112. D. Rosenthal and C. C. Woolsey, Welding, 5(31):475 (1952).
113. O'Neil, J. Iron and Steel Inst., 113:417 (1926).
114. R. Gahn, J. Inst. Metals, 493(7):1627 (1955).
115. N. N. Davidenkov and G. N. Chuchman, in: Research on Heat-Resistant Metals [in Russian], Izd. Akad. Nauk SSSR, 2:9 (1957).
116. D. Mannox, J. Metal Ind., 104(19):624 (1964).
117. J. D. Nisbet, in: Molybdenum [Russian translation], IL, p. 118 (1962).
118. S. I. Gubkin, Theory of Metal Working by Pressure [in Russian], Metallurgizdat (1947).
119. E. M. Savitskii, Dokl. Akad. Nauk SSSR, 89:1 (1958).
120. H. A. Johnson, K. S. Hartley, and M. D. Macnabb, in: Refractory Metals and Their Alloys [Russian translation], IL, p. 152 (1962).
121. E. M. Savitskii, Rare Metals and Alloys [in Russian], Izd. Domka Tekhnika, Moscow (1959).
122. E. M. Savitskii, M. A. Tylkina, and A. I. Tiranskaya, Dokl. Akad. Nauk SSSR, 101(5):857 (1955).
123. E. M. Savitskii, V. F. Terekhova, and A. V. Kolopov, Dokl. Akad. Nauk SSSR, 109(4):794 (1956).
124. E. M. Savitskii, V. V. Baron, and K. N. Ivanova, Dokl. Akad. Nauk SSSR, 113(5):1070 (1957).
125. E. M. Savitskii, M. A. Tylkina, and K. B. Povarova, Dokl. Akad. Nauk SSSR, 119(2):274 (1958).
126. M. A. Tylkina, R. V. Kirilenko, and E. M. Savitskii, Dokl. Akad. Nauk SSSR, 127(2):310 (1959).
127. E. M. Savitskii, V. V. Baron, and K. N. Ivanova, Dokl. Akad. Nauk SSSR, 126(4):771 (1959).
128. E. M. Savitskii, M. A. Tylkina, A. I. Pekarev, M. I. Gavrilyuk, and A. P. Zabavnova, Dokl. Akad. Nauk SSSR, 140(6):1301 (1961).
129. E. M. Savitskii, V. V. Baron, and Yu. V. Efimov, Dokl. Akad. Nauk SSSR, 145(3):612 (1962).
130. E. M. Savitskii and V. F. Terekhova, Transactions of the A. A. Baikov Inst. Metals [in Russian], No. 3, p. 181, Izd. Akad. Nauk SSSR (1958).
131. E. M. Savitskii and M. A. Tylkina, in: Research on Heat-Resistant Alloys [in Russian], Izd. Akad. Nauk SSSR, Vol. 4, p. 218 (1959).
132. E. M. Savitskii, Dokl. Akad. Nauk SSSR, 73:5 (1950).
133. E. M. Savitskii, Metallovedenie i Termoobrabotka Metallov, No. 7, p. 1 (1965).
134. E. M. Savitskii, M. A. Tylkina, and V. F. Terekhova, Transactions of the A. A. Baikov Institute of Metallurgy, No. 11, p. 133, Izd. Akad. Nauk SSSR (1962).
135. W. R. Rostoker, Refractory Metals and Alloys, Vol. 17, Chicago, p. 379 (1962).
136. F. A. Foyle, High Temperature Materials, 18:109 (1963).
137. M. Semchyshen and R. Q. Barr, High Temperature Materials, 18:599 (1963).
138. N. N. Morgunova, Heat-Resistant Alloys Based on Molybdenum [in Russian], VINITI (1959).
139. S. Spechner and W. R. Rostoker, Problemy Sovremennoi Metallurgii, No. 5, p. 129 (1958).
140. M. V. Mal'tsev, A. I. Baikov, and V. Ya. Solov'ev, Technology of the Production of Niobium and Its Alloys [in Russian], Izd. "Metallurgiya" (1966).
141. M. Schaufuss, Problemy Sovremennoi Metallurgii, No. 2, p. 108 (1960).
142. M. Shusslep, A. L. Mincher, J. L. Wilson, and H. E. Dufendach, in: Columbium and Tantalum, Wiley, New York (1963).
143. G. W. King and H. G. Seel, Trans. Met. Soc. AIME, 233(6):1104-1113 (1965).
144. R. I. Jaffee, W. S. Harris, and N. E. Promysel, J. Less-Common Metals, 2:95 (1960).
145. J. A. Belk et al., in: Properties and Treatment of Refractory Metals and Alloys (Transactions of an International Conference on Refractory Metals and Alloys) [Russian translation], IL, p. 26 (1961).

146. F. Henderson, S. Kvaas, and H. Wein, Problemy Sovremennoi Metallurgii, No. 3, p. 62 (1955).
147. R. I. Jaffee, W. S. Harris, and N. E. Promysel, Problemy Sovremennoi Metallurgii, No. 3, p. 108 (1961).
148. A. M. Gaudin, Lecture at the Meeting of the Metallurgical Society AIME (February 1961).
149. G. L. Kolbe, J. Electrochem. Soc., 101:601 (1954).
150. J. D. Nisbet: in: Molybdenum [Russian translation], IL, p. 118 (1962).
151. E. M. Savitskii et al., in: Research on Heat Resistant Alloys [in Russian], Vol. 9, p. 124, Izd. Akad. Nauk SSSR (1962).
152. Ya. B. Gurevich, Izv. Akad. Nauk SSSR, Metally, No. 8, p. 25 (1965).
153. K. Agte and I. Vatsek, Tungsten and Molybdenum [in Russian], Izd. "Energiya" (1964).
154. E. M. Savitskii, G. S. Burkhanov, Ch. V. Kopitskii, and G. E. Chuprikov, in: Trans actions of a Conference on the Growth of Crystals [in Russian], Vol. 6, p. 308, Izd. "Nauka" (1965).
155. E. M. Savitskii, G. S. Burkhanov, Ch. V. Kopetskii, N. N. Bokhareva, and V. G. Kardashevskaya, in: Properties and Applications of Heat-Resistant Alloys [in Russian], Izd. "Nauka" (1966).
156. E. M. Savitskii, G. E. Chuprikov, and A. S. Grokhochinskii, Obmen Opytom v Radioelektronnoi Promyshlennosti, No. 10, p. 1 (1964).
157. E. M. Savitskii, M. A. Tylkina, L. L. Zhdanova, V. N. Starkov, A. G. Fokin, L. S. Petrova, and T. I. Arkusha, in: Research on Heat-Resistant Alloys [in Russian], Vol. 9, p. 194, Izd. Akad. Nauk SSSR (1962).
158. E. M. Savitskii, V. V. Baron, V. Ya. Pakhomov, M. I. Bychkova, and V. R. Karasik, Author's Certificate No. 168250, Byull. Izobretenii i Tovarnykh Znakov, No. 4 (1965).
159. In: Metallurgy and Metal Physics of Superconductors [in Russian], Izd. Akad. Nauk SSSR (1965).
160. Express Information, Metallovedenie i Termoobrabotka, VINITI, No. 25 (Ref. 13) (1965).
161. R. V. Pikhtovnikov and V. I. Zav'yalova, Shaping Sheet Material by Explosion [in Russian], Izd. "Mashinostroenie" (1964).
162. Express Information, Metallovedenie i Termoobrabotka, VINITI, No. 28 (Ref. 14) (1965).
163. Express Information, Metallovedenie i Termoobrabotka, VINITI, No. 28 (Ref. 147) (1964).
164. J. H. Port, Metallwerk Plansee A. G. Reutte/Tyrol, p. 613 (1965).
165. D. J. Maykuth and H. R. Ogden, High-Temperature Materials, Interscience, New York, p. 81 (1963).
166. W. H. Kohl, Metal Progress, No. 85, p. 83 (1964).
167. V. S. Ivanova and L. K. Gordienko, New Methods of Increasing the Strength of Metals [in Russian], Izd. "Nauka" (1964).
168. S. T. Konobeevskii and I. Z. Miter, Kristallografiya, No. 81, p. 69 (1932).
169. A. A. Bochvar, Physical Metallurgy [in Russian], Metallurgizdat (1956).
170. I. I. Novikov and M. V. Zakharov, Heat Treatment of Metals and Alloys [in Russian], Metallurgizdat (1962).
171. G. S. Speisel and J. Wulf, in: Molybdenum [Russian translation], IL, p. 176 (1962).
172. R. C. Koo, Trans. Met. Soc. AIME, 227(1):280 (1963).
173. R. I. Jaffee, Rapporteur's Interpretive Statement, Battelle Memorial Institute (1964).
174. W. H. Chang, Trans. AIME, 218:254 (1960).
175. Ref. Zh. "Metallovedenie i Termoobrabotka," No. 8, p. 83 (1965).
176. H. Braun, M. Semchyshen, and R. Q. Barr, Metallwerk Plansee A. G. Reutte/Tyrol, p. 351 (1965).
177. W. H. Chang and I. Perlmutter, High-Temperature Materials, 18:347 (1963).
178. G. V. Zakharova and L. P. Zhorova, Tsvetnye Metally, No. 5, p. 53 (1963).
179. E. Gebhardt and R. Rothenbacher, Z. Metallkunde, 54(11):623 (1963).

180. R. W. Shussler and G. R. McKinsey, in: Columbium and Tantalum, Wiley, New York (1963).
181. V. F. Sheeley and J. L. Wilson, Some Problems of Refractory Metals and Alloys [Russian translation], p. 120 (1963).
182. R. T. Begley and W. N. Platte, Wright Air Development Center Techn. Rept. WADC-TR-St-344, Pt-4, p. 131 (1960).
183. B. Clement and I. Perlmutter, in: Properties and Treatment of Refractory Metals and Alloys [Russian translation], IL, p. 52 (1961).
184. R. T. Begley, T. L. Godshall, and R. Sticker, Metallwerk Plansee A. G. Reutte/Tyrol, p. 401 (1965).
185. M. Gensamer, E. B. Pearsall, W. S. Pellini, and J. R. Lovy, Trans. AIME, 30:983 (1942).
186. R. V. Guard, in: Mechanism of Strengthening Solids [in Russian], Izd. "Metallurgiya," p. 20 (1965).
187. R. I. Jaffee, D. J. Maykuth, and R. U. Douglass, Rhenium and Refractory Metals of the Platinum Group [Russian translation], IL (1963).
188. N. F. Eaton, Problemy Sovremennoi Metallurgii, No. 1, p. 117 (1961).
189. R. H. Gassner, Some Problems of Refractory Metals and Alloys [Russian translation], IL, p. 11 (1963).
190. V. V. D'yachenko, B. P. Morozov, M. A. Tylkina, and E. M. Savitskii, Svarochnoe Proizvodstvo, No. 7, p. 1 (1960).
191. E. M. Savitskii, M. A. Tylkina, K. B. Povarova, and V. P. Vinokurov, Proizvodsvenno-tekhn. Byull., No. 5, p. 3 (1962).
192. M. A. Tylkina and E. M. Savitskii, in: Rhenium, Trans. 2nd All-Union Congress on Rhenium Problems [in Russian], Izd. "Nauka," p. 138 (1964).
193. N. R. Lashko and S. V. Lashko-Avakyan, Soldering of Metals [in Russian], Mashgiz (1959).
194. R. E. Monroe, in: Molybdenum [Russian translation], IL, p. 129 (1962).
195. I. V. Polyakov, I. V. Mikhailev, and P. A. Rebinder, Dokl. Akad. Nauk SSSR, 167(4):873 (1966).
196. V. M. Amonenko, A. S. Tron', et al., Collective Works of the Ukrainian Scientific Research Institute of Metallurgy [in Russian], No. 10, Metallurgizdat (1964).
197. E. M. Savitskii, V. P. Polyakova, and M. A. Tylkina, Palladium Alloys [in Russian], Izd. "Nauka" (1966).
198. A. V. Krupin, I. M. Pavlov, et al., Rolling of Refractory Metals and Bimetals in a Vacuum [in Russian], "Tsvetmetinformatsiya" (1966).
199. V. I. Dobatkin et al., Ingots of Titanium Alloys [in Russian], Izd. "Metallurgiya" (1966).
200. E. M. Savitskii and O. B. Dashevskaya, Izv. Akad. Nauk SSSR, Metally, No. 3, p. 151 (1967).
201. N. I. Korneev, S. B. Pevzner, E. I. Razuvaev, and I. G. Skugarev, Pressure Working of Refractory Metals and Alloys [in Russian], Izd. "Metallurgiya," Moscow (1967).

# CONCLUSION

We shall review briefly the present position of the problem of the scientific and industrial utilization of refractory metals, rare metals, and their alloys.

Refractory metals occupy a significant place in the new technology. They are exerting an ever-increasing influence on technical progress.

Progress in the development of new metallic materials is connected to a considerable degree with the general position and level of the physical chemistry of metals, metal physics, and physical metallurgy. This refers primarily to research into the electronic structure and the nature of the interatomic bond in metallic crystals, and to the method of its quantitative measurement. The work of physical chemists and metallurgical physicists is based on a foundation of modern science – data on the structure of matter, quantum mechanics, Mendeleev's periodic law, thermodynamics, inorganic and physical chemistry, solid state physics, crystallography. From the point of view of methodology, the physical chemistry of metals makes use of the achievements of chemical engineering, metallurgy, foundary work, working of metals by pressure, and instrument manufacture. Physical chemists, physical metallurgists, and metal physicists have achieved some success in the study of the specific features of the interaction of atoms in metals and alloys; the fundamental part played by the periodic law, physicochemical analysis of metallic systems and the Kurnakov laws (variation of properties with composition), and the zone theory and the criteria of alloy formation (structural, dimensional, and electrochemical factors) have been elucidated; attempts have been made by x-ray and neutron diffraction analysis to determine the distribution of electron density in the crystals of some metals and compounds; experimental methods have been developed for the determination of the Fermi surface for a number of metals; filamentary crystals (whiskers) having theoretical strength values have been grown; concepts have been developed on the distribution and part played by impurities and defects in metallic crystals; progress has been made in research on the influence of high and low temperatures, ambient medium, and penetrating radiations on the properties of single-crystal and polycrystalline metals; pure crystals of compounds have been produced with required electrophysical properties (semiconductors and superconductors); theories of the alloying, deformability, strength, and heat resistance of metals are being developed; general concepts have been advanced on the influence of temperature on the mechanical properties of metallic compounds, polymorphic metals, etc. This, however, is as yet only the start on the road to the complete calculation of alloys. The formulation of a quantitative solid state theory and primarily a theory of metal alloys should be one of the most urgent problems of physics and physical chemistry. With regard to the experimental determination of the physical properties of refractory metals and alloys, this has already been studied, but primarily as applied to metals of commercial purity. In the forefront is the problem of acquiring the most varied data on the overall combination of the properties of refractory metals and rare metals in a state of high purity.

For elucidating the general relationships of changes in properties, the construction of constitution–property diagrams should be intensified. Currently, practically all binary phase-equilibrium diagrams involving refractory metals and rare metals are known. Ternary and multicomponent systems are being investigated, but the rate of research on multicomponent systems involving refractory metals, as theoretical bases for the development of alloys, is inadequate. The attention of research workers must be directed to the development of methods for the theoretical calculation of phase diagrams and also to the mechanization and automation of the construction of diagrams using experimental methods.

The possibilities of alloying group IVa, Va, and VIa metals in the development of materials having a predetermined combination of properties have been investigated completely, but for group Va and particularly group VIa metals these possibilties are not being fully applied owing to inadequate purity of these metals and their cold shortness. With the accumulation of experience and the production of apparatus for the purification of metals, and also their treatment by pressure, the possibilities offered by alloying will be utilized on an increasing scale. This also applies, however, to questions of the improvement of properties by thermomechanical treatment, the potentialities of which, as applied to refractory metals, are only beginning to be utilized. For controlling the structure and properties of refractory metals and alloys, it is extremely important to obtain quantitative data on the solubility limits and the form of the occurrence of interstitial impurities in them (for each impurity separately and for concurrent presence of the impurities).

The level of the chemical-metallurgical processing of crude refractory metals and rare metals may be considered to be satisfactory as regards group IVa and Va metals, and comparatively low for group VIa metals. A considerable proportion of semifinished products of tungsten and molybdenum and their alloys are produced by cermet methods. It is now the turn to profit by the indisputable advantages provided by vacuum melting for improving the quality of semifinished products and articles of refractory metals and alloys.

Electric-arc and electron-beam vacuum melting of all refractory metals has been mastered and suitable equipment devised. Currently, the production of large ingots of refractory metals and alloys presents no fundamental difficulties. The methods of introducing comparatively fusible alloy additions into the electrodes for arc melting or into electron-beam furnaces during the melting process calls for further improvement.

Vacuum zone refining of refractory metals and alloys as applied to rods of comparatively small diameter has been mastered. Single crystals of metals and alloys of predetermined crystallographic orientation are being produced. Next in turn is the problem of producing single crystals and zone-refined ingots of refractory metals and alloys not only of large size but also of different geometrical shapes.

Pressure working and heat treatment of refractory metals and alloys in a vacuum and protective atmospheres are not faced with any great difficulties, but these processes are held back by lack of the necessary equipment, especially for the treatment of ingots and semifinished products of large cross section.

No difficulties are encountered also in the chemical analysis of refractory and rare metals. In many cases, methods have already been developed for such analysis. Physical methods for the accurate determination of interstitial impurity content are insufficiently developed.

Welding, soldering, and adhesive securing of parts and semifinished products of refractory metals have been mastered to a considerable extent. Further research is, however, still necessary into brittleness of welds caused by inadequate freedom of the metals from interstitial impurities.

The current application of refractory metals (with the exception of titanium) is not in line with their potentialities. Among the principal causes of this delay must be included the high cost of semifinished products of large cross section, occurrence of brittleness in welds, inadequate technology, and inconstancy of the properties, associated with the comparatively small volume of output.

Some of the most important technical problems and projects cannot be solved without the use of refractory metals and alloys. As demand increases and production is extended and improved in all stages of the chemical, metallurgical and technological cycles, and the promotion of production to assembly-line levels, the above-mentioned deficiencies will disappear and refractory metals and alloys will occupy their rightful position as metals of technical progress.

MIX
Papier aus verantwortungsvollen Quellen
Paper from responsible sources
FSC® C105338
www.fsc.org

If you have any concerns about our products,
you can contact us on
**ProductSafety@springernature.com**

In case Publisher is established outside the EU,
the EU authorized representative is:
**Springer Nature Customer Service Center GmbH**
**Europaplatz 3, 69115 Heidelberg, Germany**

Printed by Libri Plureos GmbH
in Hamburg, Germany